BlackBerry Torch
Made Simple

For the BlackBerry Torch
and all 9800 Series Smartphones

Martin Trautschold

Gary Mazo

MadeSimple™
LEARNING

BlackBerry Torch Made Simple

Copyright © 2011 by Martin Trautschold and Gary Mazo

All rights reserved. No part of this work may be reproduced or transmitted in any form or by any means, electronic or mechanical, including photocopying, recording, or by any information storage or retrieval system, without the prior written permission of the copyright owner and the publisher.

ISBN-13: 978-1456576431

ISBN-10: 1456576437

Printed and bound in the United States of America 9 8 7 6 5 4 3 2 1

Trademarked names, logos, and images may appear in this book. Rather than use a trademark symbol with every occurrence of a trademarked name, logo, or image we use the names, logos, and images only in an editorial fashion and to the benefit of the trademark owner, with no intention of infringement of the trademark.

The use in this publication of trade names, trademarks, service marks, and similar terms, even if they are not identified as such, is not to be taken as an expression of opinion as to whether or not they are subject to proprietary rights.

    CEO: Martin Trautschold
    VP: Gary Mazo

Distributed to the book trade worldwide by Made Simple Learning 25 Forest View Way, Ormond Beach, FL 32174, USA. +1-386-506-8224 – www.madesimplelearning.com.

For information on translations, please e-mail info@madesimplelearning.com, or visit www.madesimplelearning.com.

The information in this book is distributed on an "as is" basis, without warranty. Although every precaution has been taken in the preparation of this work, neither the author(s) nor Made Simple Learning shall have any liability to any person or entity with respect to any loss or damage caused or alleged to be caused directly or indirectly by the information contained in this work.

From Martin and Gary:

This book is dedicated to our families—to our wives, Julie and Gloria, and to our kids, Sophie, Olivia and Cece, and Ari, Dan, Sara, Billy, Elise and Jonah.

Without their love, support, and understanding, we could never take on projects like this one.

More Books and Videos from Made Simple

The Made Simple team have co-authored a number of books and created video tutorials. Find our books and videos at www.madesimplelearning.com, www.amazon.com and fine booksellers around the world.

Books

BlackBerry Bold 9000 *Made Simple*

BlackBerry Bold 9700 *Made Simple* (First Edition)

BlackBerry Bold *Made Simple* (Second Edition for 9700 Series running OS 5)

BlackBerry Curve 8350i *Made Simple*

BlackBerry Curve 8500 *Made Simple* (First Edition for 8500 Series running OS 5)

BlackBerry Curve *Made Simple* (Second Edition for 8500, 9300 Series OS 5)

BlackBerry Curve 8900 *Made Simple*

BlackBerry Storm 9500 *Made Simple*

BlackBerry Storm 2 *Made Simple*

BlackBerry Tour 9600 *Made Simple*

BlackBerry Pearl 'Flip' 8200 *Made Simple*

BlackBerry Pearl *Made Simple* for 8100 Series BlackBerry smartphones

BlackBerry 8800 & 8300 Curve *Made Simple*

BlackBerry *Made Simple* for Full Keyboard BlackBerry smartphones (87xx, 77xx, 75xx, 72xx, 6xxx Series)

BlackBerry *Made Simple* for 7100 Series BlackBerry smartphones (7100, 7130, 71xx Series)

CrackBerry: True Tales of BlackBerry Use and Abuse (with Kevin Michaluk)

DROIDS *Made Simple* (with Marziah Karch)

iPad *Made Simple*

iPhone 3G *Made Simple* (For iOS 3 devices)

iPhone 4 *Made Simple* (For iOS 4 devices)

iPod touch *Made Simple* (First Edition for iOS 3 devices)

iPod touch *Made Simple* (Second Edition for iOS 4 devices)

Palm Pre *Made Simple* (Including Made Simple for WebOS on-device app)

Video Tutorials

Short video tutorials that help you or your entire organization learn how to be more productive with your BlackBerry.

These can be viewed on your computer (Windows or Mac) or on your BlackBerry.

Corporate users can deploy the videos on their Intranet (e.g. Web Server, Learning Management System, or SharePoint server) or choose to have them hosted by Made Simple Learning.

We offer a full library of 3-minute video training tutorials for all popular BlackBerry models with new videos issued frequently to keep up with the latest devices.

Individuals:

Visit www.madesimplelearning.com to get started.

Corporate Training:

Contact info@madesimplelearning.com to get started.

Contents at a Glance

Contents at a Glance	vi
Contents	viii
Part 1: Quick Start Guide	**1**
Getting Around Quickly	3
Part 2: Introduction	**30**
Introduction	31
A Full Day With Your Torch	37
Part 3: You and Your Torch	**40**
Chapter 1: Getting Started	41
Chapter 2: Windows PC Setup	94
Chapter 3: Windows PC Media and File Transfer	117
Chapter 4: Apple Mac Setup	132
Chapter 5: Apple Mac Media and File Transfer	145
Chapter 6: Typing, Spelling, Search, and Help	160
Chapter 7: Personalize Your Torch	187
Chapter 8: Phone and Voice Dialing	220
Chapter 9: Email Like a Pro	255
Chapter 10: Your Contact List	278
Chapter 11: Manage Your Calendar	296
Chapter 12: Text and MMS Messaging	318
Chapter 13: BlackBerry Messenger, PIN Messaging and More	328
Chapter 14: Social Networking	345
Chapter 15: Tasks and Notes	369
Chapter 16: Your Music Player	382
Chapter 17: Snapping Pictures	399

Chapter 18: Fun with Videos	**412**
Chapter 19: Web Browser	**420**
Chapter 20: BlackBerry App World	**437**
Chapter 21: Traveling: Maps & More	**454**
Chapter 22: Other Applications	**474**
Chapter 23: Troubleshooting	**484**
Part 4: Keyboard Shortcuts	**496**
Index	**503**

Contents

Contents at a Glance	vi
Contents	viii
Part 1: Quick Start Guide	**1**
Getting Around Quickly	**3**
Learning Your Way Around	4
Overview of Your Torch	4
Inside Your Torch	6
The Lock Key	7
The Mute Key (Pause/Play)	8
The Green Phone Key	8
The Menu Key (BlackBerry Button)	8
Multitask or Switch Applications with the Menu key	8
The Trackpad	9
The Escape/Back Key	9
The Red Phone Key	9
Touch Screen Gestures	10
Getting Around the Home Screens	11
Seeing More Icons	11
Seeing the Five Home Screens	11
The "Everything" Quick Status Bar	12
Managing Connections and Airplane Mode	12
Starting and Exiting an App	13
Opening Folders	13
Two Types of Menus: Full and Short	14
Swipe Gesture: Move to Next Day, Picture, Email	15
Scrolling Up or Down	15
Scrolling Menus	15
Showing Or Hiding Controls	15
Save Time with the Space key	16
Using Universal Search or Application Shortcuts	16
The Switch to Select Search or Shortcuts	17
Universal Search	17
Application Shortcuts ("Home Screen Hot Keys")	17
Copying, Pasting, and Multitasking	18
Mastering Your Keyboards	18
Slide Out Physical Keyboard	19
Showing and Hiding the On-Screen Virtual Keyboards	19
Choose From Three On-Screen Keyboards	20
Working with the Wireless Network	21
Reading Your Wireless Network Status	21
Wireless Network Signal Strength	22
App Reference Tables	23
Getting Set Up	23
Staying in Touch	24
Staying Organized	25
Being Productive	26
Being Entertained	27
Networking Socially	28
Personalizing Your Torch	29
Adding and Removing Apps	29
Part 2: Introduction	**30**
Introduction	**31**
Congratulations on Your Torch!	31
Unique Features on the BlackBerry Torch	31
Touch Screen	32
Built-In Social Networking	32
BlackBerry Keyboard	32
Expansion Memory Media Card	32
Always on, Always Connected	32
Things This Powerful Are Not Always Easy to Grasp—at First	33
Getting the Most out of This Book	33
How This Book Is Organized	34
Part 1: Quick Start Guide	34
Part 2: Introduction	35
Part 3: You and Your Torch	35
Part 4: Keyboard Shortcuts	35
Quickly Locating Notes, Tips, and Cautions	35
Email Tips	35
Video Tutorials	36
A Full Day With Your Torch	**37**

Part 3: You and Your Torch 40
Chapter 1: Getting Started 41

The Setup App 42
   Setup Email Accounts 43
      Setup Internet Mail Accounts 43
      Hiding Extra Email Account Icons 48
         Maintaining Your Internet Email Accounts and Signatures 48
         Setup Enterprise Accounts ("Enterprise Activation") 50
         How Can I Tell If I Am Activated on the BlackBerry Server? 52
   Wireless Email Reconciliation 53
      Disable or Enable Wireless Email Reconciliation 53
   Syncing Google Contacts Using Email Setup 54
   Troubleshooting Your Email Accounts 54
      Unable to Login to Email Accounts 54
      Verifying That Your Email Server Is POP3 or IMAP 55
      Verifying Your Email Server Settings (Advanced Settings) 56
      Solving a Gmail Enabled IMAP Error Message 56
      Why Is Some Email Missing from My BlackBerry? 57
         Why Does This Happen? 57
         How to Fix This 57
   BlackBerry Enterprise Server Express 60
      How to Get the BES Express Software 61
   Connecting with Bluetooth 61
      Bluetooth Security Tips 61
      Supported Devices 62
      Ways to Get to Bluetooth Connections Screen 62
      Pairing with a Bluetooth Device 63
      Answering and Making Calls with a Bluetooth Headset 65
         Option 1: Answering Directly From the Headset 66
         Option #2: Transferring the Caller to the Headset 66
      Voice Dialing from the Bluetooth Headset 67
      The Bluetooth Connections Menu Commands 67
      Sending and Receiving Files via Bluetooth 68
         Transfer Contacts via Bluetooth 68
         Sending and Receiving Media 68
   Troubleshooting Bluetooth 69
      Problem 1: The Device Refuses Your Passkey 69
      Problem 2: The Device Won't Pair Even with the Correct Passkey 69
      Problem 3: You Cannot Share Your Address Book 70
   Boosting Memory with a Media Card 70
      Installing Your Memory Card / Media Card 70
      Verifying the Memory Card Installation and Free Memory 71
      Formatting or Repairing Media Card 72
   Transferring Content to your Torch Using USB Drive Mode (for Mac and Windows) 72
   Connecting with Wi-Fi 75
      Understanding Wi-Fi on Your BlackBerry 75
      The Wi-Fi Advantage 76
      Setting up Wi-Fi on Your BlackBerry 76
         Connecting to Visible Wi-Fi Net. 77
      Wi-Fi Protected Push-Button Setup 78
      Manually Connect to a Network 79
         Connecting to Free Wi-Fi Networks Requiring Web Browser Login or Agreements 81
      Move, Change, or Delete Saved Wi-Fi Networks 82
      Troubleshoot Using Wi-Fi Diagnostics 83
   Using your BlackBerry as a Tethered Modem 83
      Connecting Your Laptop to the Internet with Your BlackBerry 84
      Tethering (Usually) Costs Extra 84
      Your Tethering Options 84
   Securing Your BlackBerry Data 84
      Preparing for the Worst Case 85
         Step 1: Back up or Sync Your Data 85
         Step 2: Turn on Password Security 86
         Step 3: Set the Message on Lock Screen 87
      Email Security Tips 89
      Web Browsing Security Tips 89

What to Do if You Lose your BlackBerry	89
Turning Off Password Security	89
Battery Life Tips	90
Charging your BlackBerry	90
Low Battery Warnings	90
Places to Recharge Your BlackBerry	91
Recharging from the Power Outlet	91
Recharging in a Car	91
Recharging from Your Computer	92
Extending Your Battery Life	92

Chapter 2: Windows PC Setup 94

Downloading Desktop Software for Windows	95
The Disk from the BlackBerry Box	95
Checking Your Current Version	95
Getting the Latest Version of Desktop Software	96
Installing BlackBerry Desktop Software	97
Overview of BlackBerry Desktop Software	98
Entering Your Device Password	99
Switch Devices (Old Phone to New Torch)	100
Synchronizing Organizer Data (Contacts, Calendar, Tasks and MemoPad)	101
Setting Up the Sync	101
Running the Sync	105
Automating the Sync	107
Accepting or Rejecting Sync Changes	107
Troubleshooting the Organizer Sync	108
Message That Default Calendar Service Has Changed	108
Closing and Restarting Desktop Software	109
Removing and Reconnecting your BlackBerry	109
Getting More Help for Desktop Software Issues	109
Backup Data	109
Automating the Backup Process	111
Restore Data	112
Delete Data	113
Working with Applications	114
Updating the BlackBerry System Software	115
Getting an Email when Software Updates are Available	115
Checking the BlackBerry Site for System Software Updates	115
Installing the BlackBerry System Software Update	116

Chapter 3: Windows PC Media and File Transfer 117

Verify Your Memory Card is Installed	117
USB Drive File Transfer Method	118
Sync Media with BlackBerry Desktop Software	118
Confirm Your Media Sync Options	118
Syncing Music	120
Some Music Cannot Be Synced	121
Wi-Fi Wireless Music Sync	122
Set Up Step 1: Installing the Music Sync App on Your BlackBerry	122
Set Up Step 2: Starting Wi-Fi Music Sync on your Computer	123
Using Wi-Fi Wireless Music Sync	125
Some Songs Cannot Be Wirelessly Synced	126
Working with Pictures	126
Working with Videos	129
Video Conversion and Quality	129
Troubleshooting Desktop Software	130

Chapter 4: Apple Mac Setup 132

BlackBerry Desktop Software for Mac	132
The Disk from the BlackBerry Box	133
Downloading and Installing Desktop Software for Mac	133
Starting Desktop Software – Device Options Screen	134
Main View in Desktop Software	135
Using Desktop Software for Mac	136
Backup Options	136
Setting Up Your Sync Options	137
Advanced Settings	139
Syncing Contacts, Calendar, Notes, and Tasks	140
Back Up and Restore	140
Using Backup	140
Restoring from Backup	141
Adding and Removing Applications	142
Automating the Sync	144

Chapter 5: Apple Mac Media and File Transfer — 145

- Setting Your Media Sync Options — 145
- Syncing Music — 147
- Syncing Pictures and Videos — 149
  - Syncing Pictures — 149
  - Syncing Videos — 150
  - Why can some videos and songs not be synced? (DRM) — 151
- Importing Pictures and Videos from Your BlackBerry — 152
- Using Wireless Music Sync — 152
  - Setting up Wi-Fi Music Sync — 152
  - Using Wi-Fi Wireless Music Sync on Your BlackBerry — 153
  - Some Songs Cannot Be Wirelessly Synced — 155
- USB Drive Mode Transfer for Your Media Card — 155
- Using Your BlackBerry in USB Drive Mode — 156
  - Exploring the Drive — 157
  - Copying Files Using USB Drive Mode — 158

Chapter 6: Typing, Spelling, Search, and Help — 160

- When to Use Touch Screen or the Trackpad — 160
  - Clicking versus Tapping — 161
  - Trackpad Sound and Sensitivity — 161
- Learning Three On-Screen Keyboards — 161
  - Landscape or Portrait - Full — 162
  - Portrait – Multitap Keyboard — 163
  - Portrait – SureType Keyboard — 163
  - Wait to Select Corrections in SureType Mode — 164
- Typing Tips for Your Torch — 164
  - Press and Hold for Automatic Capitalization — 164
  - Caps Lock and Num/Alt Lock — 165
  - Automatic Period and Cap at End of Sentence — 165
  - Typing Symbols — 165
    - On the Physical Keyboard — 165
    - On the Virtual Keyboards — 166
  - Quickly Typing Accented Letters and Other Symbols — 167
    - On the Physical Keyboard — 167
    - On the Virtual Keyboards — 168
  - Editing Text — 168
  - The Mighty Space key — 168
    - Using the Space key While Typing an Email Address — 169
  - Quickly Navigating Drop-Down Lists — 169
  - Jump to First Letter Trick — 169
  - Using Number Keys to Type Dates and Times — 169
- Fine-Tuning Your Keyboards — 170
- Fine-Tuning Your Typing, Spelling and More — 171
- Selecting Text to Copy, Cut or Delete — 172
- Save Time with Word Substitution — 173
  - Standard Word Substitutions — 173
  - Creating Custom Word Substitutions — 174
  - Edit or Delete a Word Substitution — 175
  - Advanced Features – Macros – Time Stamp — 176
- Using Your Spell Checker — 177
  - Using the Spelling Custom Dictionary — 178
    - Adding a Word to the Custom Dictionary — 178
    - Edit or Delete Words in the Custom Dictionary — 179
- Universal Search — 179
- Understanding How Search Works — 180
- Examples of Searches — 181
- Improving Your Search Quality — 181
  - Fine-Tuning Your Search — 182
- Using the Torch's Built-in Help — 183
  - Two Ways to Get to Help — 183
  - Using the Help Menus — 184
  - Help Tips and Tricks — 185

Chapter 7: Personalize Your Torch — 187

- Setting Your Home Screen Preferences — 187
- Using the Navigation Bar — 187
  - Changing Your Wallpaper — 188
  - Changing the Location of Your Downloads — 189
  - Resetting Your Home Screen Preferences — 190
- Organizing Your Icons — 191
  - Picking Favorites — 191
  - Moving Your Icons within a Folder — 192

Hiding and Showing Icons 192
  How to Hide Icons 192
  How to Show Hidden Icons 193
How do I know when I'm in a Folder? 194
Moving Your Icons between Folders 194
Moving Important Icons in the Top Row of the Main "All" Folder 195
Working with Folders 195
  Creating a New Folder 196
  Editing a Folder 197
  Deleting a Folder 197
Setting the Date, Time, and Time Zone 197
Changing Your Font Size and Type 199
Changing Your Theme 200
  Downloading New Themes 201
  Download from App World 201
  Download Themes, Wallpaper, and Ringtones from Other Web Sites 204
Changing Your Convenience Key 204
The Blinking Light - Repeat Notification 206
  Red Flashing Message or Alert LED 206
  Bluetooth Flashing Blue LED 207
  Coverage Flashing Green LED 208
Sounds: Ring and Vibrate 209
Preloaded Sound Profiles 209
  Selecting a Different Preloaded Sound Profile 209
Customizing a Sound Profile 210
Changing Your Phone Ring Tone 213
  Downloading a New Ring Tone 213
Setting Different Ring Tones for Contacts 215
  Using the Sound Profile App 215
  Using the Contact List 217
Accessibility Options 218

Chapter 8: Phone and Voice Dialing   220

Three Main Phone Screens 220
Working with Your Phone 221
  Placing a Call 221
    Dialing a Phone Number 222
    Dial from Call Log 223
    Dialing a Contact by Name 224
  Answering a Call 224
  Ignoring Phone Calls 225

Using the Mute Button to Turn Off the Ringing Phone 226
Dialing Numbers, Taking Notes, and Jumping to Other Apps 227
Taking Notes While On a Call 228
  View, Edit, Delete, or Send Call History and Notes 229
Adjusting the Volume on Calls 230
What's My Phone Number? 230
Changing Your Ring Tone 231
Calling Voice Mail 232
  When Voice Mail Doesn't Work 232
Using Your Call Logs 233
  Checking Your Call Logs 233
  Both Names and Numbers in Call Logs 234
  Add New Contact Entries from Call Logs 234
  Copy and Paste Phone Numbers 236
  To Show Your Call Logs in the Messages App 236
  Benefits of Adding People to Your Contact List/Address Book 237
Speed Dial on Your Torch 237
  Use On-Screen Numbers and Keyboard Letters 238
  Set Up Speed Dial from Call Logs 238
  Set Up Speed Dial from Dial Pad 239
  Set Up Speed Dial from Slide-out Keyboard 239
  Set Up from Speed Dial Icon 240
  Edit, Move or Delete a Speed Dial Number 240
  Using Your Speed Dial Entries 241
Voice Dialing Basics 241
  Voice Dialing a Contact 241
  Voice Dialing a Number 242
  Other Voice Dialing Commands 242
  Changing Your Voice Dialing Options 242
  Adapt Voice in Voice Dialing 243
  Voice Dialing Tips and Tricks 243
    Make Voice Dialing Calls More Quickly 243
    Give your Contacts Nicknames 244
    Changing Your Voice Dialing Language 244
Setting your Phone Ring Tone 245
  Changing Your Ring Tone from the Phone 245

Changing Your Ring Tone from Other Places	246
Set a Custom Ring Tone for a Caller	247
Call Waiting – Handling a Second Caller	247
Join a Conference Call or Swap	248
Call Forwarding	248
Conference Calling	250
To End or Leave a Conference Call	252
Advanced Dialing (Letters, Pauses, and Waits)	252
Dialing Letters on a Phone Call	252
Dialing Phone Numbers with Letters in your Contact List	253
Adding Pauses and Waits in Phone Numbers	253
More Phone Tips and Tricks	254

Chapter 9: Email Like a Pro 255

Composing Email	255
Send Email from Your Messages App	255
Send Email from Your Contacts App	257
Selecting a Different Email Address to Send From	258
See the Email Address	258
Replying To Messages	259
Navigating Around and Other Tidbits	259
Email Shortcut Keys	260
Email Soft Keys	260
Swipe to Navigate Your Inbox	261
Getting Rid of Blank Spots in Emails You Receive	261
Setting the Importance Level of the Email	262
Spell Checking Your Email Messages	262
Flag for Follow Up	263
Setting a New Flag	263
Changing or Editing a Flag	265
Finding Flagged Items	265
Flag Alarms	266
Attaching Contacts, Files, and Pictures	266
Attaching a Contact Entry	266
Attaching a File, Picture, Video, or Voice Note	267
Working with Email Attachments	268
Supported Email Attachment Formats	268
Knowing When You Have an Attachment	269
Opening Attachments	269
Editing Attachments with Documents to Go	270
Using the Standard Viewer	272
Using Sheet to Go, Slideshow to Go	272
To Open a Picture	273
Searching for Messages (Email, SMS, MMS)	275
The General Messages Search Command	276
Search Sender or Search Recipient	276
Search Subject	276
Search Sender or Recipient Menu Command	277

Chapter 10: Your Contact List 278

The Heart of Your BlackBerry	278
Ways to Get Your Addresses on Your BlackBerry	278
When Your Contact List Most Useful	279
Our Recommendations	279
Transfer SIM Card Contacts into Your Contact List	279
How to Easily Add New Addresses	280
Option #1: Type an Address into Contacts	280
Up to Three Email Addresses	282
Phone Numbers with Letters	282
Option #2: Add Contacts from Email Addresses	283
Option #3: Add Contacts from Phone Call Logs	283
Option #4: Add Contacts from Underlined Numbers, Email Addresses and Street Addresses	284
Option #5: Perform a Remote Lookup (If your BlackBerry is tied to an Enterprise Server)	284
How to Easily Find Names	286
Using the Find Feature in Contacts	286
Finding and Calling Someone in the Phone	287
Managing Your Contacts	287
Editing Contacts	288
For Facebook Users	289
Adding a Picture to the Contact for Caller ID	290
Changing Way Contacts Are Sorted	291
Using Categories	291

Filtering Your Contacts by Category 292
Knowing when Your Contacts Are Filtered 293
Un-Filtering Your Contacts 293
Using Groups as Mailing Lists 293
    Examples 293
Creating and Using a Group Mailing or SMS List 294
Sending an Email to the Group 295

Chapter 11: Manage Your Calendar 296

Organizing Your Life with Your Calendar 296
    Ways to Get Your Calendar on Your BlackBerry 296
Switching Views and Days 297
    Swiping to Move Between Days 298
    Using the Calendar Shortcuts 298
    Using the On-Screen Soft Keys 299
Scheduling Appointments 299
    Quick Scheduling 300
    Detailed Scheduling 300
Customizing Your Calendar with Options 303
    Changing Your Start and End Time on Day View 303
    Changing Your Initial View 304
    Default Reminder (Alarm) and Snooze Times 304
Scheduling Conference Calls 304
Copying and Pasting Information into Your Calendar 305
Dialing a Scheduled Conference Call 309
    Changing the Calendar Alarm Sound 309
    Snoozing a Ringing Calendar or Task Alarm 310
Working with Meeting Invitations 311
    Inviting to Attend a Meeting 311
    Viewing Availability of Invitees 312
    Respond to a Meeting Invitation 312
    Changing the List of Participants for a Meeting 313
    Sending an Email to Invitees 313
Using Google Sync for Contacts and Calendar 313
    Getting Started with Gmail and Google Calendar 314
    Installing the Google Sync Program 314
    How The Sync Looks 317

Chapter 12: Text and MMS Messaging 318

Text and Multimedia Messaging 318
SMS Text Messaging 319
    Composing Text Messages 319
        Sending from the Text Messages or Messages App 319
        Sending from the Contact List 321
    Text Menu Commands 321
Opening and Replying to SMS Messages 323
MMS Messaging (Picture Messaging) 323
    Sending MMS from the Message List 324
    Sending a Media File from the Media Icon 325
    Advanced MMS Commands 326
Messaging Troubleshooting 326
    Host Routing Table "Register Now" 326
    Perform a Soft Reset (Reboot) 327
    Perform a Hard Reset (Battery Pull) 327

Chapter 13: BlackBerry Messenger, PIN Messaging and More 328

BlackBerry Messenger 328
    Setting Up BlackBerry Messenger (BBM) 329
    Adding Contacts to BlackBerry Messenger 329
    Invite Using Barcodes 330
    Responding to BlackBerry Messenger Invitations 331
    BlackBerry Messenger Menu Commands 331
    BlackBerry Messenger Options and Backup, Restore 332
    Starting or Continuing Conversations and Emoticons 333
    Sending Files to a Message Buddy 335
    Pinging a Contact 335
    Set Availability to Others (My Status) 336
    Multi-person Chat with BlackBerry Messenger 337
    Using Groups 338
        Create a New BBM Group 339

Group Chats, Pictures, Lists and Calendar 340
PIN Messaging 341
Replying to a PIN Message 342
Adding Someone's PIN to Your Address Book 342
Using AIM, Yahoo, and Google Talk Messaging 343
Even More Instant Messenger Apps 344

Chapter 14: Social Networking 345

Downloading Social Networking Apps 345
Logging into the Apps 346
Facebook 347
   Facebook Setup Wizard 348
   Status Update and News Feed 349
   Top Bar Icons 349
   Communicating with Facebook Friends 350
   Uploading Pictures to Facebook 351
Twitter 352
   Create a Twitter Account 352
   Using Twitter for BlackBerry 353
      Sending Out Tweets 354
   Twitter Icons 354
      Mentions 354
      My Lists 354
      My Profile 355
      Direct Messages 355
      Find People 355
      Search 356
      Popular Topics 356
   Twitter Options 357
LinkedIn 358
   Download and Sign In - LinkedIn 358
   Navigating around the LinkedIn App 358
      Icons 359
      Search 359
      Connections 359
      Invitation 360
      Messages 360
      Reconnect 361
   LinkedIn Options 361
YouTube 362
   The YouTube App 362
   YouTube on the Web 363
   YouTube Favorites 364

Social Feeds 364
   Getting Started with Social Feeds 364
   Adding Feeds 365
   Adding RSS Feeds 365
   Using the Social Feeds App 366
   Posting to All Social Apps at Once 367
   Managing RSS Feeds 368

Chapter 15: Tasks & Notes 369

The Task Icon 369
How Do You Get Your Tasks from Your Computer to Your BlackBerry? 369
Viewing Tasks 370
Adding a New Task 370
   Categorizing Your Tasks 371
      Assigning a Task to a Category 371
   Finding Tasks 372
   Managing, Checking Off Your Tasks 373
Sorting Your Tasks and Task Options 374
The MemoPad – Virtual Sticky Notes 374
How to Sync Your MemoPad 374
1,001 Uses for the MemoPad (Notes) Feature 375
   Common Uses for the MemoPad 375
Adding or Editing Memos on the BlackBerry 375
   Quickly Locating or Finding Memos 377
   Ordering Frequently Used Memos 377
   Add a Separate Memo for Each Store 377
Viewing Your Memos 377
Organizing Your Memos with Categories 378
Switching Applications / Multitasking 378
Forwarding Memos via Email, SMS, or BlackBerry Messenger 380
   Other Memo Menu Commands 380

Chapter 16: Music Player 382

Listening to Your Music 382
Syncing Your Music and Playlists 382
Playing Your Music 383
   Finding and Playing a Song 384
   Doing Other Things While Listening to Music 385
Using a Song as your Phone Ring Tone 386
Playing All Your Music 386
Playing Your Playlists 387
   Creating Playlists on Your Computer 388

Creating Playlists on Your BlackBerry	388
Supported Music Types	390
Music Player Tips and Tricks	391
Streaming Internet Radio	391
Pandora Internet Radio	392
Installing Pandora	392
Starting Pandora for the First Time	392
Pandora Controls and Options	394
Using "Thumbs Up" and "Thumbs Down"	394
Slacker Radio	394
Downloading and Installing Slacker Radio	395
Creating or Logging in to a Slacker Account	395
Choosing Your Station	397
Slacker Controls	397
Slacker Menu Commands and Shortcut Keys	398
One-Key Shortcuts for Slacker	398

Chapter 17: Snapping Pictures 399

Using the Camera	399
Tips for Taking Great BlackBerry Pictures	399
Camera Features and Buttons	400
Starting the Camera Application	400
Icons in the Camera Screen	401
Sending Pictures with the Email Envelope Icon	402
Adjusting the Size of the Picture	403
Geotagging Your Pictures	403
Changing the Scene Mode	405
Adjusting the Picture Size/Quality	405
Using the Zoom	406
Managing Picture Storage	406
Selecting Where Pictures Are Stored	406
Viewing Pictures Stored in Memory	407
Option #1: Viewing from the Camera Program	407
Option #2: Viewing from the Pictures App	408
Picture Viewing Soft Keys	409
Viewing a Slide Show	409
Scrolling Through Pictures	410
Adding Pictures to Contacts for Caller ID	410
Transferring Pictures To or From Your BlackBerry	410

Chapter 18: Fun with Videos 412

Working with Videos on a BlackBerry	412
Transferring Videos to BlackBerry	412
Your Video Camera	412
Using the Video Camera	413
Converting DVDs and Videos to Play on the BlackBerry	415
Supported Video Formats on the BlackBerry	416
Viewing Videos on the BlackBerry	416
Playing a Video	417
Showing or Hiding controls	418

Chapter 19: Web Browser 420

Web Browsing on Your BlackBerry	420
Starting the Web Browser and the Start Page	420
Using Tabs for Browsing	421
Opening and Using Tabs	421
Opening a New Tab	422
Using Browser Keyboard Shortcuts	422
Using the Browser Menu	423
Exploring the Browser Menu Options	423
Using Your Address Bar	425
Using the Globe Shortcut Menu	428
Zoom In and Out of Web Pages	428
Copying or Sending a Web Page or Link	429
Adding Bookmarks	430
Bookmark Naming Tips	430
Viewing the Start Page or Home Page	432
Using Your Bookmarks to Browse	433
Searching with Google	435
Viewing a Google Search Location	436
Finding Places with Google Maps	436

Chapter 20: BlackBerry App World 437

The App World Concept	437
Downloading App World	437
Starting App World for the First Time	438
BlackBerry ID, Credit Card, Direct Bill or PayPal Account	438
Downloading Themes - App World	439
Featured Programs	439
Categories, Top Downloads, and Search	439

Categories	440
Top 25	441
Top 25 Paid Apps	442
Search App World	442
Downloading Apps	443
Changing Payment Options	443
Downloading and Purchasing an App	444
Using the My World Area	446
The My World Menu Commands	446
Archiving or Deleting Programs	447
Adding or Removing Apps	447
Downloading New Software Outside of App World	448
Changing Your Default Downloads Location	450
Finding More Software	450
Web Stores	450
Reviews of Software, Services, and More	451
Official BlackBerry Sites	451
Removing Software from Your BlackBerry	451
Option 1: Deleting Apps from App World	452
Option 2: Deleting Apps from the Home Screen	452
Option 3: Deleting Apps from the Options Icon	452
Option 4: Deleting Apps with the Desktop Software	453

Chapter 21: Traveling: Maps & More 454

International Travel: Things to Do Before You Go	454
Avoiding a Shockingly Large Bill	454
Before Your International Trip	455
Call Your Phone Company	455
Ask About Using a Foreign SIM Card	455
Airplane Travel: Getting into Airplane Mode	456
Things to Do When Abroad	456
Getting Your BlackBerry Ready	456
Step 1: Make Sure the Time Zone Is Correct	456
Step 2: Turn Off Data Roaming If It's Too Expensive	457
Step 3: Register with the Local Network	457
Step 4: Look for Free Wi-Fi Networks	458
Returning Home	458
Step 1: Check Your Time zone	458
Step 2: Reset Your Data Services	458
Step 3: Register Your BlackBerry on the Local Network	458
Step 4: Turn Off Your Special International Plan	458
BlackBerry Maps, Google Maps, and Bluetooth GPS	459
Enabling GPS on Your BlackBerry	459
Using BlackBerry Maps	459
Viewing a Contact's Map	460
Getting Directions with BlackBerry Maps	462
BlackBerry Maps Menu Commands	463
Google Maps	465
Google Map Commands and On-Screen Icons	466
Google Maps Search - Finding an Address or Business	467
Keyboard Shortcut Keys	469
Getting Directions	469
Google Latitude	470
Layers – Finding More Things Nearby	470
Layers – Seeing Transit Lines	472
Viewing Current Traffic	472

Chapter 22: Other Apps 474

The Calculator – Tips and Conversions	474
The Clock – Alarm, Stopwatch, Timer	476
Bedside Mode:	477
Voice Notes Recorder	478
Password Keeper	479
Podcasts App	480
Browse and Search for Podcasts	481
Subscribe to Podcasts	482
Enjoy Podcasts	483
Podcast Options	483

Chapter 23: Troubleshooting 484

Solving Connection Issues	484
Low Signal Strength	484
Managing Connections Manually (Airplane Mode)	485
Register - Host Routing Table	485
Managing Your Applications	486
Viewing the Diagnostic Help Me!	488
Clearing the Event Log	489

Saving Battery Life	490
Resetting Your BlackBerry	491
Performing a Soft Reset	491
Performing a Hard Reset	492
Sending Internet Account Service Books	492
If the Problem Persists...	495

Part 4: Keyboard Shortcuts 496

Home Screen Hotkeys / Application Shortcuts	497
The Switch to Select Search or Shortcuts	497
List of Home Screen Hotkeys	498
Email Messages Shortcuts	499
Web Browser Shortcuts	500
Calendar Shortcuts	501
Media Player Shortcuts	502

Index   503

About the Authors

Martin Trautschold is the founder and CEO of Made Simple Learning, a leading provider of Apple iPad, iPhone, iPod touch, BlackBerry, Android, and Palm webOS books and video tutorials. He has been a successful entrepreneur in the mobile device training and software business since 2001. With Made Simple Learning, he helped to train thousands of Smartphone users with short, to-the-point video tutorials. Martin has now co-authored twenty "Made Simple" guide books. He also co-founded, ran for 3 years, and then sold a mobile device software company. Prior to this, Martin spent 15 years in technology and business consulting in the US and Japan. He holds an engineering degree from Princeton University and an MBA from the Kellogg School at Northwestern University. Martin and his wife, Julia, have three daughters. He enjoys rowing with the Halifax Rowing Association in Daytona Beach, Florida and cycling with friends. Martin can be reached at martin@madesimplelearning.com.

Gary Mazo is Vice President of Made Simple Learning. Gary joined Made Simple Learning in 200g and has co-authored the last nineteen books in the Made Simple series. Along with Martin, and Kevin Michaluk from CrackBerry.com, Gary co-wrote *CrackBerry: True Tales of BlackBerry Use and Abuse*—a book about BlackBerry addiction and how to get a grip on one's BlackBerry use. Gary also teaches writing, philosophy, technical writing, and more at the University of Phoenix. He holds a BA in anthropology from Brandeis University. Gary earned his M.A.H.L (Masters in Hebrew Letters) as well as ordination as Rabbi from the Hebrew Union College-Jewish Institute of Religion in Cincinnati, Ohio. He has served congregations in Dayton, Ohio, Cherry Hill, New Jersey and Cape Cod, Massachusetts. When not writing or teaching, Gary enjoys cycling and playing the piano. Gary is married to Gloria Schwartz Mazo; they have six children. Gary can be reached at: gary@madesimplelearning.com.

Acknowledgments

A book like this takes quite a bit of time successfully complete. We would like to thank AT&T for lending us devices to use for this book.

We would like to thank our spouses and families for allowing us to have the time needed to complete this project.

We would like to thank our Made Simple Learning customers for supporting our endeavors and keeping us on our toes with great questions and suggestions.

Part 1:

Quick Start Guide

In your hands is the first ever BlackBerry to combine a highly responsive touch screen with a slide out physical keyboard, the BlackBerry Torch. You have the best of both worlds - you can use the beautiful touch screen to do all your viewing and browsing, then when you really need to crank out some fast typing (such as a text message, email or editing a Word document), just slide out the keyboard and type away. This Quick Start Guide will help get you and your new Torch up and running in a hurry. You'll learn all about the keyboard, buttons, and ports, and how to use the responsive touch screen to help you get around. Our **App Reference** tables will introduce you to the apps on your Torch and serve as a quick way to find out how to accomplish a task.

Getting Around Quickly

This Quick Start is meant to be just that—a tool that can help you jump right in and find information in this book—and learn the basics of how to get around and enjoy your Torch right away.

We start with the nuts and bolts in the "Learning Your Way Around" section—what all the keys, buttons, switches, and symbols mean and do on your Torch. You will learn how to get inside the back of your Torch to remove and replace the battery, SIM card, and media card. You'll also learn how to use the **Green Phone** key, **Red Phone** key, **Menu** key, **Trackpad**, and **Escape** key.

Then we move on to the "Touch Screen Gestures" section, where we show you all the best tips and tricks for getting around using the touch screen, including how to start apps, get into folders, and perform some time-saving tasks.

In "Using Universal Search or Application Shortcuts" we show you how to switch between the search or shortcut hot keys when you start typing from your Home screen – both features can be very useful. Play with both options and get a feel for which is more useful to you.

In the "Copying, Pasting, and Multitasking" section, we cover the very useful copy-and-paste function and how to jump between apps on your Torch.

You'll want to know how to type on your Torch, so in the "Mastering Your Keyboards" section, we include some useful tips and tricks for both the virtual keyboards and the slide out physical keyboard.

In "Working with the Wireless Network," we help you understand how the letters, numbers, and symbols at the top of your Torch screen tell you that you can make phone calls, send SMS text messages, send and receive email, or browse the Web. We also show you how to handle your Torch on an airplane, when you might need to turn off the radios.

In "App Reference Tables," we've organized the app icons into general categories so you can quickly browse the icons and jump to a section in the book to learn more about the app a particular icon represents. Here are the tables:

- Getting Set Up
- Staying in Touch
- Staying Organized
- Being Productive
- Being Entertained

- Networking Socially
- Personalizing Your Torch
- Adding and Removing Apps

Learning Your Way Around

To help you get comfortable with your Torch, we start with the basics—what the keys, buttons, ports, and symbols do, and how to open up the back cover to get at your battery, media, and SIM card. Then we move into how you start apps and navigate the menus. We end this section with a number of very useful time-saving tips and tricks for getting around the touch screen.

Overview of Your Torch

Figures 1 and 2 show many of the things you can do with the buttons and ports on your Torch. Go ahead and try out a few things to see what happens. Your phone multitasks (that is it can do multiple tasks at once). Try pressing the **Menu** key for 2 seconds, press and hold the **Green Phone** key (to dial from the address book), press the side **Convenience** key, tap an icon, and then press the **Escape** key to see what happens. Have some fun getting acquainted with your device.

QUICK START GUIDE

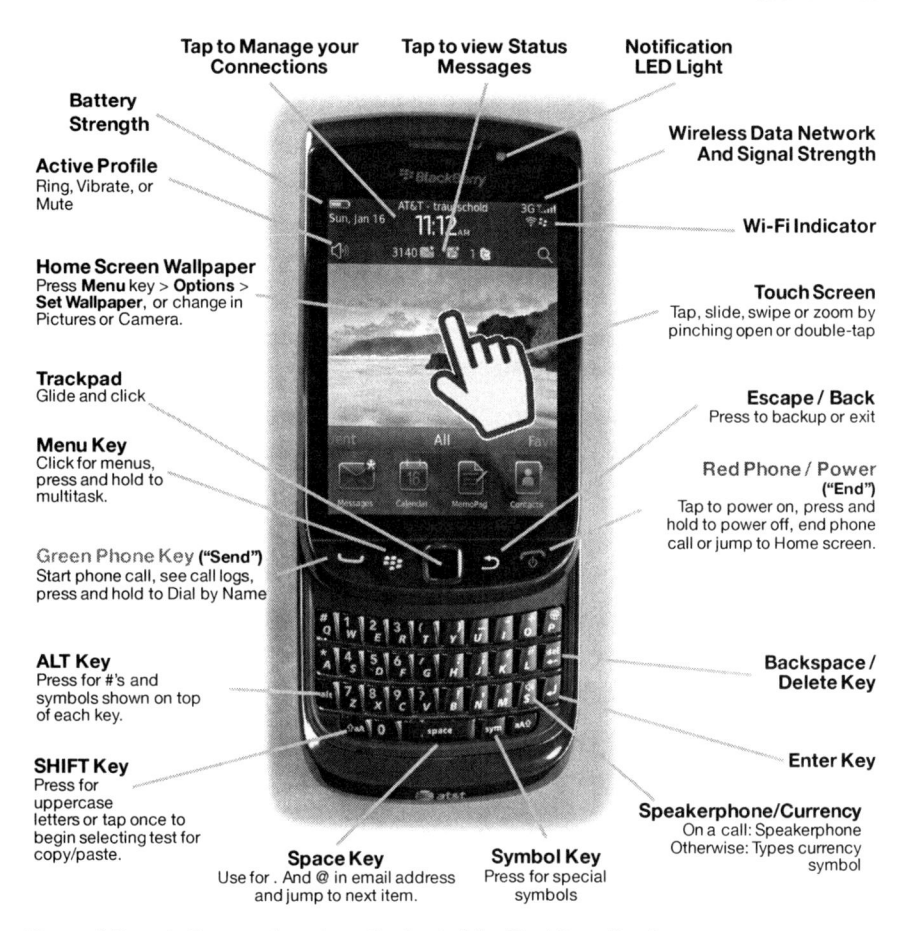

Tap to Manage your Connections

Tap to view Status Messages

Notification LED Light

Battery Strength

Active Profile
Ring, Vibrate, or Mute

Wireless Data Network And Signal Strength

Wi-Fi Indicator

Home Screen Wallpaper
Press **Menu** key > **Options** > **Set Wallpaper**, or change in Pictures or Camera.

Touch Screen
Tap, slide, swipe or zoom by pinching open or double-tap

Trackpad
Glide and click

Escape / Back
Press to backup or exit

Menu Key
Click for menus, press and hold to multitask.

Red Phone / Power ("End")
Tap to power on, press and hold to power off, end phone call or jump to Home screen.

Green Phone Key ("Send")
Start phone call, see call logs, press and hold to Dial by Name

ALT Key
Press for #'s and symbols shown on top of each key.

Backspace / Delete Key

SHIFT Key
Press for uppercase letters or tap once to begin selecting test for copy/paste.

Enter Key

Speakerphone/Currency
On a call: Speakerphone
Otherwise: Types currency symbol

Space Key
Use for . And @ in email address and jump to next item.

Symbol Key
Press for special symbols

Figure 1. Keys, buttons, and, ports on the front of the BlackBerry Torch

Top Edge

Keyboard Lock
Press to lock/unlock the screen and keyboard.

Mute Key
Press to mute, will also pause or play music and videos.

3.5 mm Headset Jack

Volume Up & Down Keys

Micro USB Port
Plug into USB cable for computer and charger

Convenience Key

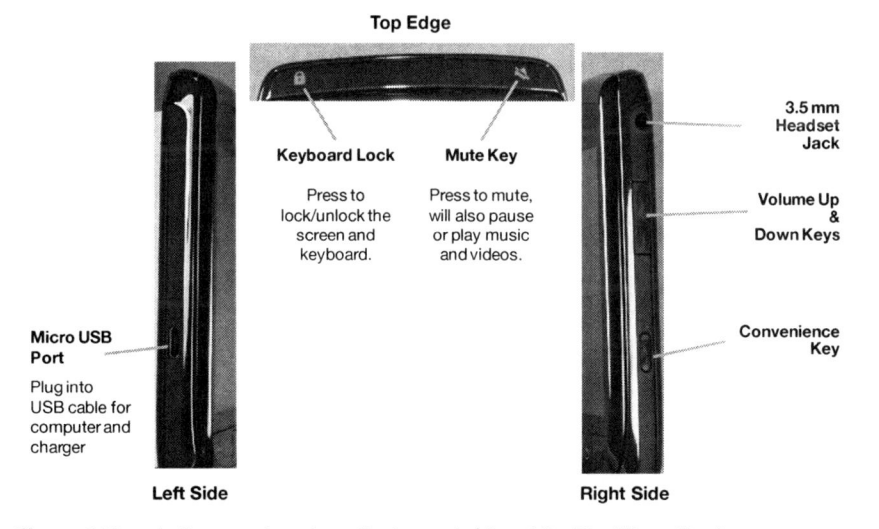

Left Side

Right Side

Figure 2. Keys, buttons, and, ports on the top and sides of the BlackBerry Torch

View Video Tutorials and Free Tips at www.MadeSimpleLearning.com

5

Inside Your Torch

You have to get inside your Torch to access your battery, SIM card slot, and media card slot. The following instructions and Figure 3 show you how.

> **TIP**: To remove the back cover, press with two thumbs and slide the back down and off the BlackBerry.

To insert a memory card (MicroSD format), you do not need to remove the battery. Just perform the following steps:

1. Lift up the black plastic flap and gently place the media card with the metal contacts facing down and the notch toward the bottom left.
2. Then slide it completely down into the media card slot near the bottom right of the device.

To remove the memory card, reverse the procedure: lift the black plastic flap and slide the card up out of the holder and lift it out.

To remove or replace the battery, do the following:

1. Gently put your fingernail next to the top edges of the battery (the semicircles) and pry it up and out. You may need to turn over the BlackBerry gently tap the open back in your palm to remove a stubborn or stuck battery.
2. To replace it, insert it from the bottom edge and then press the top edge down.

To insert or remove a SIM card, do the following:

1. Remove the battery, and then place the SIM card on the left edge with the notch in the upper-left corner.
2. Slide it completely into the SIM card slot (see below).

To replace the back cover, place it so that it almost completely covers the back, then slide it up to the top until it clicks or locks into place.

QUICK START GUIDE

Camera Lens
Keep clean for highest quality pictures.

3.5 mm Headset Jack

Volume keys

Convenience key

Camera Flash
Keep clean for the brightest flash.

Battery Slot
Remove by prying up at the gray circles here with your fingernail and lift out from top edge.

SIM Card
Battery needs to be removed to insert or remove the SIM Card.

Lift black flap to insert or remove Media Card.

Media Card
MicroSD Format
(Supports up to 32 GB size)

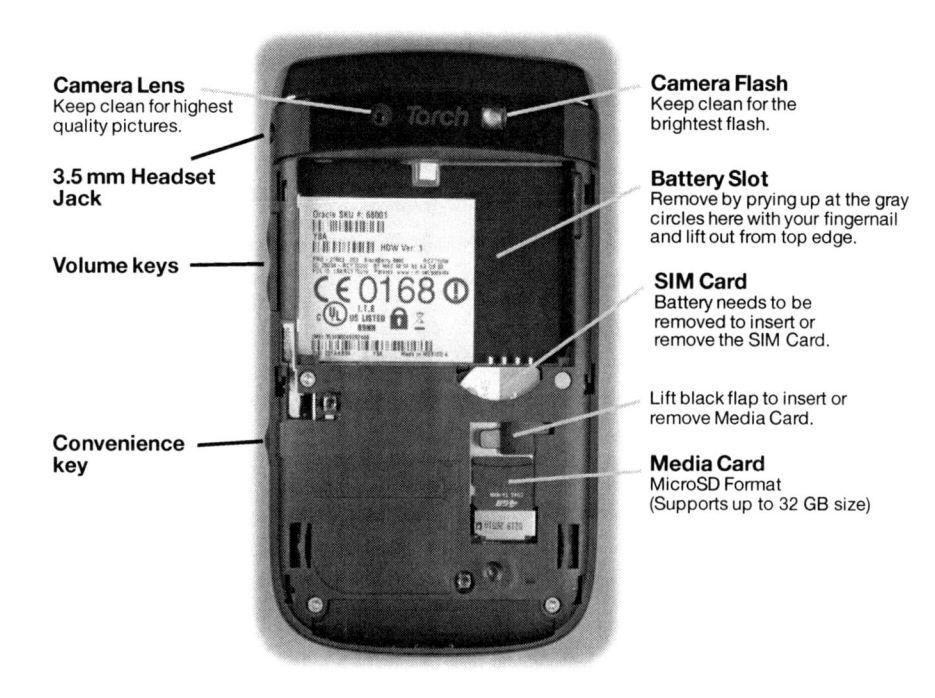

Figure 3. Inside your Torch— the battery, media card, and SIM card slots

The Lock Key

At the top left of your BlackBerry is your **Lock** key. Simply tap it to lock the BlackBerry screen and keyboard, and tap again to unlock.

CAUTION: The **Lock** key will <u>not</u> password lock your BlackBerry. If you want to require a password be entered to unlock your device, you first need to enable it and then use the **Password Lock** icon. To learn how, see the "Password Security" section of Chapter 1: "Getting Started."

The Mute Key (Pause/Play)

On the top-right edge of your BlackBerry is the **Mute and Pause/Play** key. Tap this key to mute a ringing phone call, mute yourself on a phone call, and pause any playing media (e.g., a video, song, or audio book). Tap it again to un-mute or play the media.

The Green Phone Key

Start up your phone by pressing this **Green Phone** key from any location.

> **TIP:** Call Any Underlined Phone Number. Pressing the Green Phone key when you see any underlined phone number (e.g., 313-555-1212 in an email signature, web site, note, etc.) will start a call to that number. You can learn about your phone in Chapter 8: "Phone and Voice Dialing."

The Menu Key (BlackBerry Button)

The **Menu** key—that key with the BlackBerry logo—is the doorway to all the possibilities of your device. In every application, pressing this key brings up a menu with all of the options for the app. From the basic home screen, pressing this key brings up all the other icons.

> **TIP:** Pressing and holding the **Menu** key will allow you to Multitask. See below for details.

While viewing a contact, pressing the **Menu** key allows you to call, email, SMS, or communicate in other ways with that particular contact. Try pressing it in every program you open to see the myriad of possibilities.

Multitask or Switch Applications with the Menu key

You can multitask using the **Menu** key. Press and hold it to bring up the multitasking pop-up window, which will allow you to multitask by taking the following steps.

1. Press and hold the **Menu** key to see the pop-up window of running apps.

2. Swipe up or down to the app you want to start and click it.

3. Click the **Home** app if you don't see the icon you want to start. Then click the icon you want to start from the home screen.

4. Repeat the procedure to return to the app you started in.

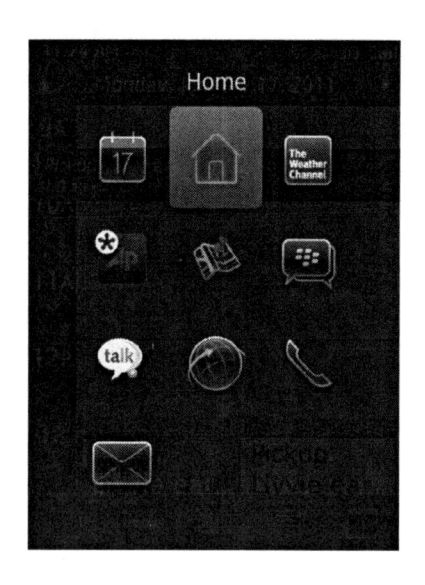

The Trackpad

The **Trackpad** allows you to move around the screen and select items by clicking it or pressing it in. It can be quite helpful when you need to carefully move the cursor or click on something small on the screen such as a list of small text items (BlackBerry Help screens or web pages). It can be more accurate than trying to touch something small with your fingers.

The Escape/Back Key

This key does just what its name suggests—it goes back to where you were before. It also prompts you to save, delete, or discard things you might be working on. If you do something wrong or find yourself wanting to get back to where you were a second ago, just press this key.

The Red Phone Key

This is equivalent to the "end" key that you may be familiar with from your old phone. You can end your calls, ignore calls, and perform other familiar actions with the **Red Phone** key.

BlackBerry Torch Made Simple

You can also multitask using the **Red Phone** key. Just press it (when not on a call) and you'll jump right to the **Home** screen.

> **TIP:** Knowing when your information is saved
>
> When you use the **Red Phone** key or press and hold the **Menu** key to jump out of an app everything you were doing in the app you are leaving is saved. For example, if you are in the middle of typing an email message and need to check your **Calendar**. Using the **Red Phone** key or pressing and holding the **Menu** key to jump to the **Calendar** will leave your email message exactly as it was. This allows you to jump back to the **Messages** app and finish your email.

Touch Screen Gestures

The BlackBerry Torch has a very responsive touch screen. Understanding the correct gestures will help get the most out of your Torch. The great thing is that because of the slide-out keyboard, you can type on the keyboard instead of the screen. Some people prefer the physical keys instead of "typing on glass."

You can pretty much do anything on your Torch by using a combination of the following:

- Touch screen gestures
- Pressing soft keys on the screen
- Using the Trackpad and four navigation buttons on the bottom of the BlackBerry
- Typing on the slide out keyboard
- Pressing the **Convenience** key on the side

Here are a few of the basic touch screen gestures:

- **Touch or Tap:** To start an app, tap its icon. To select a menu item, tap it gently.
- **Long press**: For many items including icons, contacts, phone numbers and other things if you touch and hold the item you will see a pop-up menu showing you various commands related to the item you touched.
- **Swipe**: To move to the next screen on your **Home** screen, picture, email message or day in your calendar swipe your finger in one direction to drag items on the screen in that direction

- **Scroll up/down**: To see what is above the top or below the bottom of the screen lightly touch and drag your finger up or down.

- **Pinch to Zoom**: Place your fingers on the screen and pinch open to zoom in. Pinch closed to zoom out.

- **Double-tap to Zoom**: You can also double-tap the screen to zoom in on a web page, picture, map or other item. Double-tap again to zoom out.

- **Multi-touch**: To cut or copy text, tap the beginning and end of the text that you want to cut or copy with two fingers simultaneously.

In the sections that follow, we will show you exactly how to master each of these gestures. Read the following pages for a graphical step-by-step description of how to get around and master your BlackBerry Torch.

Getting Around the Home Screens

Go ahead and play around with your new Torch as shown in Figure 4. Try touching, then press and physically click the glass down on an icon to start it then press the **Escape** key to back out. Try just hovering your finger over an icon. Press the **Menu** key and then swipe up to see all your icons. Press the **Escape** key to get back to your home screen.

Seeing More Icons

To see all the icons in a particular area of the home screen, tap the bar above the icons or swipe your finger up as shown in Figure 4.

You then see all icons in the **All** section of the **Home** screen.

Swipe up to see the rest of the icons.

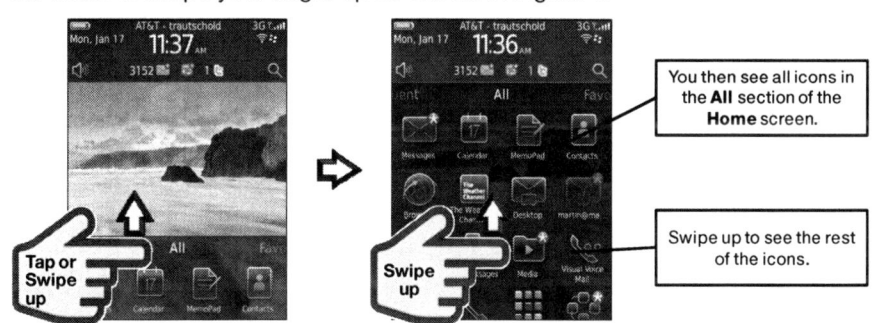

Figure 4. Touch screen gestures

Seeing the Five Home Screens

The **Home** screen is actually grouped into five areas: **All**, **Favorites**, **Media**, **Downloads** and **Frequent**. You see each area by swiping left or right. Once you swipe right from **Frequent** you will end up back in the **All** section. **All** has all your icons, you can assign icons to **Favorites** by long-pressing them, **Media** has all your music, videos, pictures and related items, **Downloads** are new apps you download and **Frequent** are those apps you use most often (see Figure 5).

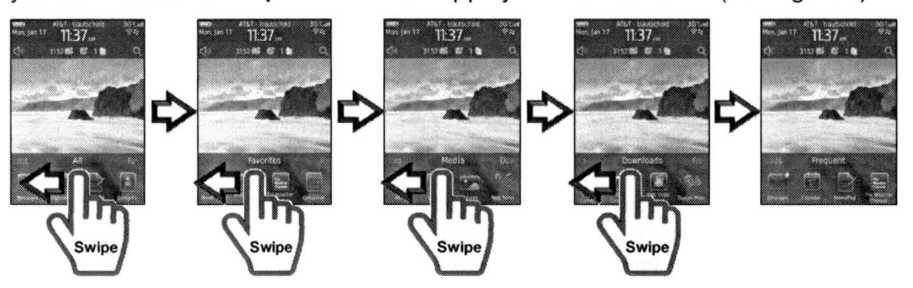

*Figure 5. The Five Home Screens: **All**, **Favorites**, **Media**, **Downloads** and **Frequent**.*

The "Everything" Quick Status Bar

If you tap the screen just under the time as shown in Figure 6, you can see a quick summary of the latest happenings in your world. You see recent **Messages** (email, instant, and related), upcoming **Calendar** events, missed **Phone** calls, **BlackBerry Messenger** alerts, **Social Feeds**, **Twitter** updates and more.

Figure 6. Tap to see a Quick Summary of all Status Messages.

Managing Connections and Airplane Mode

Tap anywhere in the very top bar from the battery to time to the status icons to bring up the **Manage Connections** app. This allows you to go into **Airplane Mode** by turning all your connections off, or selectively turn on or off your **Mobile Network (cell)**, **Wi-Fi**, or **Bluetooth** connections. You can also setup all these connections from this screen (see Figure 7). Tap **Turn All Connections Off** to go into **Airplane mode**.

QUICK START GUIDE

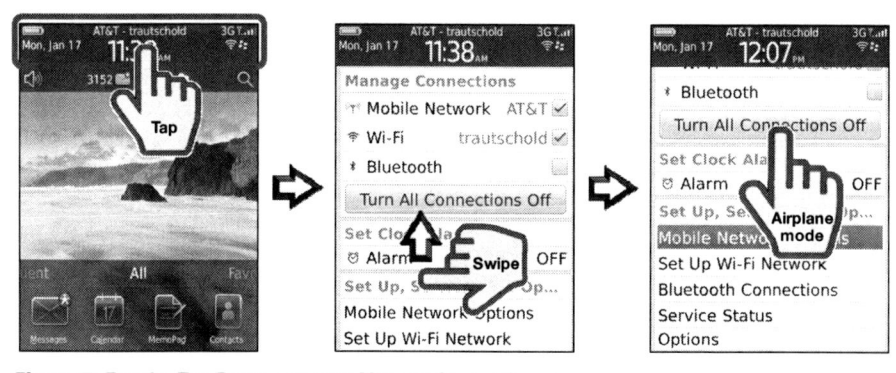

Figure 7. Tap the Top Bar to see your Manage Connections app.

Starting and Exiting an App

Tap any icon to start the app. For example in Figure 8, we tapped the **MemoPad** icon to start the **MemoPad** app. Then to exit, we pressed the **Escape** key.

Tap any icon to start the app.

Press the **Escape** key to exit.

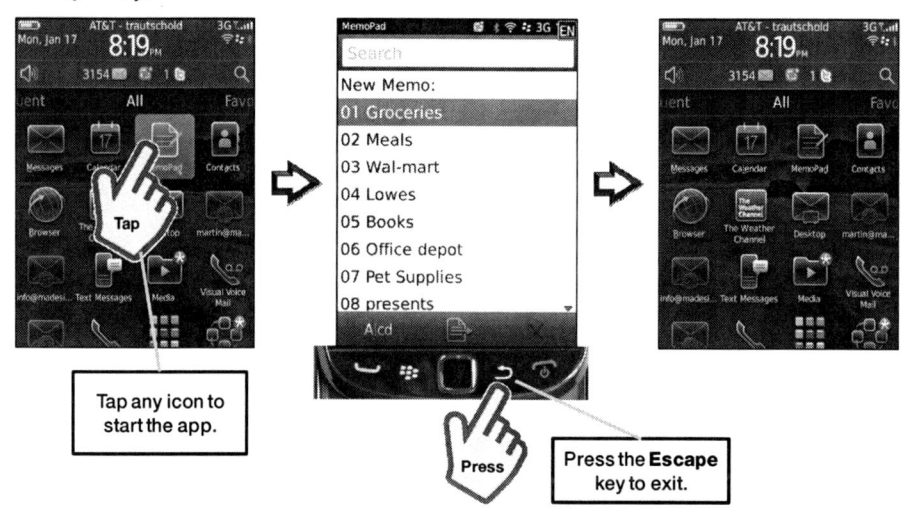

Figure 8. How to start and exit apps.

Opening Folders

You will notice some icons look like folders. You can use these folders to organize sets of icons into logical groups. There will be a few folders already on your Torch such as Applications, Media and Instant Messaging, but you can create more folders, as shown in Chapter 7: "Personalize Your Torch." Tap any folder to open it and press the **Escape** key to back out and close the folder (see Figure 9).

View Video Tutorials and Free Tips at www.MadeSimpleLearning.com   13

BlackBerry Torch Made Simple

Tap the **Applications** folder to open it.

Press the **Escape** key to exit the folder.

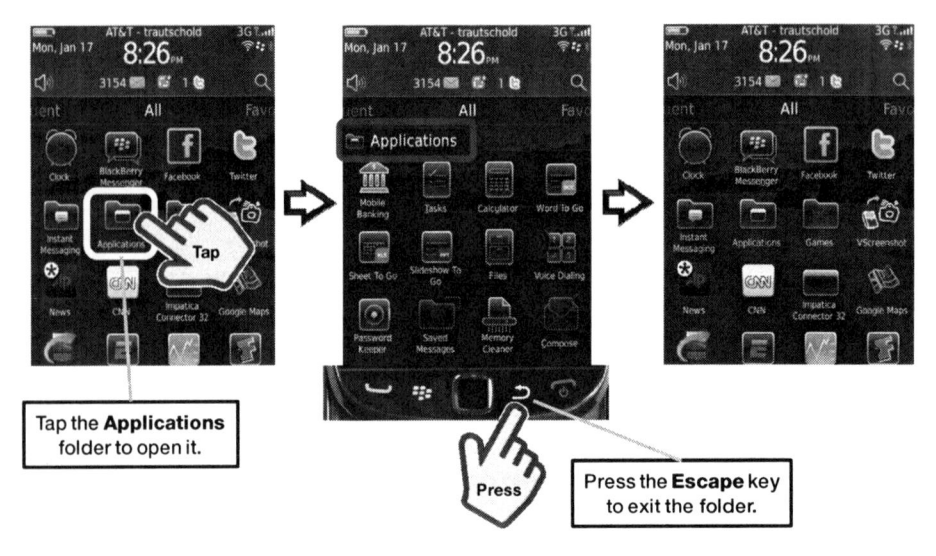

Figure 9. Opening and closing folders where you store and organize icons.

> **NOTE: Icon versus App** --In this book we use both icon and app. We thought we should clarify here. The icon is simply the picture on the **Home** screen that starts the app. For example, you click on the **Messages** icon to start the **Messages** app.

Two Types of Menus: Full and Short

You will see two types of menus on your Torch (see Figure 10). The **Short Menu** that pops up in the middle of the screen after you press and hold an item. The **Full Menu** appears when you press the **Menu** key.

Press and hold any item to see the short pop-up menu.

Short Pop-Up Menu

Full Menu

Press the **Menu** key to see the full menu.

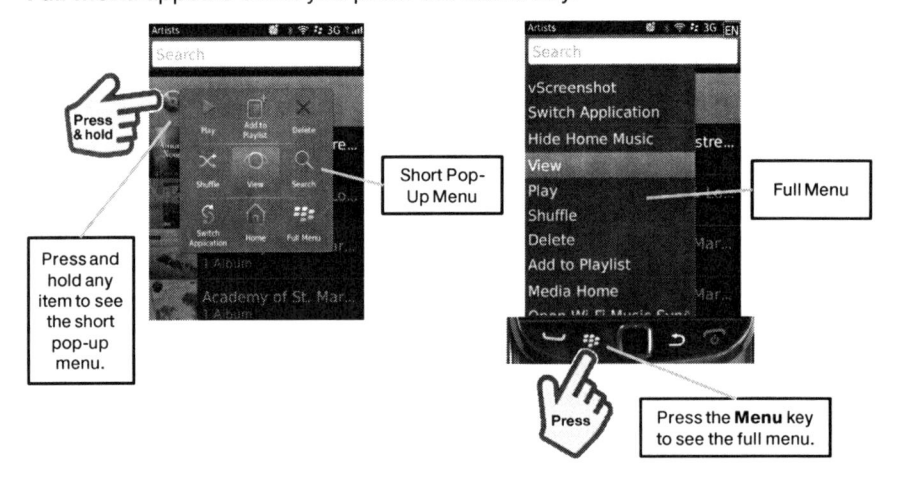

Figure 10. Short Menus and Full Menus.

14

QUICK START GUIDE

Swipe Gesture: Move to Next Day, Picture, Email

The Torch gives you the ability to swipe through items such as screens of icons, days in your calendar, email messages or pictures. To swipe, gently touch and slide your finger, and the app will move to the next/previous screen (calendar day, email, etc.).

Scrolling Up or Down

If you want to scroll, simply place your finger on the screen and drag it up or down on the screen. You can swipe it up or down quickly to move faster.

Scrolling Menus

You can use the same scroll movement to view more menu items off the bottom or top of the screen (see Figure 11).

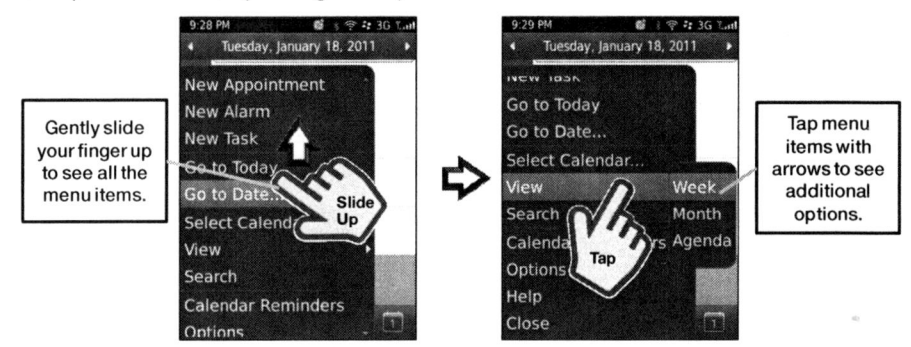

Figure 11. Scrolling up or down menus and showing additional menu levels.

Showing Or Hiding Controls

Also, when listening to music or watching a video, the playback controls will disappear after a second or so. To bring them back, simply tap the screen (see Figure 12). If you tap the screen again, you can hide them.

View Video Tutorials and Free Tips at www.MadeSimpleLearning.com

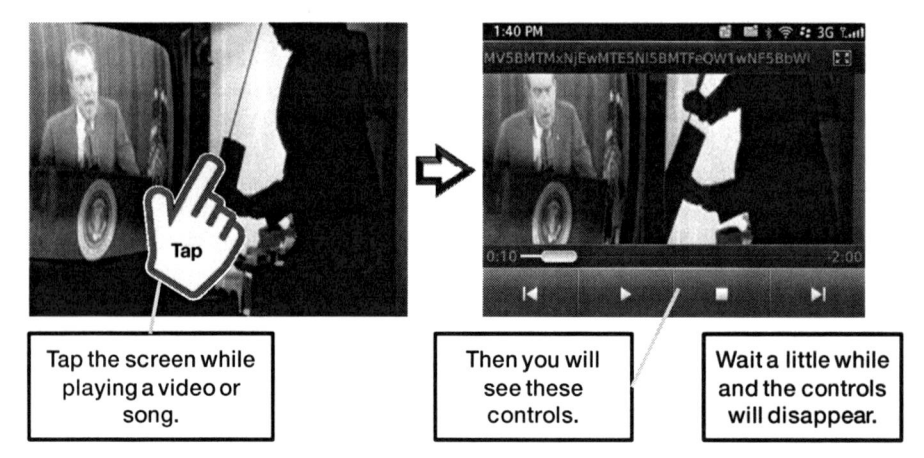

Figure 12. Tap the screen to view video or music controls.

Save Time with the Space key

You can use the **space** key to save time when you are typing email addresses, web addresses, and typing the ends of sentences. You can use the **space** key to simplify typing in an email address. For instance, typing:

susan **space** company **space** com generates susan@company.com

Similarly, to get the "dots" in a web address, typing:

www **space** google **space** com yields www.google.com

In addition, pressing the **space** key twice will enter an automatic period and make the next letter you type uppercase.

Using Universal Search or Application Shortcuts

You can access either the **Universal Search** or **Applications Shortcuts** by typing letters from your **Home** screen. We describe both features below and give you much more detail on Shortcuts of all sorts in Part IV of this book.

The Switch to Select Search or Shortcuts

By default, typing on the **Home** screen will start the **Universal Search**. To change this to **Application Shortcuts**, follow these steps:

1. Press the **Menu** key from the **Home** screen and select **Options**.

2. Change the Launch by Typing to Application Shortcuts.

3. Press the **Menu** key and **Save**.

Universal Search

Once you have enabled **Universal Search** in the **Launch by Typing** part of the options shown above, you can simply type your search while viewing the **Home** screen without having to tap the **Search** icon first.

After you type a few letters such as the name of a contact, app icon, or even a web search string, you should see these letters appear in the Search window and the icons below the search bar change as you type more letters. Tap any of the icons to select those search results.

Check out the "Universal Search" section of Chapter 6 for more details.

Application Shortcuts ("Home Screen Hot Keys")

If you select Application Shortcuts in the Launch by Typing area of the Home screen options, then you can start many of the popular icons by pressing a single letter on the keyboard.

View Video Tutorials and Free Tips at www.MadeSimpleLearning.com   17

Some of the more common shortcuts are: **B = Browser, M = Messages, L = Calendar, C = Contacts, N = BlackBerry Messenger,** and **D = MemoPad.**

Check out the "Home Screen Hot Keys" section of Part IV of this book for a complete list.

Copying, Pasting, and Multitasking

A few very useful tricks on your Torch are copying, pasting, and multitasking (i.e., jumping between applications—also referred to as application switching).

You may find times when you want to cut/copy and paste text from one app into another. One example might be copying directions for a trip from an email message into a calendar event so that you have them available when you need them (see Figure 13).

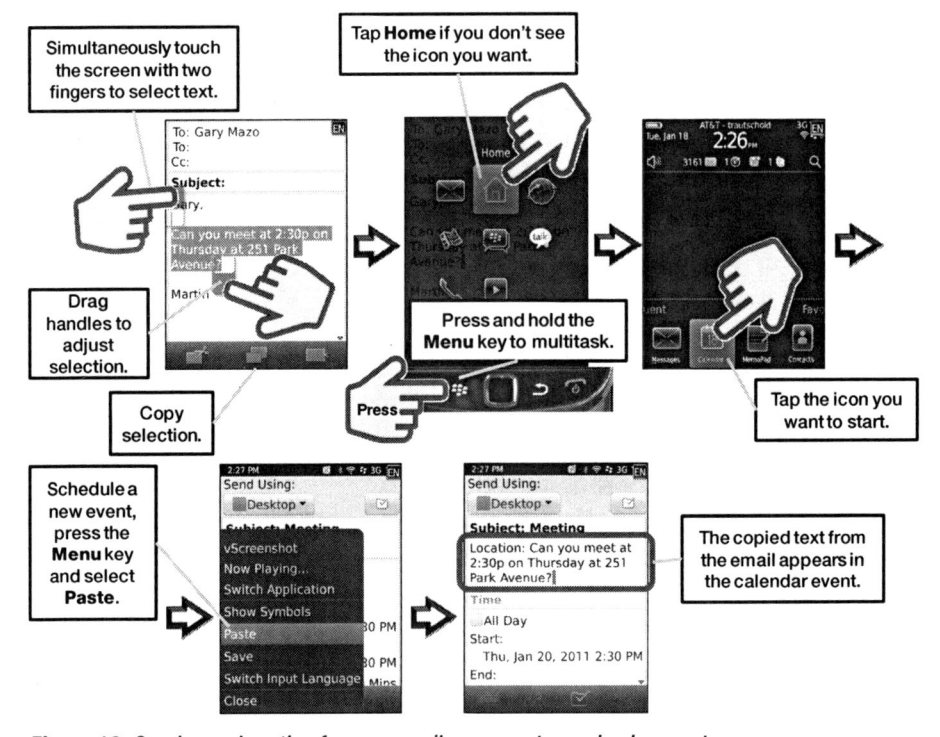

Figure 13. Copying and pasting from an email message to a calendar event.

Mastering Your Keyboards

Typing on your Torch is quite flexible. With three virtual keyboards on the screen and the slide out physical keyboard, you should be able to find exactly what you need at any time. If you are just typing a very quick reply, or are very comfortable typing on the screen, you might prefer the on-screen keyboards. If

QUICK START GUIDE

you prefer the feedback you get from typing on a physical keyboard, just slide it out and use it.

Slide Out Physical Keyboard

Probably the easiest keyboard to use is the physical one – just slide it out the bottom of the BlackBerry and start typing. Figure 14 shows a few special keys an how to use them on your physical keyboard. Chapter 6: "Typing, AutoText, Spelling, Search and Help" shows you more typing tips.

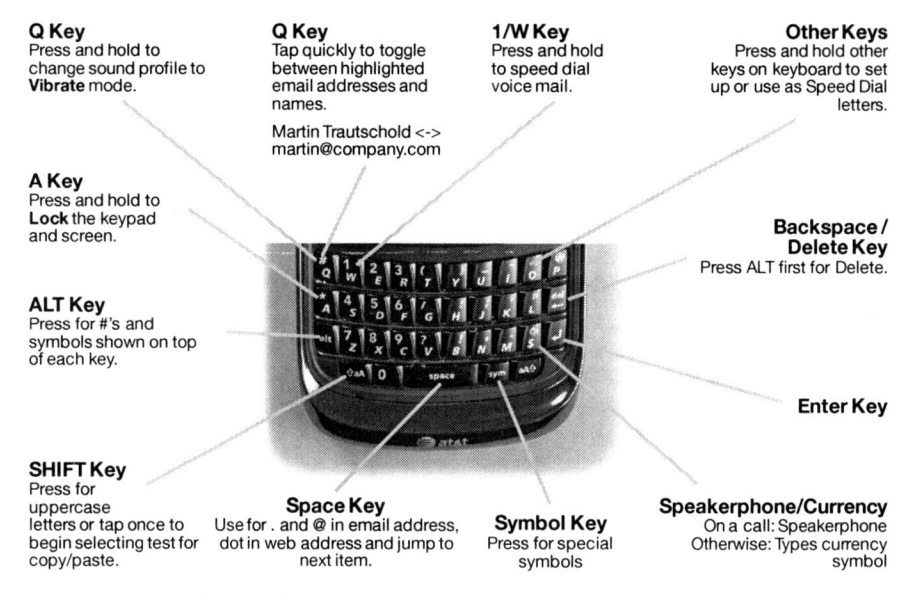

Q Key
Press and hold to change sound profile to **Vibrate** mode.

Q Key
Tap quickly to toggle between highlighted email addresses and names.
Martin Trautschold <-> martin@company.com

1/W Key
Press and hold to speed dial voice mail.

Other Keys
Press and hold other keys on keyboard to set up or use as Speed Dial letters.

A Key
Press and hold to **Lock** the keypad and screen.

ALT Key
Press for #'s and symbols shown on top of each key.

Backspace / Delete Key
Press ALT first for Delete.

Enter Key

SHIFT Key
Press for uppercase letters or tap once to begin selecting test for copy/paste.

Space Key
Use for . and @ in email address, dot in web address and jump to next item.

Symbol Key
Press for special symbols

Speakerphone/Currency
On a call: Speakerphone
Otherwise: Types currency symbol

Figure 14. Tips and Tricks with the slide out physical keyboard.

Showing and Hiding the On-Screen Virtual Keyboards

One thing you will quickly figure out is that when your physical keyboard is open, you cannot access your on-screen virtual keyboards. This is by design. It can be a little disconcerting when you are browsing the web and turn your BlackBerry sideways (landscape) without closing the slide out keyboard. In order to see the on-screen keyboard, simply close the slide out one.

Once you close the slide out keyboard, then you will be able to see all your on-screen keyboards.

If you want to see the onscreen keyboard when it's not visible and the physical keyboard is closed, tap the **Keyboard** soft key (usually the bottom left soft key).

To hide the on-screen keyboard; simply swipe your finger down from the top of the virtual keyboard (see Figure 15). You can also show and hide the on-screen keyboards from the menus.

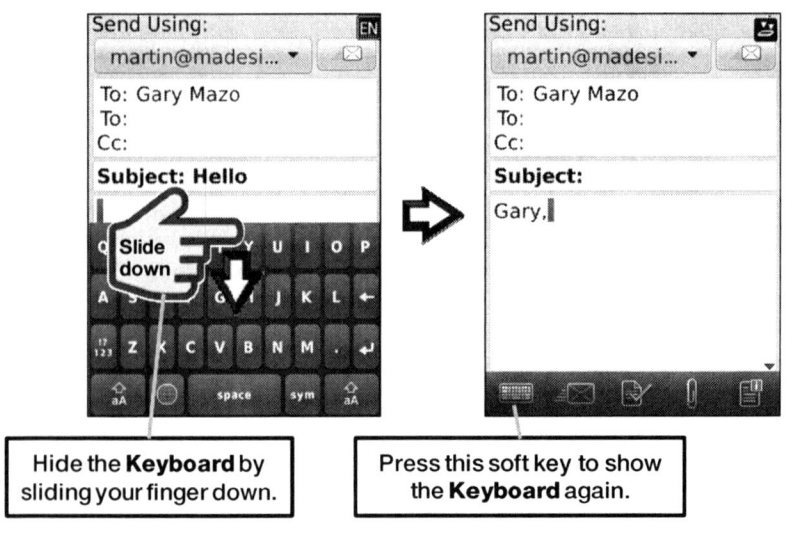

Figure 15. Hiding and showing the virtual keyboard

Choose From Three On-Screen Keyboards

You have three types of virtual keyboards from which to choose on your Torch. Use the menus to switch between the various keyboards.

> **NOTE:** In order to see the **SureType** or **MultiTap** keyboards, you first have to press the **Menu** key and select **Enable Reduced Keyboards** from the menu.

- The **Full** keyboard gives you a single letter per key and is available when you hold your Torch in both vertical and horizontal orientations.

- The **SureType** keyboard has larger keys with two letters per key and is only available in portrait mode.

- The **MultiTap** keyboard, also only available in portrait mode looks like an old-fashioned mobile phone keyboard. There are large numbers and three to four letters on each key, and you press the key one, two, three, or four times for each letter (see Figure 16).

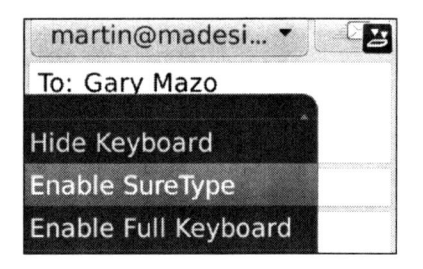

20

The **Full** keyboard.	The **SureType** keyboard.	The **MultiTap** keyboard.

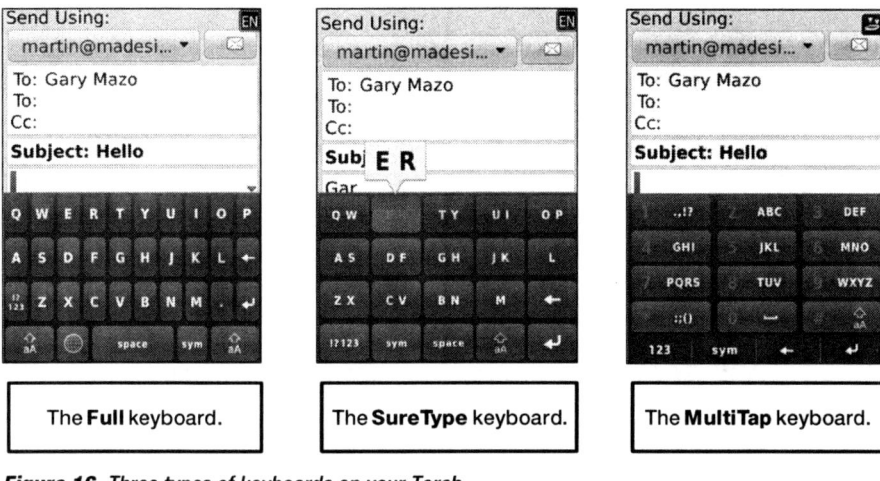

Figure 16. *Three types of keyboards on your Torch*

Working with the Wireless Network

Since most of the functions on your Torch work only when you are connected to the internet (**Messages**, **Browser**, **App World**, **Facebook**, etc.), you need to know when you're connected. Understanding how to read the status bar can save you time and frustration. You will also want to know how to quickly turn off your wireless radio or other radios when you get on an airplane.

Reading Your Wireless Network Status

In Table 1, we show you how to read the status icons at the top of your Torch screen so you can save time and stay connected. Check marks indicate that a connection is active and an X indicates an inactive connection. Check out Chapter 24: "Troubleshooting" for help with getting this working.

Table 1. Reading Your Wireless Network Status

In the Upper-Right Corner, If You See Letters and Symbols	Email and Web	Phone Calls & SMS Text	Speed of Data Connection
3G (3G with logo)	✓ = Active	✓	High
EDGE, GPRS	✓	✓	Medium, Slow
(with any letters shown)	✓	✓	High
1X, EDGE, GPRS, GSM, or 3G	✗ = Inactive	✓	None
OFF, X	✗	✗	None

Wireless Network Signal Strength

The following list shows the various wireless signal strength icons that will appear (signal strength varies between one and five bars):

Strong signal:	📶	4-5 bars is a strong signal – fast downloads, web browsing and email should, good voice quality
Weak signal:	📶	1-2 bars is a weak signal – all data functions will be slower and possibly voice calls may be broken up.
No signal:	X	Your radio is on, but no wireless signal is available.
Radio off:	OFF	Your radio is turned off. From your **Home** screen, tap the top status bar where your time is shown and select **Turn All Connections On.**

QUICK START GUIDE

App Reference Tables

This section groups the apps on your Torch, as well as other apps you can download, into handy reference tables. Each table gives you a brief description of the app and tells where to find more information in this book.

Getting Set Up

In Table 2 are some apps and quick links to help you get your email, Bluetooth, contacts, calendar, and more loaded onto your device. You can even boost the memory by adding a memory card, which is important to install if you'd like to have more pictures, videos, and music on your Torch.

Table 2. Getting Set Up

To Do This	Use This	Where to Learn More
Set up email, Bluetooth, date/time, fonts, Wi-Fi, and personalize.	**Setup App**	Chapter 1: Getting Started Chapter 7: Personalize
Set up or change your internet email, change email signatures	**Setup App > Email Accounts**	Chapter 1: Getting Started.
Connect to Bluetooth devices	**Setup App > Bluetooth**	Chapter 1: Getting Started.
Connect to Wi-Fi networks	**Setup App > Wi-Fi**	Chapter 1: Getting Started.
Set up Social Networks (Facebook, Twitter, more)	**Setup App > Social Networking**	Chapter 1: Getting Started.
Share addresses, calendar, tasks, and notes with your computer	**BlackBerry Desktop Software (for Windows or Apple Mac)**	Windows: Chapter 2. Mac: Chapter 4.

View Video Tutorials and Free Tips at www.MadeSimpleLearning.com   23

Add memory to store your music, videos, and pictures		Media card Image Courtesy SanDisk Corp.	Chapter 1: Getting Started.
Load up your music, pictures, and videos		USB Drive Mode Desktop Software Media Sync	USB Mode: Chapter 1. Windows: Chapter 3. Mac: Chapter 5.

Staying in Touch

Getting familiar with the apps in Table 3 will help you stay in touch with your friends and colleagues. Whether you prefer calling, emailing, texting, or using instant messaging, your BlackBerry has many options. Use the browser to stay up to date with the latest happenings on the Web.

Table 3. Staying in Touch

To Do This	Use This		Where to Learn More
Read and reply to Email		Messages	Chapter 9: Email.
AP News Reader – Stay Informed		AP News	Chapter 20: App World.
Keep up on the Latest Weather		The Weather Channel	Chapter 20: App World.
BlackBerry Messenger		BlackBerry Messenger ("BBM")	Chapter 13: BlackBerry Messenger
Send and read Text and Picture messages		Text Messages	Chapter 12: Text and MMS Messaging.

24

QUICK START GUIDE

To Do This		Use This	Where to Learn More
Get on the internet/browse the Web		**Browser**	Chapter 19: Web Browser.
Call voicemail		**The 1 key**	Chapter 8: Phone and Voice Dialing.
Start a call, dial by name, and view call logs		**Green Phone key**	Chapter 8: Phone and Voice Dialing.
Dial by voice		**Voice Dialing**	Chapter 8: Phone and Voice Dialing.
Turn off the radio (important when flying on an airplane)		**Manage Connections**	Quick Start Guide.
Maximize your battery life to talk, message, and play more		**Battery Life Tips**	Chapter 1: Getting Started.

Staying Organized

From organizing and finding your contacts to managing your calendar, taking written or voice notes, and calculating a tip using your built-in calculator, your BlackBerry can help you do it all. See the apps in Table 4.

Table 4. Staying Organized

To Do This	Use This		Where to Learn More
Manage your contact names and numbers		**Contacts**	Chapter 10: Contacts.
Manage your calendar		**Calendar**	Chapter 11: Calendar.

View Video Tutorials and Free Tips at www.MadeSimpleLearning.com   25

Set a wakeup alarm, or use a countdown timer or stopwatch		**Clock**	Chapter 22: Utilities.
Store all your important passwords		**Password Keeper**	Chapter 22: Utilities.
Find lost names, email, calendar entries, and more		**Universal Search**	Chapter 6: Typing & Search.

Being Productive

Sometimes you need to get work done on your Torch. Use the apps shown in Table 5 to get things done on your BlackBerry.

Table 5. Being Productive

To Do This	Use This		Where to Learn More
Manage your to-do list		**Tasks**	Chapter 15: Tasks and Memos.
Find Things, Get Directions, See Traffic		**Google Maps**	Chapter 21: Maps.
Take notes, store your grocery list, and more		**MemoPad**	Chapter 15: Tasks and Memos.
Leave yourself a voice note		**Voice Note Recorder**	Chapter 22: Other Applications

QUICK START GUIDE

View and edit Microsoft Office Word, Excel, and PowerPoint		**Word to Go, Sheet to Go, Slideshow to Go**	Chapter 9: Email
Use the built-in text-based help		**Help**	Chapter 6: Typing & Search.
Calculate your MPG (miles per gallon) or meal tips, or convert units		**Calculator**	Chapter 22: Other Applications.

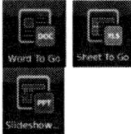

Being Entertained

Use the apps in Table 6 to have fun with your BlackBerry Torch.

Table 6. Being Entertained

To Do This	Use This		Where to Learn More
Quickly get to all your music		**Music**	Chapter 16: Your Music Player.
Play videos		**Videos**	Chapter 18: Fun with Videos.
View your pictures		**Pictures**	Chapter 17: Snapping Pictures.
Enjoy podcasts		**Podcasts**	Chapter 22: Other Applications
Take a photo		**Camera**	Chapter 17: Snapping Pictures.

View Video Tutorials and Free Tips at www.MadeSimpleLearning.com

Capture video	**Video Camera**	Chapter 18: Fun with Videos.
Play a game	**Games folder**	Chapter 20: App World.
Enjoy free internet radio with your own favorite music	**Slacker**	Chapter 16: Your Music Player.
Enjoy free internet radio with your own favorite music	**Pandora**	Chapter 16: Your Music Player.

Networking Socially

Connect and stay up to date with friends, colleagues, and professional networks using the social networking tools on your Torch (see Table 7).

Table 7. Networking Socially

To Do This	**Use This**	**Where to Learn More**
Connect with Facebook friends	**Facebook**	Chapter 14: Social Networking and Social Feeds.
Setup and View RSS Feeds	**Social Feeds**	Chapter 14: Social Networking and Social Feeds.
Follow people and tweet using Twitter	**Twitter**	Chapter 14: Social Networking and Social Feeds.
Connect with colleagues	**LinkedIn**	Chapter 14: Social Networking and Social Feeds.

QUICK START GUIDE

Personalizing Your Torch

Use the apps in Table 8 to personalize the look and feel of your Torch.

Table 8. Personalizing Your Torch

To Do This	Use This		Where to Learn More
Mute or Vibrate your Phone and other Sounds	(Normal) Sound and A...	Sounds	Chapter 7: Personalize.
Change your background home screen picture	Wallpaper	Setup > Wallpaper	Chapter 7: Personalize.
Change your font size	A Fonts	Setup > Fonts	Chapter 7: Personalize.
Change your programmable **Convenience** key	Convenience Key	Setup > Convenience Key	Chapter 7: Personalize.

Adding and Removing Apps

Use the apps shown in Table 9 to add apps to your BlackBerry.

Table 9. Adding and Removing Apps

To Do This	Use This		Where to Learn More
Add and remove apps on your Torch	App World	**BlackBerry App World**	Chapter 20: App World.
Find all the apps you've downloaded	Downloads	**Downloads folder**	Chapter 20: App World.
Manage Applications		**Options > Manage Applications**	Chapter 23: Troubleshooting

View Video Tutorials and Free Tips at www.MadeSimpleLearning.com

Part 2:

Introduction

Welcome to your new BlackBerry Torch. In this section we will introduce you to how the book is organized and where to find useful information. Also at the end of this section we include our "Day in the Life" section, where we give you some scenarios to describe how you can use your Torch for work and play. We even show you how to find some great tips and tricks sent right to your Torch as well as how to view Video Tutorials either on your computer or load up videos on your Torch.

Introduction

Congratulations on Your Torch!

The BlackBerry Torch is a unique offering for BlackBerry with both a touch screen and the famous BlackBerry keyboard. The screen sensitivity, processor, and memory have all been upgraded and enhanced for a better and faster experience. With this guide, we hope to help you tap the power of this great smartphone.

In 2001 the BlackBerry name came into the marketplace. One popular story is that the keys on the very early devices looked like seeds to some of the creators. The creators looked at various seeded fruit and decided that "BlackBerry" would be a friendly and inviting name for the device.

However, the origin of the name is less important than understanding the philosophy. One thing to understand from the outset is that this is not just a phone. The BlackBerry is a computer—a sophisticated messaging device that does a bunch of things at the same time—and a phone.

The BlackBerry takes most of the major needs that we have—information, communication, constant contact, media, accessibility, and more—and puts them in one device that can do just about everything.

Unique Features on the BlackBerry Torch

Your BlackBerry Torch has many shared features with the BlackBerry family and some unique features as well. Following are some of the key features (some of which are described in greater detail in the following subsections):

- Unique touch screen design
- BlackBerry OS 6.0
- Built-in social networking
- BlackBerry keyboard
- Camera (5.0 megapixel) with flash and auto focus
- Media player (for pictures, video, and audio)
- Video recording
- Memory Expansion with MicroSD Media Card
- 3G/Wi-Fi capability

Touch Screen

Having a touch screen is nothing new these days. What separates the Torch from the rest is that RIM has made a touch screen that supports different types of gestures and touches.

Built-In Social Networking

Use your Torch for accessing all your Social Feeds; your Facebook page, Twitter, YouTube, Flickr and virtually every IM program. Keep in touch with everyone in the ways you like most.

BlackBerry Keyboard

The BlackBerry design allows you to quickly type your messages using the familiar BlackBerry keyboard as well as an on-screen keyboard which is very handy when in landscape mode.

Expansion Memory with Media Card

Your Torch may come bundled with a 4 or 8 GB MicroSD card (which can be expanded further) that allows you to store large numbers of pictures, songs, and videos.

Always on, Always Connected

Perhaps nothing sums up what a BlackBerry can do better than the fact that it always keeps you connected (which virtually no other device can do). Your BlackBerry will push your email (up to ten different accounts) right into your hand—all the time, day and night. (Now, you can put limitations on that—but the reason it is sometimes called a CrackBerry is that once you experience this, you might not want to limit it at all!)

You may be used to turning your old cell phone on and off and only checking it to see if you missed a call. Your BlackBerry stays on all the time—even when it is in standby mode, it is still on. You will find yourself looking at it not only to see if you missed a call, but to see what emails have come in, who is sending you an instant message, who just posted a note on your Facebook page, what time your next appointment is, what you need to pick up at the store, and so on. In short, your life can be managed from your BlackBerry.

Things This Powerful Are Not Always Easy to Grasp—at First

Your BlackBerry is grouped into the category of things called "smartphones" by many. A BlackBerry, however, is really . . . a BlackBerry—more than a smartphone by any other name, because it does so much so well.

The pros of that are clear—in your hand is probably the most capable and most complicated technology available today.

The cons are that the BlackBerry, especially a touch screen model, is not always intuitive at first use. For some, this is frustrating—they want their BlackBerry to do what their old phone did—in just the same way.

Like a computer, your BlackBerry has a unique operating system (OS) that is proprietary and only found on other BlackBerry devices. Your Torch takes the OS one step further and adds an innovative new touch technology that takes a little getting used to.

Take your time—this is not a device to pick up for an hour and then throw down in frustration—there is a lot to learn here. Remember when you got your first Windows or Mac computer? When we did, we didn't know what a window was, let alone where to find things or how to type a letter—it took time.

With the BlackBerry—like most other things in our lives—the more we invest in learning, the more we will ultimately get out of it.

Getting the Most out of This Book

You can read this book cover to cover, but you can also peruse it in a modular fashion, by chapter or topic. Maybe you just want to check out BlackBerry App World, try the web browser, and get set up with your email and contacts; or you might just want to load up your music. You can do all this and more with our book.

You will soon realize that your Torch is a very powerful device. There are, however, many secrets locked inside, which we help you unlock throughout this book.

Take your time—this book can help you on your way to learn how to best use, work, and have fun with your new Torch. Use this book to get up to speed and learn all the best tips and tricks more quickly.

Remember that devices this powerful are not always easy to grasp—at first. You will get the most out of your Torch if you can read a section and then try out what you read. For most people, reading about and then doing an activity gives us a much higher retention rate than simply reading alone. So, in order to learn and remember what you learn, we recommend the following: *read a little, try a little on your Torch, and repeat!*

How This Book Is Organized

Knowing how this book is organized will help you more quickly locate things important to you. Here we show you the main organization of this book. Remember to take advantage of our abridged table of contents, detailed table of contents, and comprehensive index to help you quickly pinpoint items of interest.

Part 1: Quick Start Guide

The Quick Start Guide covers the following topics:

Learning your way around: Learn about the keys, buttons, and ports on your Torch, as well as how to get inside and change the battery, SIM card, and memory card. Then learn many time-saving tips about getting around quickly, as well as how to multitask.

Touch screen gestures: We show you all the best tips and tricks for getting around the touch screen. We cover how to start apps and get into folders, as well as some time-saving techniques.

Using Universal Search or Application Shortcuts: Learn how to switch between the search or shortcut hot keys when you start typing from your **Home** screen – both features can be very useful.

Copying, pasting, and multitasking: We cover the very useful copy-and-paste function and how to jump between apps on your Torch.

Mastering your keyboards: We explore the virtual (on-screen) and physical (slide-out) keyboards and provide some useful tips about how to get more out of each one.

Working with the wireless network: Learn how to read the letters, numbers, and symbols at the top of your Torch screen so you know when you can make phone calls, send SMS text messages, send and receive email, and browse the Web. Also learn how to handle your Torch on an airplane.

App references: This section includes a series of reference tables that allow you to quickly peruse the icons or apps in particular categories. Get a thumbnail of what all the apps do on your Torch and chapters to jump right to the details of how to get the most out of each app in this book.

Part 2: Introduction

Here we describe the layout of the book.

A Full Day With Your Torch

Located right after this Introduction chapter, this is an excellent section full of examples of the myriad ways you can use your Torch along with cross-reference chapters. So if you see something you want to learn, simply thumb to that chapter and learn about it—all in just a few minutes.

Part 3: You and Your Torch

This is the meat of the book, organized in easy-to-understand chapters packed with loads of pictures to guide you every step of the way.

Part 4: Keyboard Shortcuts

This is where we have gathered all the best keyboard shortcuts to help you speed your day with your BlackBerry Torch.

Quickly Locating Notes, Tips, and Cautions

If you flip through this book, you can instantly see these items based on their formatting. For example, if you wanted to find all the calendar tips, you would flip to the Calendar chapter –and quickly find them.

> **NOTE:** Notes, tips, and cautions are all formatted like this, with a gray background, to help you see them more quickly.

Email Tips

Check out the authors' web site at www.madesimplelearning.com for a series of very useful BlackBerry tips and tricks. On this site, we have taken a selection of great tips out of this book and even added a few new ones. Click the Free Tips section and register for your tips in order to receive a tip right in your Torch inbox about once a week. Learning in small chunks is a great way to master your Torch.

Video Tutorials

The authors also offer video tutorials on their web site that cover much of the material presented in this book and some new things. Whenever there is a video tutorial available for the material covered in the book we add the graphic shown here.

How to get the videos? Visit www.madesimplelearning.com from your computer and click on the BlackBerry Torch.

Some of the videos are free, for example the entire set of videos showing you how to use **Desktop Software** for Windows.

In addition, there are over 80 video tutorials that can be purchased to show you how to use your Torch.

You can choose to watch on your computer or load these videos up and take with you to view on your BlackBerry, anytime, anywhere.

A Full Day With Your Torch

Here we give you a brief glimpse into the possibilities of what you can do you're your BlackBerry. With all the apps, there are so many things you can do.

Time	What I do with my BlackBerry Torch...	Learn more...
6:00 AM	My BlackBerry Alarm wakes me.	Alarm Clock – Ch.23
6:10 AM	I connect my BlackBerry up to my stereo Bluetooth docking station in the bathroom to listen to my music free Pandora or Slacker Internet radio.	Play Music – Ch. 16 Bluetooth – Ch. 1
6:45 AM	At the breakfast table, I check my emails to get a jump on my day before I get to the office. I also browse the New York Times and other web sites from apps I've installed.	Email - Ch. 9 App World - Ch. 20
6:55 AM	I check traffic using Google Maps and find out I need to take my alternate route to work today.	Traffic – Ch. 21
7:00 AM	I do a little social networking and check out the latest about my friends on Facebook, my colleagues on LinkedIn, and post a couple of Tweets on Twitter.	Facebook – Ch. 14 LinkedIn – Ch. 14 Twitter – Ch. 14
8:30 AM	Waiting for a meeting, I snap a picture and send a Multi-Media Message (MMS) and post the picture on Facebook.	MMS – Ch. 12 Facebook – Ch. 14
9:00 AM	I take pictures of my new client's art work with my BlackBerry to send to a prospective buyer. I select a few and email the best snapshots.	Camera - Ch. 17 Email Pictures - Ch. 9
10:00 AM	I find out I have to fly from New York to Vancouver tonight to visit a client - so I check travel web sites on my BlackBerry Browser and book my airplane and rental car.	Web Browser – Ch. 19

Time	Activity	Reference
10:05 AM	I received an email message from my client with the location of our meeting in Vancouver. I map the address and then add this as a Favorite in Google Maps.	Email – Ch. 9 Google Maps – Ch. 21
10:10 AM	I need to find hotels around my Client's address so I just type "hotels" into Google Maps search window and quickly find all hotels in the area. I call to reserve a room right from Google Maps.	Search Business – Ch. 21
10:25 AM	I add a bookmark for Vancouver, BC, Canada on www.weather.com to my Bookmark list and see that it will be really cool and rainy. Now I know how I need to pack.	Add Bookmark –Ch. 19
11:00 AM	An alert pops up on my BlackBerry reminding me to deal with the email message I received yesterday about the meeting. (Flag for Follow-up feature) I then add the event to my Calendar.	Follow-up – Ch. 9 Calendar – Ch. 11
11:05 AM	I look up the phone number for the restaurant and call them using Google Maps to reserve a table.	Google Maps – Ch. 21
11:07 AM	I reply to the text message and say "Sure – we're all set for 1PM - see you there!"	SMS Text – Ch. 12
1:30 PM	While at lunch, I show my spouse my cool new pictures I took on the BlackBerry in the Pictures app.	Pictures – Ch. 17
2:00 PM	Since I am traveling internationally, I check out the Travel Tips to make sure I'm not surprised by a huge data or voice roaming bill when I return.	Travel Tips – Ch. 21
3:45 PM	The buyer emails me back and says she wants to buy some of the art pieces. I click on her phone number in her email signature and give her a call right away. I then conference call the artist and we seal the deal right there. I then make sure to add her to my Contacts app.	Call Number – Ch. 8 Contacts – Ch. 10
4:30 PM	Another quick check of Google Maps for traffic for the commute home and again for the trip to the airport. Armed with the latest information, I'm set.	Google Traffic – Ch. 21
5:00 PM	I turn on Pandora Internet radio and listen to my favorite Bruce Springsteen station on the commute home. It sounds great because I stream the music through my Bluetooth car stereo system.	Pandora – Ch. 16 Bluetooth – Ch. 1

8:00 PM	Sitting at the gate in the airport, I browse BlackBerry App World for a fun game or two to play on the airplane.	App World – Ch. 20
8:50 PM	I settle into my seat and make sure to turn off my wireless radio so I can use my BlackBerry without causing trouble with the pilots. I rest assured that my time zone will update automatically when I land on the East Coast with the auto-update time zone feature.	Airplane Mode - Ch. 1 Time zone - Ch. 7
9:00 PM	Sitting back in my seat, I plug in my headset and play some of my new games. Another great day with my BlackBerry Torch!	App World - Ch. 20

Part 3:

You and Your Torch

This is the heart of *BlackBerry Torch Simple*. In this section, you'll find clearly labeled chapters—each explaining the key features of your Torch. You'll see that most chapters focus on an individual app or a specific type of application. Many of the chapters discuss applications that come with your Torch, but we also include some fun and useful apps you can download from BlackBerry App World. Sure, the Torch can help you get work done, but it's for a whole lot more, too. We finish with some handy troubleshooting tips that can help if your Torch isn't working quite right.

Chapter 1: Getting Started

This chapter is about setting up your BlackBerry. We cover everything from email setup (both Internet and Enterprise), connecting to Bluetooth and Wi-Fi, security tips and how to prolong your battery life.

Your BlackBerry supports a wide array of email accounts including (Google, Comcast, Yahoo, MSN, or more generic POP3/IMAP accounts as well as Microsoft Exchange and Lotus Domino corporate accounts). The various types of accounts will have different wireless sync capapbilites for contacts and calendars in addition to email. We show you all the differences and help you get each type set up in this chapter.

Sometimes email setup can cause problems; we show you how to fix some common errors and make sure you receive all your email on both your computer and your BlackBerry. While you can adjust email setup from your BlackBerry, you may also want to adjust your email settings from your computer; we cover that in this chapter.

If your BlackBerry is tied to a BES (BlackBerry Enterprise Server), we show you how to get connected or 'activated' on the server. And, more recently, RIM, BlackBerry's maker, announced a free version of the BES software. This is a fantastic value if your organization uses Microsoft Exchange and would like to take advantage of the BES features without additional software licensing and, in most cases, without additional hardware costs.

> **CAUTION:** Before you read any further, or use the **Setup** App or anything else, please take a few minutes to check out the Quick Start Guide earlier in this book, if you haven't already. It is meant to help you find lots of useful things in this book, as well as give you some great beginning and advanced time-saving tips and tricks so you can get up and running quickly.

The Setup App

The first time you turn on your BlackBerry, you will most likely be placed in the **Setup** app. Most of the basic setup (email, instant messaging, social networking, Bluetooth and Wi-Fi) as well as personalization or customization can be accessed from **Setup**.

If you want to start the **Setup** app at any time, tap the **Setup** icon which is found in your **All** folder on your **Home** screen.

To see all of your icons, swipe your finger up or press the **Menu** key and select **Open Tray** then swipe up in the **All** folder.

You will be presented with screens similar to the ones that follow.

NOTE: These Setup screens may vary based on your carrier and can be updated by RIM, BlackBerry's maker, so you might see them in a different order or see some different icons.

Swipe your finger up and down to see all the icons grouped into the following categories: **Setup**, **Personalization** and **Help & Tutorials** (see Figure 1-1).

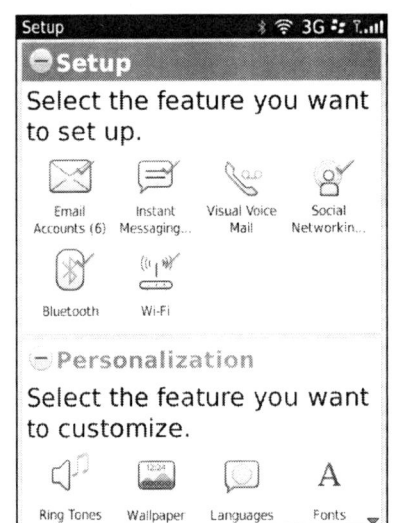

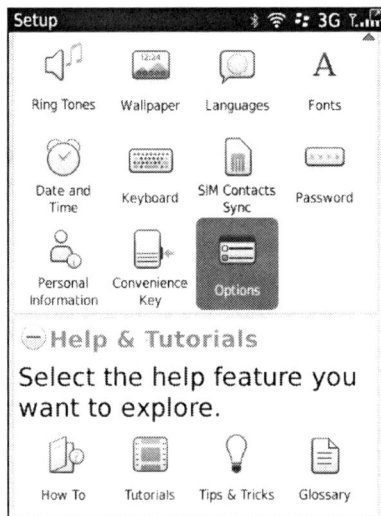

Figure 1-1. Setup app screens

We cover the many of the various set up options shown as icons in Figure 1-1 in this and subsequent chapters. Below is a handy list showing you where to find more information.

CHAPTER 1: Getting Started

Email Accounts	See "Setup Internet Email" or "Setup Enterprise Email" in this chapter.
Instant Messaging	See Chapter 12: "SMS Text and MMS" and Chapter 13: "BlackBerry Messenger"
Social Networking	See Chapter 14: "Social Networking and Social Feeds"
Bluetooth	See "Connecting With Bluetooth" in this chapter.
Wi-Fi	See "Connecting With Wi-Fi" in this chapter.
Personalization	See Chapter 7: "Personalize your Torch" and Chapter 8: "Phone and Voice Dialing."

The nice thing about the icon layout inside the **Setup** app is that you can choose to setup as little of or as much of your BlackBerry as you want at any single time. You are not forced into setting everything up at once. If you want to setup **Email** today, click the **Email Accounts** icon. You can come back into the **Setup** app at a later time and complete **Instant Messaging**, **Social Networking** or **Bluetooth** setup when you need them.

In the next sections, we assume you are already inside the **Setup** app so you can click the icon shown inside **Setup** to get started.

Setup Email Accounts

You can setup both **Internet Mail** (e.g. Gmail, Yahoo, AOL, or other) and **Enterprise Account** (e.g. your work email) accounts from this **Email Accounts** icon.

Email Accounts (6)

> **NOTE:** You need to have an enterprise email service plan from your wireless carrier to be able to setup Enterprise Accounts, skip to the "Setup Enterprise Account" section to setup your work email account.

Setup Internet Mail Accounts

As of publishing time, you can setup a maximum of 10 Internet Mail accounts on your BlackBerry using this process. If you are not able to successfully login to your Internet mail account using the steps in this section, then please see the "Troubleshooting Email Settings" section for help. Sometimes email accounts require that you enter a few advanced settings to work correctly with your BlackBerry.

View Video Tutorials and Free Tips at www.MadeSimpleLearning.com   43

1. Click the **Email Accounts** icon inside the **Setup** app.
2. If you see an option for **Internet Mail Account**, click it.
3. If you do not have an Internet Email Setup account yet, you will be asked to create one. After you have created your account, you will see the login screen shown.
4. Enter your **User name** and **Password**, click **Remember password** so you don't need to re-type your password every time.
5. Click **Continue** to login.

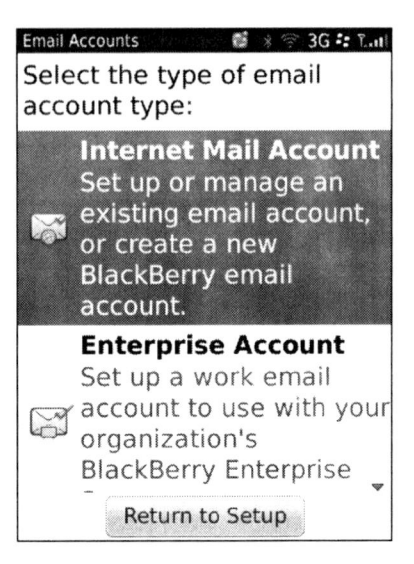

CHAPTER 1: Getting Started

6. If you see a screen similar to the one shown to the right, then you had another BlackBerry device setup previously and can simply move the email accounts to your new BlackBerry. Tap **Move** and re-enter the password for each of the accounts listed. Then you are done with email setup. Only if you want to add more accounts do you need to continue below.

7. The first time you login, or when you choose to setup another email account, you will see the screen shown to the right allowing you to setup various types of email accounts.

8. Click **Yahoo!**, **Gmail**, **AOL**, or **Windows Live** to setup one of those types of accounts. If you have another type, click **Other**.

NOTE: Only click **Create New Address** if you want to create a brand new email address that is associated only with your BlackBerry. You don't click this item if you want to integrate one of your existing email accounts.

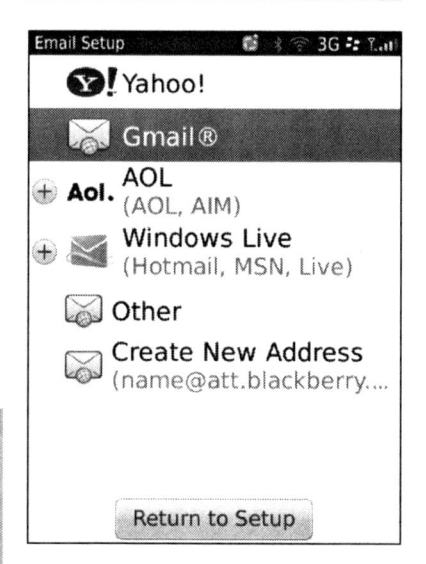

View Video Tutorials and Free Tips at www.MadeSimpleLearning.com   45

9. If you clicked **Yahoo!, Gmail, AOL, Windows Life** or **Other**, you will see a screen asking you for your **Email address** and **Password**. Click **Continue** when done.

TIP: Click the **Show Password** checkbox to see the letters typed in your password. This can be very helpful if you password contains numbers or symbols.

10. If the BlackBerry is able to successfully login to your email account you will see a screen similar to this one.

NOTE: If the BlackBerry is not able to login to your email account and you have verified your login settings, then skip to the "Troubleshooting Email Settings" section for help.

11. Click **Continue** to setup another email account or click **Return to Setup** to get back to the **Setup** app.

TIP: You can adjust your email signature and other settings by clicking the **Change Settings** button. This is the same as editing the email account which we show you in the "Edit Internet Email Accounts and Signature" section below.

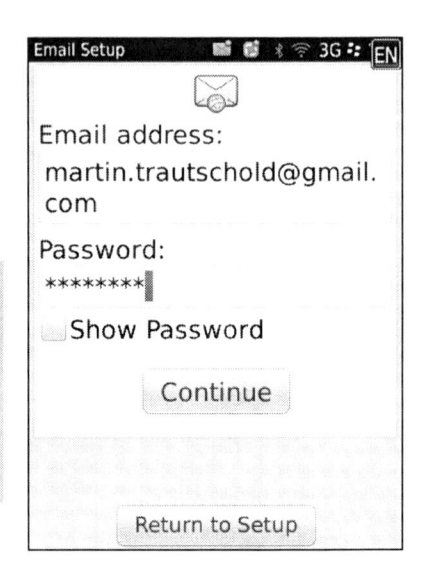

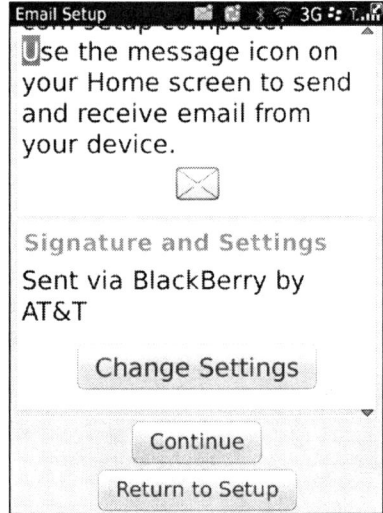

CHAPTER 1: Getting Started

12. Each email account you setup will be listed at the top of the **Email Accounts** screen as shown to the right.

13. Tap the **Set up another email account** button if you need to setup more accounts. Then repeat steps 7 – 11.

14. When you are done setting up accounts, press the **Red Phone** key to exit to your **Home** screen. You will see a new icon for each email account you just setup. Tap **Messages** to see all your accounts in one unified inbox. Email will start flowing to each account after about 20 minutes as each Activation email message shows.

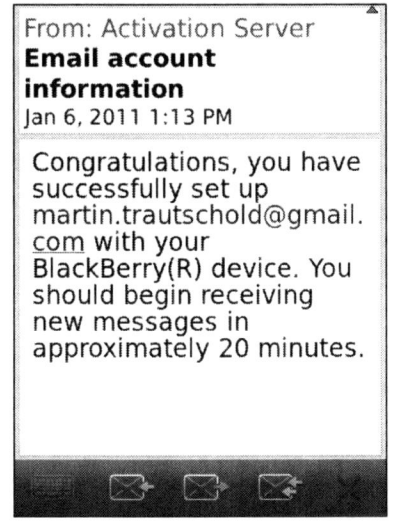

View Video Tutorials and Free Tips at www.MadeSimpleLearning.com

Hiding Extra Email Account Icons

As you set up each email account, a new icon will appear on your home screen tied to that particular account. If you like having individual icons, you can leave them alone. However, since all your email goes into your main **Messages** inbox, you can hide these icons to clean up your **Home** screen. Follow these steps to hide these extra icons:

1. Touch and hold the icon you want to hide until you see the pop-up menu as shown.
2. Select **Hide**.

If you want to get the icon back, then follow the steps to shown in the "Hiding and Showing Icons" section of Chapter 7.

TIP: Press the **space** key to get the @ (at) and . (dot) symbols whenever typing an email. For example, for sara@company.com, type, **Sara** *space* **company** *space* **com**.

Maintaining Your Internet Email Accounts and Signatures

You can add more accounts, delete or edit your email accounts including your signature and other settings on your Internet Email accounts from the **Email Accounts** icon.

CHAPTER 1: Getting Started

1. Tap the **Email Accounts** icon (and tap **Internet Mail Account**, if asked) within the **Setup** app.

2. After logging into your account, you should see all your email accounts listed as shown.

3. Tap the email account you wish to adjust to see the pop-up menu shown to the right.

   - Tap **Delete** to remove this account and confirm your selection on the next screen.

   - Tap **Edit** to make changes to the account and follow the steps below.

   - Tap **Add Email Account** to add more accounts.

4. Scroll down the page a little to the Signature section and type your new email signature.

   Your name:
   Made Simple Learning

   Signature:
   Martin Trautschold
   Made Simple Learning
   www.MadeSimpleLearning.com

5. To adjust your username or password for this email account, scroll down and tap Login information to enter the new information.

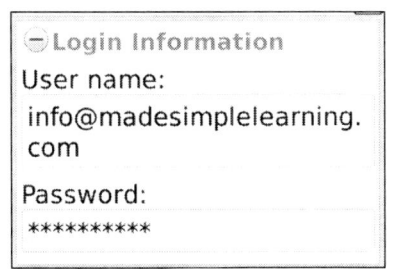

   Login Information
   User name:
   info@madesimplelearning.com
   Password:
   ****

View Video Tutorials and Free Tips at www.MadeSimpleLearning.com   49

6. To have a blind carbon copy (BCC) of every email you send from your BlackBerry sent to another email address, enter that in the Delivery Options section. This might be good for record keeping purposes.

7. If you do not want to synchronize deleted items (or Calendar and Contacts, if shown) between your BlackBerry and your main mail account, then uncheck the boxes in the Synchronization Options section.

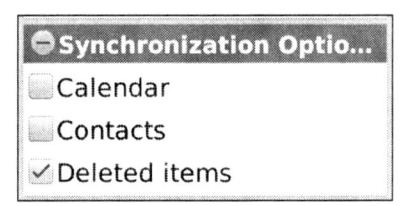

8. To force an SSL (Secure Socket Layer) connection, make sure the Use SSL box is checked in the **Advanced Options** section at the bottom. (You may not see this section for all account types.)

9. Tap the **Save** button at the bottom to save all your settings.

Setup Enterprise Accounts ("Enterprise Activation")

If you have an enterprise service plan from your carrier, and have received an Activation Password from your email administrator, you are ready to activate or connect your BlackBerry to the secure Enterprise Server.

CHAPTER 1: Getting Started

NOTE: If you have not received your Activation Password, then you need to ask your help desk or technology support department for that password before you can complete this process.

This will allow you to securely send and receive corporate email, securely browse the corporate intranet and receive wireless updates of your Contacts, Calendar, Tasks and MemoPad items on your BlackBerry.

1. Click the **Email Accounts** icon inside the **Setup** app.
2. Tap the **Enterprise Account** option.

NOTE: You must see this selection screen in order to be able to activate your BlackBerry on the server. If you don't see this screen, you will need to contact your IT Help Desk to let them know that your BlackBerry may not have the correct enterprise service plan.

3. Now enter your work email address in the **Email** field and the **Activation Password** you received from your administrator. Usually, the administrator will email you the password so check your work email account on your computer. The other thing to keep in mind is that these passwords expire within 24 – 48 hours.

NOTE: The **Activation Password** is not the same as your regular email login password.

View Video Tutorials and Free Tips at www.MadeSimpleLearning.com   51

4. If you entered the **Email** and **Activation Password** correctly, you should then see a series of status screens showing messages similar to the screen shown to the right.

TIP: You can tap the **Hide** button and continue using your BlackBerry while the synchronization is taking place in the background. This process could take 30 minutes or more to complete depending on the size of your contact list, calendar items and related items.

How Can I Tell If I Am Activated on the BlackBerry Server?

A simple rule of thumb is that if you can send and receive email and you have names in your BlackBerry address book (which you can access through the **Contacts** icon), then it has been successfully set up (enterprise activation is complete).

To verify that your BlackBerry is successfully configured with your server, do the following:

1. Tap your BlackBerry **Contacts** icon.

2. Look at the very top, just under **New Contact**, you should see a row called **Remote Lookup** as shown to the right. You will also see a new menu item called **Lookup**.

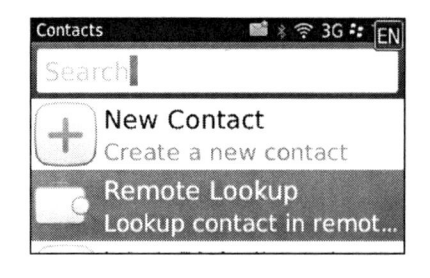

TIP: The **Remote Lookup** or **Lookup** command allows you to search your corporation's complete address list on the server (sometimes called a "Global Address List"). Then you can add them to your personal address book in the **Contacts** app on your BlackBerry.

Wireless Email Reconciliation

The BlackBerry allows you to turn on or off wireless reconciliation, which is the feature that synchronizes deletion of email between your regular mailbox and your BlackBerry. In other words, you can set it up so that if you delete an email on your BlackBerry, the same email message is also automatically deleted from your regular email account.

Disable or Enable Wireless Email Reconciliation

Usually, this is turned on by default and set to **Prompt** the user about deleting emails, but you can disable or adjust it.

> **NOTE:** If you work at an organization that supplied your BlackBerry to you, this feature may be controlled centrally by your administrator and may not be adjustable.
>
> Some wireless carriers (phone companies) do not support this feature.

1. Click the **Messages** icon.
2. Press the **Menu** key and select **Options** (or press the letter **O** to jump down to **Options**).
3. Then select **Email Reconciliation**.
4. Set Delete On to **Handheld**.
5. Uncheck the box next to **Wireless Reconcile** to disable it.

Re-Enable Reconciliation

To re-enable email reconciliation, set **Delete On to Mailbox & Handheld**, and check the box next to **Wireless Reconcile**.

On Conflicts

A conflict can occur if you work with the same email message on both your computer and your BlackBerry. Adjusting the **On Conflicts** field will give priority to either the handheld or the mailbox in case of conflict.

Purge Deleted Items

If you have turned on **Wireless Reconcile** and want to get rid of old email that you deleted from either your main email inbox or your BlackBerry, then press the **Menu** key and select **Purge Deleted Items**, and then select the email address.

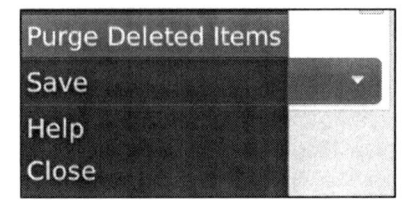

Syncing Google Contacts Using Email Setup

You can choose to sync Google Contacts and Calendar using the Email Setup app, however as of publishing time, we recommend using the separate Google Sync app and not the email setup. See the Caution below.

> **CAUTION:** As of publishing time, when we had a BlackBerry Torch on AT&T service with the enterprise service plan, we experienced that syncing contacts with the email setup shown earlier in this chapter caused all the first line of the street addressed to be erased by the sync. Therefore, we recommend you do not use the email setup to sync your Google Contacts and Calendar, instead use the **Google Sync** app. We describe how to setup **Google Sync** in the Contacts and Calendar chapters.

Troubleshooting Your Email Accounts

Sometimes your email accounts just don't work quite right, and unfortunately, sometimes email isn't as easy to set up as just shown. The following subsections provide a few tips for handling some of the more common errors.

Unable to Login to Email Accounts

With certain types of internet email accounts, entering the email address and password of your account does not work. In these cases, you need to enter some advanced settings screens shown below.

CHAPTER 1: Getting Started

1. If you have attempted to login with the correct username and password, but cannot get it to work, you will be left at this screen. Tap the link **I will provide the settings**.

> **TIP:** Make sure to check the **Show Password** box as shown so you can make sure your password is entered correctly, especially if it has numbers or characters in it.

2. Contact your email administrator to ask them for the correct **Email Server** name (usually **mail.yourdomainname.com**) and **User name** (usually your full email address) for this screen.
3. Then tap **Continue**.
4. Then, if the settings are correct, your BlackBerry should login to your email account and show you a screen similar to the one in step 9 above.

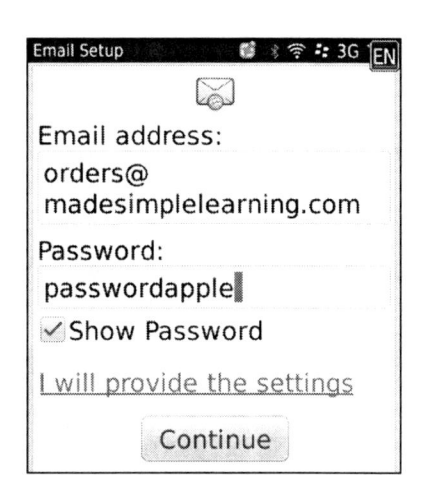

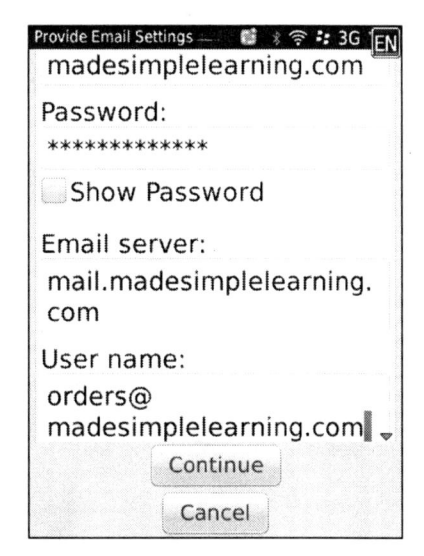

Verifying That Your Email Server Is POP3 or IMAP

Some email servers cannot be accessed by BlackBerry Internet Service, so you cannot use these types of email accounts on your Torch. Contact your email service provider, tell them you are trying to access your email from a BlackBerry smartphone, and verify that the server is of a type called POP3 or IMAP.

View Video Tutorials and Free Tips at www.MadeSimpleLearning.com   55

Verifying Your Email Server Settings (Advanced Settings)

Another setup issue might be that your server uses SSL security or has a nonstandard email server name. Contact your service provider to find out about these settings. To change these settings on your Torch, you need to follow the steps shown in the "Maintaining Your Internet Email Accounts and Signatures" section earlier in this chapter.

Solving a Gmail Enabled IMAP Error Message

If you receive an error message in your email inbox telling you to turn on IMAP settings in Gmail, you need to log into your Gmail account from your computer and follow these steps:

1. Click the **Settings** link (usually in the top-right corner).
2. Click the **Forwarding and POP/IMAP** tab.
3. Make sure that **IMAP Access** is set to **Enable IMAP**, as shown in Figure 1-2.
4. Then click the **Save Changes** button.

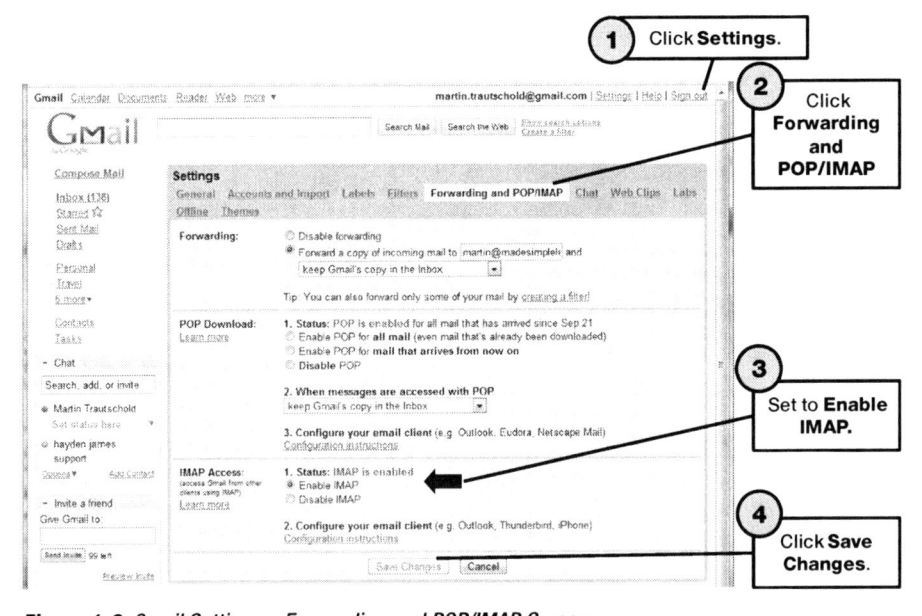

Figure 1-2. Gmail Settings – Forwarding and POP/IMAP Screen.

5. After you make this change, you will need to go back into your Email setup and validate the account.

6. Tap the invalid account and select **Validate** from the pop-up menu.

7. Type the password for your invalid email account and click **OK**.

Why Is Some Email Missing from My BlackBerry?

If you download your email messages to your computer using an email program such as **Microsoft Outlook**, **Outlook Express**, or similar, and you have your Internet Email Account set up on your BlackBerry, then you need to turn on a specific setting in your computer's email program. If you do not leave a copy of your messages on the server from your computer's email, you may end up receiving all email on your computer, but only a limited set of email on your BlackBerry.

Why Does This Happen?

By default, most email programs pull down or retrieve email from the server every 1 to 5 minutes, and then erase the retrieved messages from the server. By default, the BlackBerry Internet Email service usually pulls down email every few minutes. So, if your computer has pulled down the email every minute and erased it from the server, your BlackBerry will only receive a very limited set of messages (those that haven't yet been pulled down by your computer).

How to Fix This

The answer is to set your computer's email program to keep your messages on the server. This way, the BlackBerry will always receive every email message. Here's how:

1. In your computer email program (e.g., Microsoft Outlook), find the location where you can configure or change your email accounts. For example, Tools, Configure Accounts, Account Settings, or something similar (Figure 1-3).

2. Select or change the appropriate email account. (Sometimes you just double-click the account to edit it.)

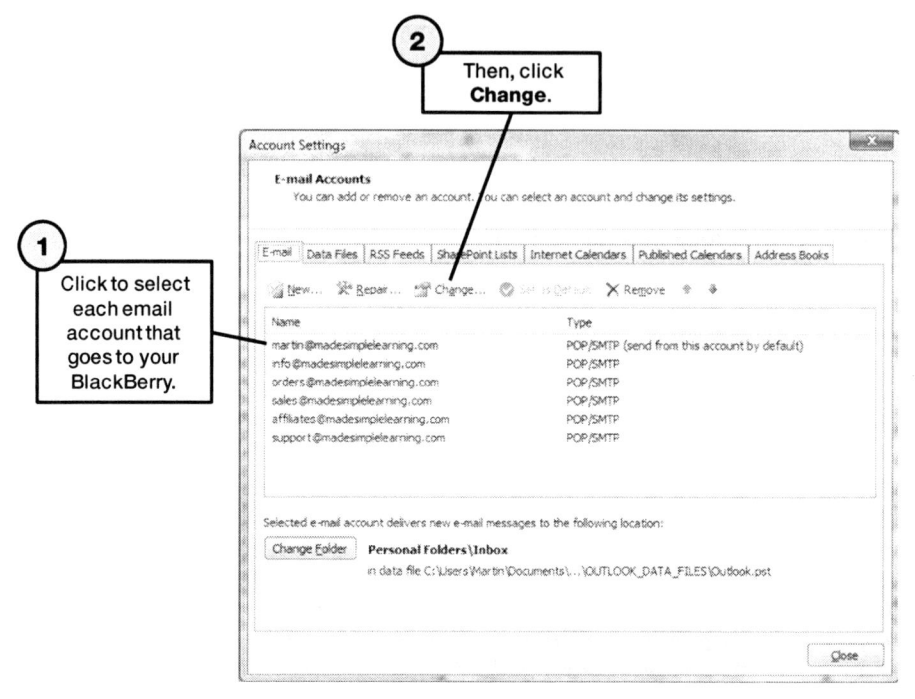

Figure 1-3. Account Settings screen in Microsoft Outlook

3. You will then usually go to an Advanced Settings area to make changes to **Leave a copy of the message on the server**. In Microsoft Outlook, click the **More Settings** tab and then the **Advanced** tab (Figure 1-4).

CHAPTER 1: Getting Started

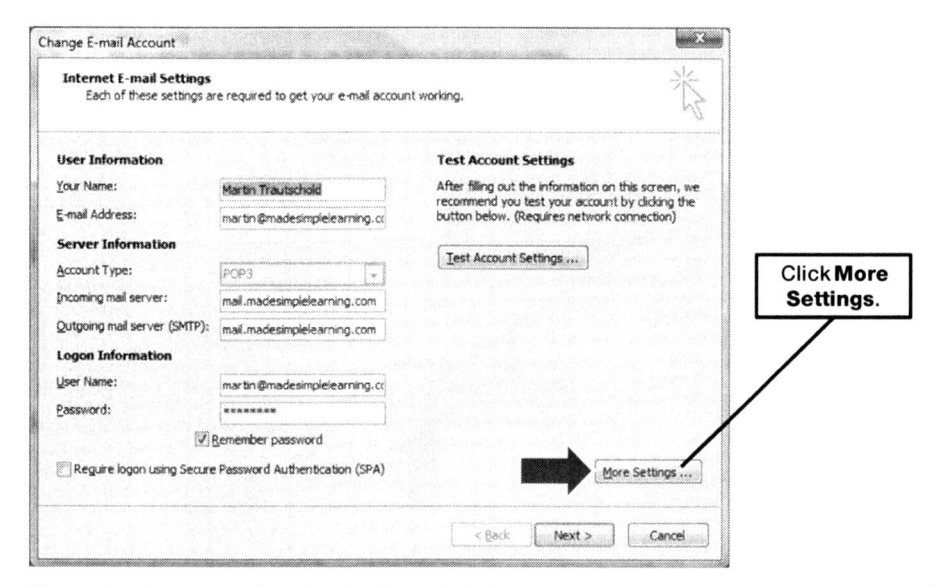

Figure 1-4. Internet email settings in Microsoft Outlook

4. At the bottom of the screen, under Delivery, check the **Leave a copy of the message on the server** box (see Figure 1-5).

5. Then check the **Remove from server after [X] days box**. We suggest changing the number to at least **10**. This allows you time to make sure that the message reaches both your BlackBerry and your PC, but doesn't clutter the server for too many days. If you make the number of days too high, you may end up with a "Mail Box Full" error and have your incoming email messages bounced back to the senders.

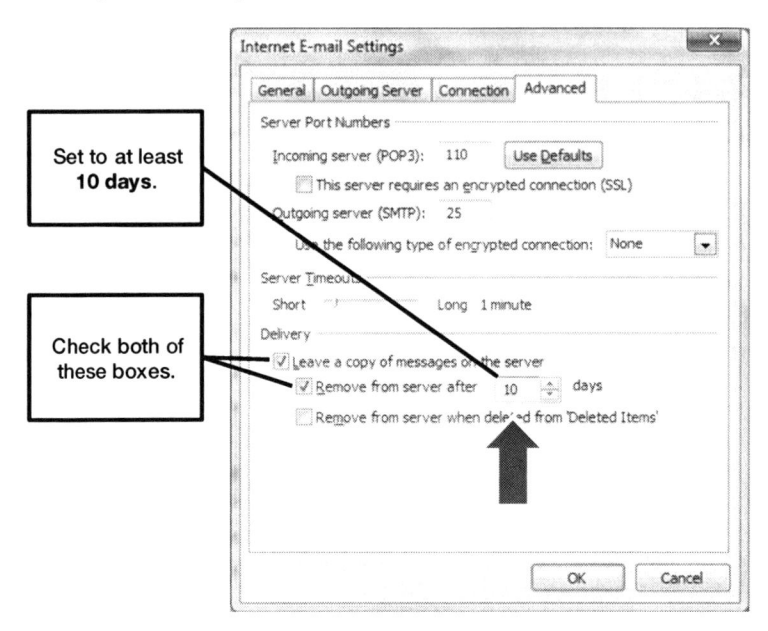

Figure 1-5. Leave a Copy of Messages On Server in Microsoft Outlook

> **NOTE:** Remember to repeat the preceding process for every email account that you have going to both your computer and your BlackBerry.

BlackBerry Enterprise Server Express

If you work at an organization that uses a **Microsoft Exchange** or **Microsoft Windows Small Business Server**, then you can now acquire the **BES Express** software for free.

A **BES** ("**BlackBerry Enterprise Server**") is a server that typically sits behind your organization's firewall and securely connects your BlackBerry to your organization's email and data, as well as wirelessly synchronizes (shares) contacts, calendar, tasks, and memo items between your corporate computer and your BlackBerry.

Image from RIM

BES allows your organization to gain all the benefits of a BlackBerry Server with no additional software costs. This saves thousands of dollars over the old pricing model from RIM. You might want to let your IT group know about this great new deal if you use BlackBerry devices at your workplace.

You should be able to support up to 75 BlackBerry users on the same box as your email server. You can support up to 2,000 users by putting the BES Express software on a separate server.

How to Get the BES Express Software

To acquire BES Express, go to www.blackberry.com and follow these steps:

1. Click the **Software** link at the top.

2. Click **BlackBerry Enterprise Server Express** under **Business Software** in the left column, and you'll see a screen similar to the one shown in (You might be able to get straight to the page by typing this link: http://us.blackberry.com/apps-software/business/server/express/.)

3. Then follow the onscreen steps to start the free download.

4. Once the software is downloaded, the administrator uses a web-based console to set up and administer all the users. The full setup instructions for **BES Express** are beyond the scope of this book, but please follow the onscreen help and tutorials found at www.blackberry.com.

Connecting with Bluetooth

Bluetooth allows your BlackBerry to communicate with devices as varied as headsets, GPS devices, and other hands-free systems with the freedom of wireless. Bluetooth is a small radio that transmits from each device. The BlackBerry gets paired, or connected, to the peripheral. Many Bluetooth devices can be used up to 30 feet away from the BlackBerry.

In this section, we will show you how to pair to a Bluetooth headset, how to connect to a Bluetooth stereo device, and how to prioritize your Bluetooth connections.

We'll start by explaining how to use Bluetooth with your Blackberry. We'll walk you through how to enable Bluetooth, to how to use it to connect to various devices, to troubleshooting some potential issues.

Bluetooth Security Tips

If you're going to use Bluetooth on your Blackberry, you need be cognizant of the security concerns around this technology. For example, here are a few security tips from a recent BlackBerry IT Newsletter. Following these will steps help you prevent hackers from getting access to your BlackBerry through Bluetooth:

- Never pair your BlackBerry when you are in a crowded public area.
- Disable the **Discoverable** setting (or use the **2 Minute** setting) after you're done pairing your BlackBerry with a device.
- Do not accept any pairing requests with unknown Bluetooth devices, and only accept connections from devices with names you recognize.
- Change the name of your BlackBerry to something other than the default value (e.g., *BlackBerry 9800*). This will help you keep hackers from easily finding your BlackBerry.

You can find this information and many other useful facts from BlackBerry's web site: www.blackberry.com.

Supported Devices

Your BlackBerry should work with most Bluetooth headsets, car kits, hands-free kits, keyboards, and GPS receivers that are compliant with Bluetooth 2.0 and earlier. At the time of writing, Bluetooth 2.1 was just entering the marketplace; you will need to check with the device manufacturer of newer devices to make sure they are compatible with your BlackBerry.

Ways to Get to Bluetooth Connections Screen

There are a couple of ways to get to Bluetooth and other wireless settings on your BlackBerry. You can get there from the **Bluetooth** icon inside the **Setup** app as shown in the left portion of Figure 1-5 or you can tap the top status bar from any **Home** screen (where the time is shown), then tap **Bluetooth Connections** (see Figure 1-6).

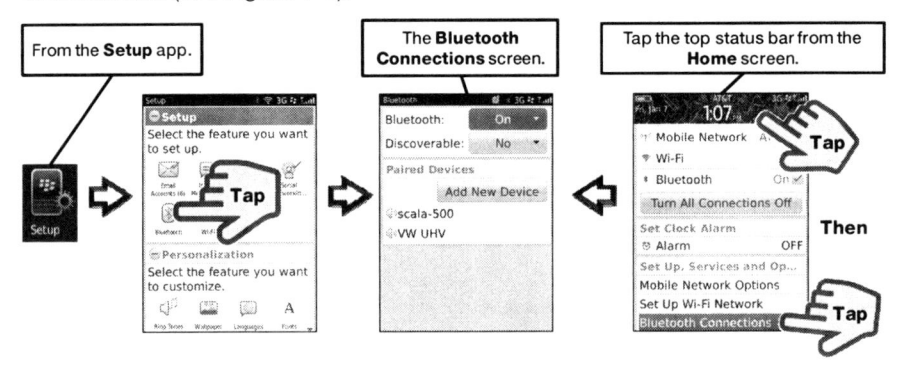

Figure 1-6. Two Ways to Get to the Set Up *Bluetooth Connections Screen*

CHAPTER 1: Getting Started

Pairing with a Bluetooth Device

You can think of *pairing* as the act of establishing a wireless connection between your BlackBerry and a peripheral, whether it's a headset, global positioning device, external keyboard, or a Windows-based or Mac computer. Pairing is dependent on entering a required *passkey* that locks your BlackBerry into a secure connection with the peripheral. Use one of the methods shown in Figure 1-6 above to get to the Bluetooth settings screen.

From the **Bluetooth** settings screen, you can pair devices and manage all your Bluetooth connections.

> **NOTE:** If you are pairing your BlackBerry with your computer, then you need to make sure that both your BlackBerry and your computer are in **Discoverable** mode. Set this in the **Bluetooth Options** screen by setting the **Discoverable** property to **Yes** or **Ask**. (The default setting is **No**, which will prevent you from pairing your BlackBerry and computer.)

1. Begin by putting your Bluetooth device (headset, computer, car stereo) in **pairing** mode, as recommended by the manufacturer. Also, have the passkey ready to enter later on in the process.

2. From the **Bluetooth** settings screen, make sure **Bluetooth** is set to **On**.

3. If you are connecting to a Bluetooth device that needs to find the BlackBerry to connect to it, set **Discoverable** to **2 Minutes**, otherwise set it to **No**. (If you are connecting a headset or keyboard that will be listening for your BlackBerry to connect, then setting **Discoverable** to **No** is fine.)

4. Tap **Add New Device** to get started.

> **NOTE:** All devices currently paired are listed at the bottom of the screen. In the image above there is a **scala-500** headset and a **VW UHV** car stereo.

View Video Tutorials and Free Tips at www.MadeSimpleLearning.com   63

5. Then, you may be prompted to change the name of your device from the default **BlackBerry** to help avoid Bluetooth hackers. Go ahead and make the name something unusual, in this example we just used **xyz**. Tap **OK**.

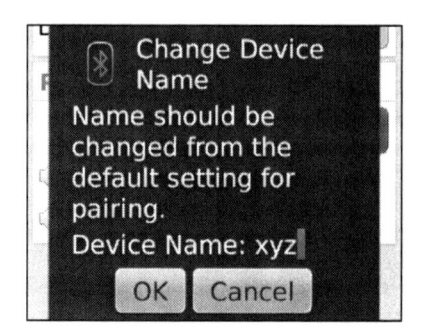

6. Click **Search** to search for Bluetooth devices in pairing mode. Click **Listen** if another device is looking for your BlackBerry.

7. Then, as long as the other device you are searching for is in pairing mode and close to your BlackBerry, it should be found as shown to the right. Click on the device name that appears.

TIP: If the BlackBerry cannot find any devices, then make sure the other Bluetooth device is truly in pairing mode and try searching again.

CHAPTER 1: Getting Started

8. You may be prompted to enter the four-digit passkey provided by the manufacturer of the Bluetooth peripheral. Enter this passkey, and then press the **Enter** key. (Many default passkeys are just *0000, 1111,* or *1234.*)

NOTE: Many newer headsets will pair automatically now without the need to enter a passkey.

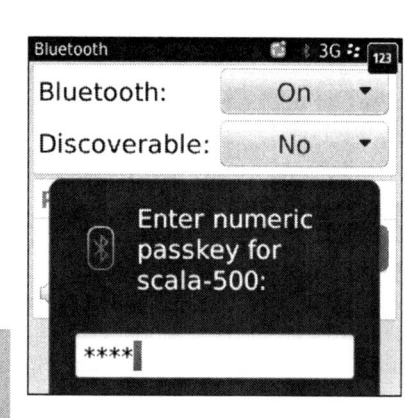

9. You will then be prompted to accept the connection from the new device; do so.

TIP: If you check the box next to **Don't display this again**, you will only have to do this once.

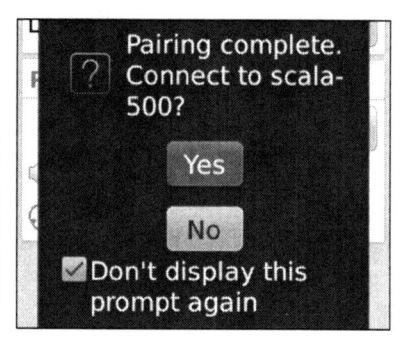

10. Now you should see your newly connected Bluetooth device shown at the bottom of the Bluetooth settings screen. In this image you can see the **scala-500** headset.
11. Repeat the above procedure to connect with any Bluetooth devices.

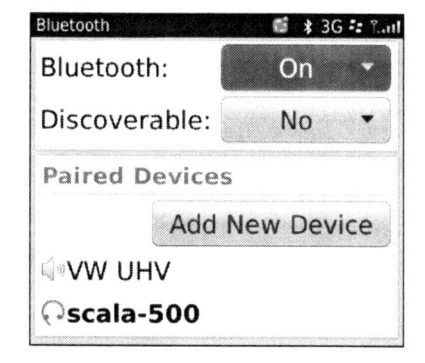

Answering and Making Calls with a Bluetooth Headset

Some Bluetooth headsets support an **Auto Answer** feature that will, as it sounds, automatically answer incoming calls and send them right to the headset. This is very helpful when driving or in other situations where you should not be looking at your phone to answer the call. Sometimes, you will need to push a button – usually just one – to answer your call from the headset. You have two options for answering and making calls with a headset, which are covered next.

Option 1: Answering Directly From the Headset

When a call comes into your BlackBerry, you should hear an audible beep in the headset. Just press the **Multi-function** button (usually the main button) on your headset to answer the call. See Figure 1-7. Press the **Multi-function** button again to hang up the phone call.

Slide up/down for **Volume** control. Press and hold for **power** or **paring mode**.

The **Multi-function** button. Use to answer, hang up or start **Voice Dialing**.

Figure 1-7. A Bluetooth Headset (Scala-500 model) showing the Multi-Function button.

Option #2: Transferring the Caller to the Headset

When a phone call comes into your BlackBerry, follow these steps to transfer the caller to your headset:

1. Press the **Menu** key.

2. Scroll to **Activate (***your Bluetooth headset's Device Name***)**, and the call will be sent to the selected headset.

In image to the right, the headset's **Device Name** is *Jawbone*.

You can switch back to the handset by pressing the **Menu** key and selecting **Activate Handset**.

Voice Dialing from the Bluetooth Headset

Pressing the Multi-function button on your Bluetooth headset will start Voice Dialing when you are not on a phone call. See Figure 1-7 above.

The Bluetooth Connections Menu Commands

There are several options available to you from the **Bluetooth Connections** menu. Learning these commands can help you to take full advantage of Bluetooth on your BlackBerry.

To get to the **Bluetooth Connections** screen, tap the top status bar of your **Home** screen to view your **Manage Connections** screen. Then scroll down and tap **Bluetooth Connections** near the bottom. (See Figure 1-6 above).

You can access the following functionality from the **Bluetooth Setup** menu:

- **Disable Bluetooth:** Gives you another way to turn off the Bluetooth radio, which can help you to save battery life if you don't need the Bluetooth active.

- **Connect / Disconnect:** (This option is only visible if you have highlighted a device in the list at the bottom of the screen before pressing the **Menu** button.) Lets you immediately connect/disconnect you to/from the highlighted Bluetooth device.

- **Add Device:** Lets you connect to a new Bluetooth peripheral.

- **Delete Device:** Removes the highlighted device from the BlackBerry.

- **Device Properties:** Lets you check whether the device is trusted or encrypted. It also lets you see whether the **Echo Control** is activated.

- **Transfer Contacts:** (This option is only visible if you have highlighted a device in the list at the bottom of the screen before pressing the **Menu** button.) Enables you to send your address book through **Bluetooth** to a PC or another Bluetooth smartphone, assuming you are paired and connected to that device.

- **MAP Options:** Shows you the Message Access Options screen where you can check or uncheck your various email accounts and SMS/MMS. If you uncheck an email account or message type, then it is not accessible via **Bluetooth** connectivity.

- **Options:** Shows the **Options** screen, which we've covered previously.

Sending and Receiving Files via Bluetooth

Once you have paired your BlackBerry with your computer, you can use Bluetooth to send and receive files. At the time of writing, these files were limited to media files (e.g, videos, music, and pictures) and address book entries; however, we suspect that you will be able to transfer more types of files in the future.

Transfer Contacts via Bluetooth

From the **Bluetooth Connections** screen, highlight the device to which you want transfer contacts to or from. Press the **Menu** button and select **Transfer Contacts**. Follow the steps on the screen to complete the transfer.

Sending and Receiving Media Files

Sending and receiving media files through Bluetooth on your BlackBerry is a straightforward process. All you need to do is follow these simple steps:

1. Tap the icon for the type of media you wish to transfer, such as **Music**, **Ringtones**, **Pictures** or **Videos**.

2. Navigate to particular item you wish to transfer.

3. Press the **Menu** button and select Send, then select **Bluetooth**.

4. Select the device to which you want to transfer the files, usually your computer.

5. From here, you will need to follow the prompts on your computer to receive the file.

NOTE: You may need to set up your computer so it can receive files through Bluetooth.

If you are receiving a file (or files) on your BlackBerry, then you need to select the **Receive via Bluetooth** option. Go to your computer and select the file or files you want to send, and then follow the commands required to **Send via Bluetooth**. You may be asked on the BlackBerry to confirm the Bluetooth transfer.

> **NOTE:** You can send (transfer) only media files that you have put onto your BlackBerry yourself. You cannot use Bluetooth to transfer the pre-loaded media files.

Troubleshooting Bluetooth

Bluetooth is still an emergent technology, and sometimes it just doesn't work as well as you might hope. If you are having difficulty, perhaps one of these suggestions will help. So in this section, we will cover common problems and possible approaches to solving them.

Problem 1: The Device Refuses Your Passkey

Sometimes a Bluetooth device just won't accept your passkey. In this case, it's possible that you have the incorrect passkey. Most Bluetooth devices use either *0000*, *1111*, or *1234* for the passkey; however, but some have unique passkeys.

If you lost your product manual for the Bluetooth device, you can often find a PDF version of the manual online using a search engine such as Google or Yahoo. Typically, you'd find this PDF at the device manufacturer's site, but it can show up on other sites, too.

Problem 2: The Device Won't Pair Even with the Correct Passkey

Sometimes you have the right passkey, but you still cannot pair the device. In this case, it is possible that the device is not compatible with your BlackBerry. One thing you can try is to turn off encryption. Follow these steps to do so:

1. Get to the **Bluetooth Connections** screen as shown above.

2. Highlight the **Name** of the problem device and then press the **Menu** button to select **Device Properties**.

3. In the **Device Properties** dialog, disable encryption for that device and try to connect again.

Problem 3: You Cannot Share Your Address Book

Sometimes you might encounter problems when trying to share your address book with a Bluetooth device. Follow these steps if you encounter this issue:

1. Inside the **Bluetooth Connections** screen, press the **Menu** key and select **Options**.

2. Double-check that you have enabled the **Contacts Transfer** field. If you haven't, do so.

NOTE: Many Bluetooth headsets and car kits do not fully support address book transfers; be sure to double-check the documentation that came with your Bluetooth device.

Boosting Your Memory with a Media Card

Your BlackBerry comes with 512 MB (Megabytes) of flash memory for programs and 4 GB of "on-board" memory for things like pictures and music, but all that won't all be available to you. The operating system and installed software take up some of that room and so will all your personal information.

Since your BlackBerry is also a very capable media device, you will probably want more room to store things like music files, videos, ring tones, and pictures.

The image of the SanDisk MicroSD card and the SanDisk logo are copyrights owned by SanDisk Corporation.

That's where the MicroSD memory card comes in. Some wireless carriers are pre-installing an 4GB or 8GB memory card in the Torch—if you have a smaller card or no card at all, you can expand the memory to 32GB MicroSD memory card if you need.

Installing Your Memory Card / Media Card

The BlackBerry actually has the MicroSD card slot inside the back cover, near the battery. Getting to it is simple:

CHAPTER 1: Getting Started

1. Turn your BlackBerry over and hold it in both hands. Press on the back cover and slide it down with both thumbs.

2. Lift up the little black plastic flap. (See image).

3. Hold the memory card with the small notch on the lower left-hand side.

4. Slide the card down into the slot until it stops.

5. Replace the back cover.

First, lift up this little black plastic flap.

Then, slide the **MicroSD memory card** in here.

CAUTION: Before you remove the media card, we recommend using the **Safely Remove Media Card** from the **Storage** screen in **Options**. To get to this command, type **Storage** in the universal search, tap the **Options** icon, tap **Storage**, press the **Menu** key and select **Safely Remove Media Card**. Then you can follow the steps above to remove the card.

Verifying the Memory Card Installation and Free Memory

Once installed, it is a good idea to double-check that the card is installed correctly and the amount of free space is available.

1. Tap the **Search** icon in the upper right corner of the **Home** screen and type **Storage**.
2. Tap the **Options** icon.
3. Tap **Storage** from the results list.
4. This image shows that a media card has been inserted (see **Media Card Storage** at the bottom).
5. The total space figure at the bottom of the screen shows that a 4.0 GB card will read about 3.6 GB Total Space (1GB equals about 1,000 MB). If you see that, all is well.

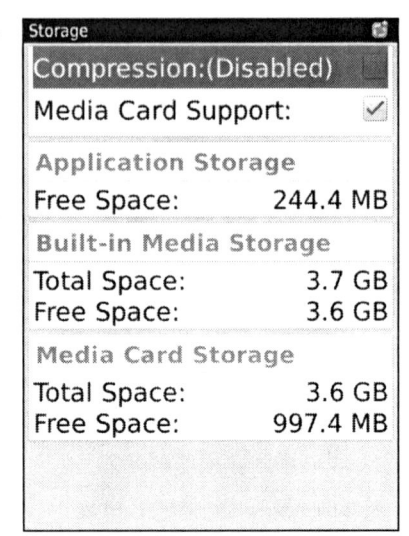

Storage	
Compression:(Disabled)	
Media Card Support:	✓
Application Storage	
Free Space:	244.4 MB
Built-in Media Storage	
Total Space:	3.7 GB
Free Space:	3.6 GB
Media Card Storage	
Total Space:	3.6 GB
Free Space:	997.4 MB

Formatting or Repairing the Media Card

You can format or repair the media card from this same **Storage** screen in **Options**.

Press the Menu button in the **Storage** screen.

- To format the media card, select **Format**.
- To repair the media card, select **Repair**.
- To remove the media card, select **Safely Remove Media Card**.

Transferring Content to your Torch Using USB Drive Mode (for Mac and Windows)

This works regardless of whether you have a Windows or a Mac computer. We will show images for the Windows computer process, and it will be fairly similar for your Mac. This transfer method assumes you have stored your media on a MicroSD media card in your BlackBerry.

CHAPTER 1: Getting Started

1. Connect your BlackBerry to your computer using the USB cable.

2. Select **USB Drive** from the options.

NOTE: Select **Sync Media** if you are syncing media (music, pictures, etc.) using BlackBerry Desktop Software.

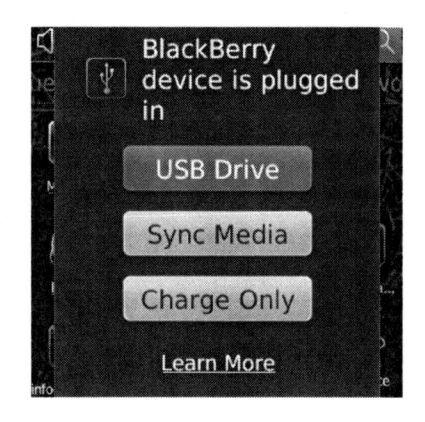

3. After your BlackBerry is connected and in USB Drive mode, just open your computer's file management software.

    a. In Windows, press the **Windows** key + **E** to open **Windows Explorer**.

    b. On your Mac, start your **Finder**. Look for another hard disk or "BlackBerry (model number)" that has been added.

You may see two separate disks:—

Removable Disk (your media card) and

BLACKBERRY1 (your BlackBerry internal memory).

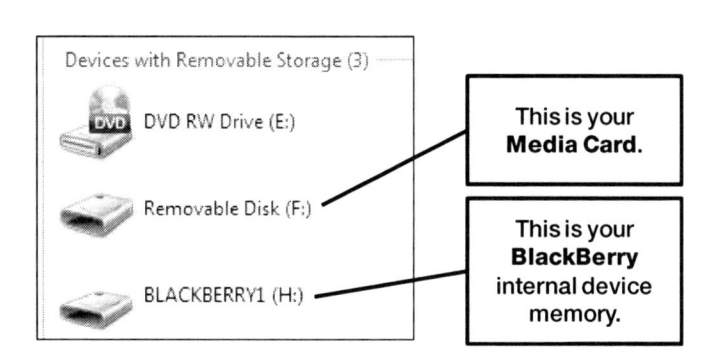

This is your **Media Card.**

This is your **BlackBerry** internal device memory.

You will notice that all your files are stored in separate folders.

View Video Tutorials and Free Tips at www.MadeSimpleLearning.com   73

In the BlackBerry internal memory (shown as **BLACKBERRY1**), the main folder where all your information is stored is called **home** then **user**. In that folder there are separate folders depending on the type of files.

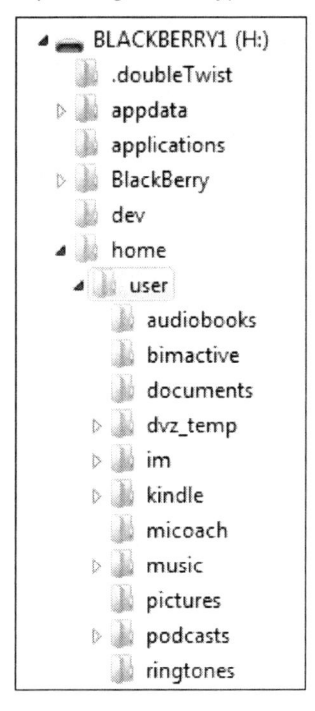

In the media card (**Removable Disk**) they are all under the main folder called **BlackBerry.** From there you will see audiobooks, camera, documents, etc. In that folder there are separate folders depending on the type of files.

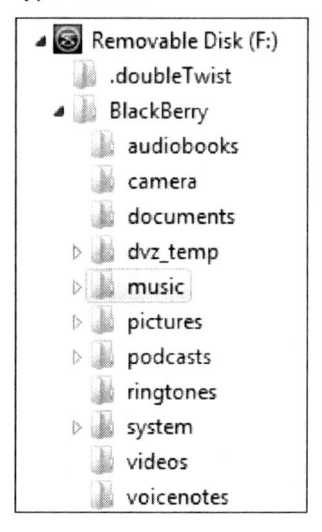

```
BLACKBERRY1 (H:)
    .doubleTwist
    appdata
    applications
    BlackBerry
    dev
    home
        user
            audiobooks
            bimactive
            documents
            dvz_temp
            im
            kindle
            micoach
            music
            pictures
            podcasts
            ringtones
```

```
Removable Disk (F:)
    .doubleTwist
    BlackBerry
        audiobooks
        camera
        documents
        dvz_temp
        music
        pictures
        podcasts
        ringtones
        system
        videos
        voicenotes
```

In the images above, you notice that different types of media are stored in different folders.

To copy a specific type of media, navigate to the correct folder on your BlackBerry or the media card and copy or paste files into that folder.

You can also drag and drop files between these folders and folders on your computer as you normally do with other files on your computer.

For example, to copy media (pictures, videos, music) from your BlackBerry, select the files from the **BlackBerry / pictures** folder. Then draw a box around the pictures, or click one and press **Ctrl+A** or **Command+A** (Mac) to select them all.

Or hold the **Ctrl** key (Windows) or **Command** key (Mac) down and click individual pictures to select them. Once selected, right-Tap (Windows) or right-Tap the Mac.

Or press **Ctrl** and click (Mac) one of the selected pictures and select **Cut** (to move) or **Copy** (to copy).

Then click any other disk/folder—like **My Documents**—and navigate to where you want to move / copy the files. Once there, right-Tap again in the right window where all the files are listed and select **Paste**.

On your Mac, click the **Finder** icon in the lower left-hand corner of the Dock.

You will see your Devices (including both BlackBerry drives) on the top and your Places (where you can copy and paste media) on the bottom.

You can also delete all the pictures / media / songs from your BlackBerry in a similar manner. Navigate to the BlackBerry / (media type) folder, like **BlackBerry / videos**. Press the key combination shown previously on your computer keyboard to select all the files, and then press the **Delete** key on your keyboard to delete all the files.

You can also copy files from your computer to your BlackBerry using a similar method. Just go to the files you want to copy, and select them (highlight them). Then right-Tap **Copy** and paste them into the correct BlackBerry / (media type) folder.

> **Important:** Not all media (videos), pictures (images), or songs will be playable or viewable on your BlackBerry. If you use desktop software such as Desktop Software for Windows (See Chapter 3: "Windows PC Media and File Transfer") for Mac (See Chapter 5: "Apple Mac Media and File Transfer") to transfer the files, most files will be automatically converted for you.

Connecting with Wi-Fi

Wi-Fi is a wireless-connection technology that enables you to connect computers, printers, game consoles, mobile devices (like your blackberry) and more to the Internet. Today we are fortunate to live in a world where Wi-Fi is increasingly ubiquitous. Indeed, it is difficult to go anywhere and not have access to a Wi-Fi network.

In this section, we will show you how to connect your BlackBerry to available Wi-Fi networks, how to prioritize and organize your networks, and how to diagnose wireless problems.

Understanding Wi-Fi on Your BlackBerry

Like other devices that use Wi-Fi, your BlackBerry can send and receive a wireless signal to and from a wireless router. If your BlackBerry is Wi-Fi equipped, you can take advantage of high-speed Web Browsing and file downloading speeds through your home or office wireless network. You can also access millions of Wi-Fi hotspots in all sorts of places, such as coffee shops and hotels, many of which are free!

> **NOTE:** The origin of the term *Wi-Fi* is disputed. According to some, Wi-Fi stands for Wireless Fidelity (IEEE 802.11 wireless networking), whereas others say that the term refers to a wireless technology brand owned by the Wi-Fi Alliance. Regardless of the term's origins, we're immensely grateful for its increasingly widespread adoption!

The Wi-Fi Advantage

Wi-Fi is a great advantage to BlackBerry users around the globe. The advantages to using a Wi-Fi connection (as opposed to a carrier data connection such as GPRS/EDGE/3G) are many:

- Web browsing speeds are much faster.
- You are not using up data from your data plan,
- Most file downloads will be faster.
- You can get often great Wi-Fi signals when you cannot get any regular cell coverage (1XEV/EDGE/3G), such as in the bottom floors of a thick-walled building.

Setting up Wi-Fi on Your BlackBerry

Before you can take advantage of the speed and convenience of using Wi-Fi on your BlackBerry, you will need to set up and configure your wireless connection.

Similar to getting to the Bluetooth screen, there are a couple of ways to get to the **Set Up Wi-Fi** screen. You can get there from the **Wi-Fi** icon inside the **Setup** app as shown in the left portion of Figure 1-8 or you can tap the top status bar from any **Home** screen (where the time is shown), then tap **Setup Wi-Fi Network**.

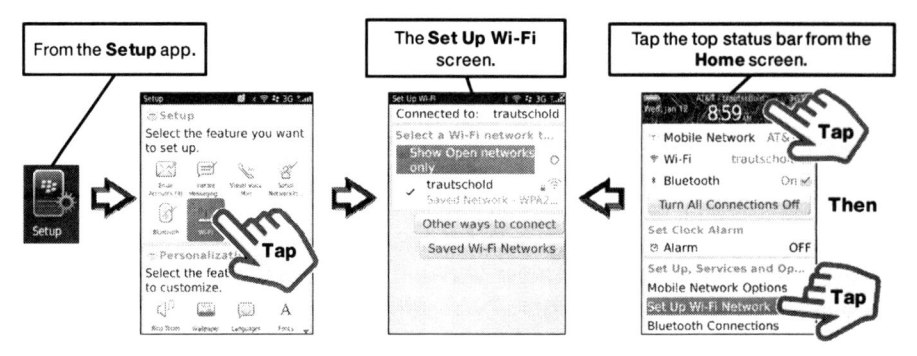

Figure 1-8. Ways to Get to the Set Up *Wi-Fi Network Screen*

Connecting to Visible Wi-Fi Networks

If the Wi-Fi network is broadcasting its network name (also known as the "ssid"), then you can see it on your list of available networks from the **Set Up Wi-Fi** screen. Connecting is as easy as clicking on the network name.

1. From the **Set Up Wi-Fi** screen, you will immediately see all Wi-Fi networks that are broadcasting their network names. In this image you can see two available networks: **trautschold** and **johnsonwifi**.

2. If you see the network to which you want to connect, tap it.

3. If the network has security enabled, you will be asked for the passphrase. Type it in and click Connect. Otherwise, if the network is unsecure, then you will be connected automatically.

4. Once you have been connected to the network, you will see a checkmark next to the network name as shown here.

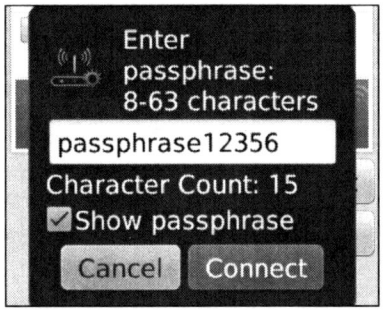

Wi-Fi Protected Push-Button Setup

If you have a newer Wi-Fi router, it will have a special button to help you quickly connect Wi-Fi devices such as your BlackBerry. You will need to have a button that looks like this on your router to use this method.

1. From the **Set Up Wi-Fi** screen, click the **Other ways to connect** button near the bottom.

2. Press **Wi-Fi Protected Setup** on this screen.

3. Here you have two choices:
   Choice 1: (Easier) Click **Press button on router** and press the button on your router. Then the BlackBerry will wait for a signal from the router to connect.

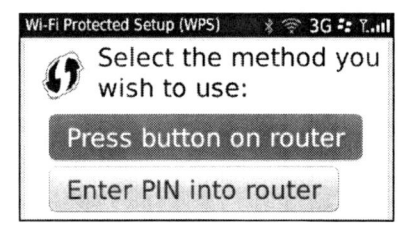

CHAPTER 1: Getting Started

4. **Choice 2:** (Harder) Click **Enter PIN into router**. Then, then if the BlackBerry finds a network you will see a screen where you press **Start**. Then you will see a screen showing a randomly generated pin number. You will need to enter that PIN into your router configuration screen (usually by typing 192.168.1.1 or similar number into your computer's web browser and logging in).

5. When successfully connected, you will see this screen with the network name in the top row with a checkmark next to it.

Manually Connect to a Network

Sometimes, you will have to manually type in the Wi-Fi network information, this could be because the network is not broadcasting its name or it has advanced security settings. Contact the Network Administrator prior to using this option, because you will have to type in detailed information about the Wi-Fi network and its security settings in order to connect to it.

1. From the **Set Up Wi-Fi** screen, click the **Other ways to connect** button near the bottom.

View Video Tutorials and Free Tips at www.MadeSimpleLearning.com   79

2. Press **Manually Connect to Network** on this screen.

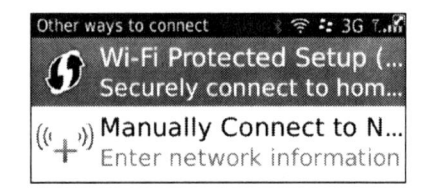

3. Type in the **SSID** (network name) and click **Next**.

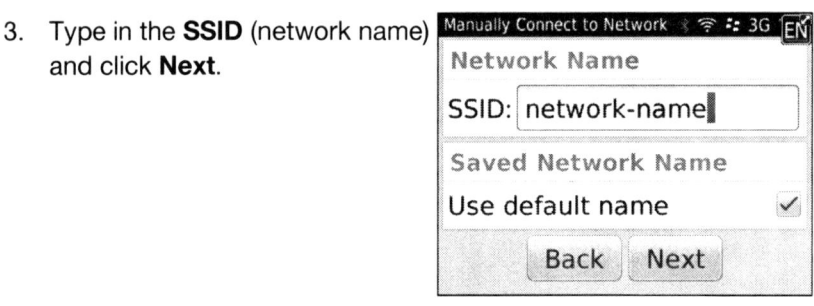

4. If the network has no security, then leave **Security Type** set to the default **None** and click **Save and Connect** at the bottom.

5. If the Network has security enabled, then follow these steps.

6. Select the **Security Type** and select one from the list: **WEP, WPA/WPA2 Personal,** or **WPA/WPA2 Enterprise.**

7. After selecting the **Security Type**, more fields will appear. Type in the requested information, such as the passphrase or other information.

8. Click the **Save and Connect** button at the bottom.

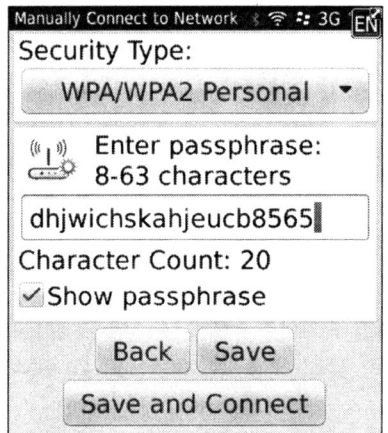

CHAPTER 1: Getting Started

9. Now the BlackBerry will try to connect to the specified network. You will see a checkmark next to the network name on the **Set Up Wi-Fi** screen.

10. If you have problems connecting, you may need to verify your information and try again.

Connecting to Free Wi-Fi Networks Requiring Web Browser Login or Agreements

Sometimes you will be able to connect to a Wi-Fi network easily as described above, but you will not have full internet access until you open your BlackBerry web browser and click a check box or enter a password.

1. Go to the Wi-Fi setup as you did above.

2. Choose a network to connect to; in this example we are connecting to the Panera Hotspot.

3. Click OK if you get a screen notifying you that there might be some legal terms or conditions to which you need to agree.

4. Agree to the terms and conditions of the hotspot in order to connect.

NOTE: In the case of this Panera Hotspot, we had to agree to a limited 30 minutes of access during peak times.

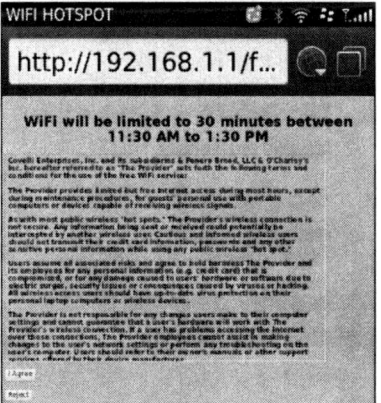

View Video Tutorials and Free Tips at www.MadeSimpleLearning.com   81

Move, Change, or Delete Saved Wi-Fi Networks

One of the nice things about Wi-Fi is that you can connect to a wireless network just about anywhere. While you may have saved your home and work networks in the steps above, there may be times you want to prioritize, change or delete networks.

Let's say you connect to your home wireless network 90% of the time. You will want to make sure that your home network is at the top of the list of networks your BlackBerry searches for.

Follow these steps to move, change or delete your saved Wi-Fi networks:

1. Access your **Wi-Fi settings**, as described previously.

2. Click the **Saved Wi-Fi Networks** button.

3. Press and hold the network you wish to move, change or delete until you see the pop-up menu as shown to the right.

4. From here you may:

    a. Tap **Move** to move the network up or down the priority list (higher networks are connected first.) Then tap a new location to move the network to that location on the list.

    b. Tap **Disable** to disconnect from this network.

    c. Tap **Delete** to remove this saved network.

    d. Tap **Edit** to change the network settings.

5. When done, press the **Escape** key and select **Save**.

Troubleshoot Using Wi-Fi Diagnostics

There may be times when your Wi-Fi Connection doesn't seem to be working well. Thankfully, your BlackBerry has powerful, built-in troubleshooting programs to help you in those instances.

> **CAUTION:** These troubleshooting tools are highly technical and should be used only by highly technical individuals or while talking to the network administrator. We have not described their detailed use because it is beyond the scope of this book.

1. Get to the **Setup Wi-Fi** screen as shown above.

2. Press the **Menu** button and select **Troubleshoot** to see the screen shown here.

3. Select **Wi-Fi Diagnostics** to see technical information about your current network.

4. Select **Site Survey** to see relative signal strengths of available Wi-Fi networks.

5. Select **DNS Lookup** to see the local name servers (used to connect you to the Internet), and

6. Select **Ping** to see how long a packet of data from your BlackBerry takes to get to a specific location on the Internet.

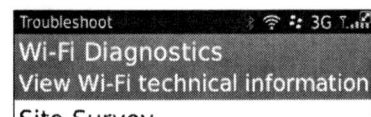

Using your BlackBerry as a Tethered Modem

Tethering is the process of connecting your BlackBerry to your computer and using the BlackBerry as a "modem" to access the internet. This is particularly useful if you are in an airport or a hotel with no internet connection on your notebook, and you need capabilities more robust than those provided by your BlackBerry.

Connecting Your Laptop to the Internet with Your BlackBerry

Depending on your wireless carrier (phone company) and what type of software you use, you should be able to use your BlackBerry to connect your laptop (PC or Mac) to the internet. This is variously called a *tethered modem*, *tethering*, or even an *IP modem*. You need a USB cable to connect your BlackBerry to a PC; however, sometimes you can use Bluetooth to connect to a Mac.

> **NOTE:** Not every BlackBerry wireless carrier supports using your BlackBerry as a modem. Please check with your carrier before attempting to use this feature. Because there are so many ways to connect using your BlackBerry and the methods vary considerably, we could not include them all in this book. Please contact your carrier or the software provider for assistance with this feature.

Tethering (Usually) Costs Extra

Most (but not all) wireless carriers charge an extra fee to allow you to use your BlackBerry as a tethered modem. You may not be able to connect using your BlackBerry as a modem, however, unless you have specifically signed up for a *BlackBerry as Modem* service or a similar data plan.

Your Tethering Options

You have several options to choose from if you decide to use your BlackBerry as a tethered modem for your laptop (PC or Mac):

- **Option 1:** Purchase third-party software. We recommend **Tether**, available at http://tether.com/
- **Option 2:** Use your wireless carrier's software (contact your carrier to ask whether it has Tethering software available).
- **Option 3:** Use the **BlackBerry Desktop Software** (this is usually only an available for you if Option 2 is not.)

Securing Your BlackBerry Data

You have a lot of valuable and highly personal information on your BlackBerry; obviously, keeping it private and personal is critically important. In this section, we will cover general security principles for your BlackBerry, including how to secure your BlackBerry data. For example, we will cover how to prepare for the possibility that you lose your BlackBerry, and the steps you should take to hedge the potential downside from such an unfortunate occurrence.

We will also show you how to enable and use password security and alert you to good email and web security tips and tricks.

Consider what would happen if your BlackBerry were lost or stolen. Would you be uncomfortable if someone found and easily accessed all the information on your BlackBerry? For most of us the answer would be, "Yes!"

On many of our BlackBerry smartphones, we store vital information about our friends and colleagues, including their names, addresses, phone numbers, confidential emails, and notes. Some of our devices may even contain Social Security Numbers, passwords, and other important information in your **Contact** notes or your **MemoPad** program.

For this reason, you will want to enable or turn on the **Password Security** feature. When you turn this on, you and anyone else, will need to enter the correct password to access and use the BlackBerry. In many larger organizations, you do not have the option to turn off your password security; it is automatically turned on by your BlackBerry Enterprise Server Administrator.

Preparing for the Worst Case Scenario

Fortunately, you can take several steps to make this worst-case scenario less painful, should it ever occur. For example, taking the following steps will help you minimize the potential impact of losing your BlackBerry or having it stolen:

- **Step 1**: Back up your BlackBerry to your computer, a detached hard drive, a USB memory stick, or some other device or service.
- **Step 2**: Turn on (**enable**) the BlackBerry's **Password Security**.
- **Step 3**: Fill out the **Owner Information** field; be sure to include an incentive for returning your BlackBerry.

In the next several sections, we will walk you through how to implement each of these steps.

Step 1: Back up or Sync Your Data

Windows and Mac users can use the data backup feature of **BlackBerry Desktop Software** described in Chapters 2 and 4.

CAUTION: If you have important media or other information (e.g., personal pictures, videos, music, and so on) stored on your media card, see Chapters 3 (Windows) and 5 (Mac) to learn how to transfer or sync that media from your BlackBerry to your computer.

Step 2: Turn on Password Security

A critical step in securing your BlackBerry is to make sure you enable **Password Security**. This is really your "Plan B" to safeguard all your data, should the device get lost or stolen. Follow these steps to enable **Password Security**:

1. Click the **Search** icon in the upper right corner of your Home screen and type the word "lock." Then, click the **Options** icon (wrench).

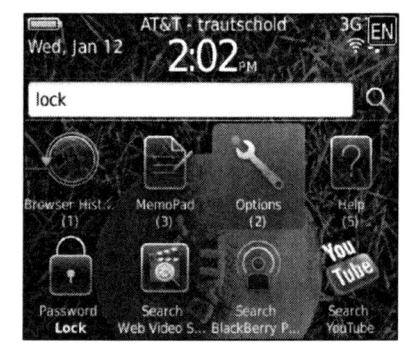

2. Click **Password** from the Options screen.

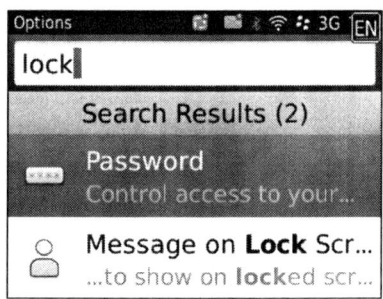

CHAPTER 1: Getting Started

3. Click **Enable** to turn on the checkmark next to it.

4. Click **Set Password** to create your password. You can adjust the other settings unless they are locked out by your BlackBerry administrator.

5. Press the **Escape** key and **Save** your settings.

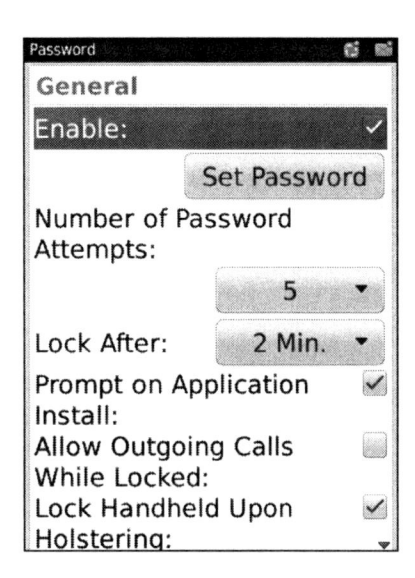

CAUTION: If you cannot remember your password, then you will lose all the data on your BlackBerry, including email, addresses, calendar events, and tasks – everything!. Once you enter more than the set number of password attempts (the default is ten attempts), the BlackBerry will automatically *wipe,* or *erase all data.*

Step 3: Set the Message on Lock Screen

We recommend including something like this in the message shown on the screen when your BlackBerry is locked: *Reward offered for safe return*. This is the text that appears on the screen if someone found your BlackBerry in a **Locked** mode.

1. Click the **Search** icon in the upper right corner of your Home screen and type the word "lock." Then, click the **Options** icon (wrench).

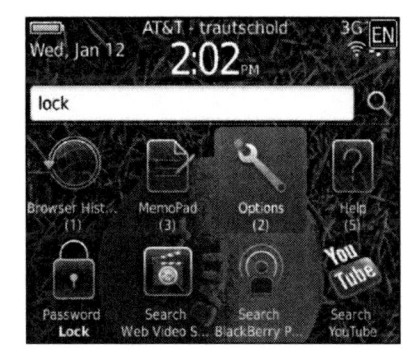

87

2. Click **Message on Lock Screen** from the Options screen.

3. Now type your name in the **Display Owner Name**. We recommend typing something that offers a reward in the **Display Information**. This may give the finder of your BlackBerry incentive to return it to you.

4. Press the **Escape** button and **Save**.

Now test what you've entered by locking your BlackBerry by pressing the **Password Lock** icon (scroll all the way to the bottom of your **All** folder on your **Home** screen).

> **CAUTION:** Just pressing the **Lock** button on the top of your BlackBerry will not Lock your device to where you need to enter the password. Instead, only the screen is locked until you press the **Lock** button again.

When your BlackBerry is locked, you should now see your owner information, as shown in the figure to the right.

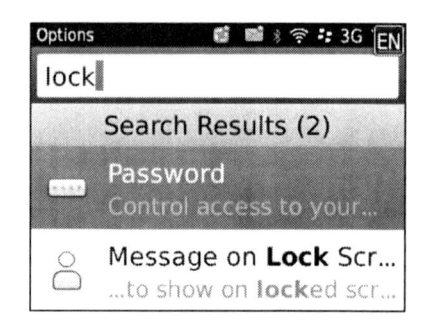

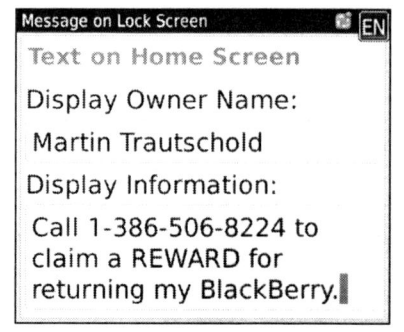

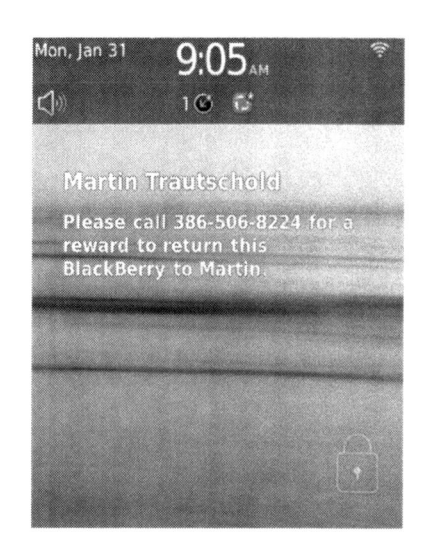

Email Security Tips

You should *never* send personal information or credit card information via email. Examples of other information you should never send through email include: credit card numbers, Social Security Numbers, your date of birth, your mother's maiden name, or sensitive passwords / PIN numbers (e.g., a bank account ATM card). If you have to transmit this information, it's best to call the trusted source or, if possible, visit that source in person.

Web Browsing Security Tips

Next, we'll cover some basic safety tips you should keep in mind when browsing the Web. If you are in an organization with a BlackBerry Enterprise Server, then all the browsing you do with the built-in Browser program is secured by the encryption from the Server. Your communications will be secure as long as you are browsing sites within your organization's intranet. Once you go outside the intranet to the internet, and you are not using an HTTPS connection, your web traffic is no longer secure.

Anyone who visits a site with HTTPS (Secure Socket Layer) connection will also have a secure connection from your BlackBerry to the web site, just as when you're using your computer's web browser.

Whenever you are browsing a regular HTTP web site on the internet (not your organization's secure intranet), then you need to always be aware that your connection is not secure, even if are using a BlackBerry Enterprise Server. In this case, please make sure that you do not to type or enter any confidential, financial, or personal information.

What to Do if You Lose your BlackBerry

If you work at an organization with a BlackBerry Enterprise Server or use a Hosted BlackBerry Enterprise Server and you lose your Blackberry, then you should immediately call your Help Desk and let it know what has happened. Most Help Desks can send an immediate **Wipe** command to erase all data stored on your BlackBerry device. You or the Help Desk should also contact the cell phone company to disable the BlackBerry's phone service.

If you are not at an organization that has a BlackBerry Enterprise Server, then you should immediately contact the cell phone company that supplied your BlackBerry and let it know what happened. Hopefully, a good Samaritan will find your BlackBerry and return it to you.

Turning Off Password Security

It's possible that you might want to turn off your BlackBerry's **Password Security** feature for some reason. Follow these steps to disable **Password Security**:

1. Follow the steps shown above to get to this **Password** screen inside **Options**.
2. Click **Enable** to turn off the checkmark next to it.
3. Press the **Escape** key and **Save** your settings.
4. You will have to enter your password one last time to complete the process.

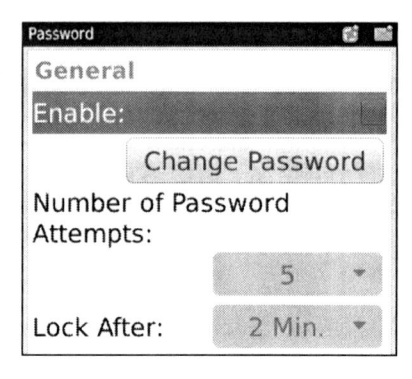

NOTE: If you work at an organization with a BlackBerry Enterprise Server, you may not be allowed to turn off your password.

Battery Life Tips

As everyone knows, your BlackBerry's battery is critical to its smooth operation. But what you may not know is what happens when it gets too low, how often you should charge it, whether there are extended-life batteries available, or even what things can cause it to be drained more quickly. In this section, we will give you an overview of battery issues that relate to using your BlackBerry.

Charging your BlackBerry

Many people wonder how frequently the BlackBerry's battery should be charged. We recommend charging the battery every night. Using your phone is the fastest way to consume your battery life; so, if you talk a lot on your BlackBerry, you will definitely want to charge it every night. Another big drain on your battery is using its speaker to play music or video sound – especially at higher volumes.

TIP: Plug in your BlackBerry charger cable wherever you set your BlackBerry down every night. Then, before setting the BlackBerry down, plug it into the charging cable.

Low Battery Warnings

Also, your BlackBerry provides displays a series of increasingly critical warnings as your battery power drops below certain percentages of a full charge. The following list describes the various battery-related warnings the BlackBerry displays:

- **Low Battery Warning**: This warning is displayed if the battery drops to only 15% of its full charge. If this happens, you will hear a beep and see a warning dialog. Also, the device's **Battery Indicator** meter will usually change to red. At this point, you will want to start looking for a way to charge your BlackBerry – either by connecting it to your computer with a USB cable or through its regular power charger.

- **Very Low Battery:** When the battery gets down to 5% of its full charge, you are in a **Very Low Battery** condition. At this point, the BlackBerry will automatically turn off your wireless radio to conserve what little remaining power there is. This will prevent you from making any calls, browsing the Web, sending/receiving email or messages of any kind at all. You should try to charge the device immediately or turn it off.

- **(Almost) Dead Battery:** When the BlackBerry senses the battery is just about to run out altogether, then it will automatically shut off the BlackBerry itself. This is a preventative measure, meant to ensure your data remains safe on the device. However, we recommend getting the BlackBerry charged as quickly as possible after you are in this condition.

Places to Recharge Your BlackBerry

You can charge your BlackBerry in many on-the-go other settings, including in your car or from your computer. The next couple sections provide more information on the various places you might be able to recharge your BlackBerry.

Recharging from the Power Outlet

This is the fastest and most effective way to charge your BlackBerry. Connect the USB cable to the power charger and plug it into the wall. If you're in a hurry and have access to an outlet, use this method.

Recharging in a Car

Obviously, if you're traveling, you will want to be able to recharge your BlackBerry in a car. The easiest way to do this is to purchase a car charger from a BlackBerry accessory store. To find a car charger online, type *blackberry car charger* in your favorite search engine.

> **NOTE:** If you are coming from an older BlackBerry, the Torch uses the new thin Micro USB charger; your older chargers may not fit this device.

The other way to charge your BlackBerry is to use what is called a *vehicle power inverter* that converts your vehicle's 12V direct current (DC) into alternating current (AC) that you can plug your BlackBerry charger into. The other option with a power inverter is to plug your laptop into it, and then connect your BlackBerry with the USB cable to the laptop (a true mobile office setup!).

Recharging from Your Computer

As outlined previously in this book, you can also recharge your BlackBerry by connecting the device to your computer with a USB cable. Your device will charge as long as your computer is plugged into a power outlet (or vehicle power inverter) and your BlackBerry is connected to your computer through the aforementioned USB cable.

Extending Your Battery Life

If you are in an area with very poor (or no) wireless coverage, you can extend your battery life by turning off your device's wireless radio through the **Manage Connections** feature. When the wireless radio is searching for the network, it consumes much more battery power.

Here are some additional tips you can follow to extend your BlackBerry's battery life:

- **Use headphones instead of the built-in speaker to listen to music or watch videos:** The built-in speaker consumes much more power than a headset does, so minimizing the use of the speaker will extend your expected battery life.

- **Use your speakerphone sparingly:** The speakerphone uses much more energy than either the regular speaker or your headset.

- **Decrease the backlight brightness or reduce the timeout:** You can control this property through your device's **Screen Display** options. (Get to this screen by typing "screen" in your **Search** icon, tap **Options** then tap **Screen Display**.)

- **If possible, set your profiles to ring and/or vibrate less often:** You might want to turn off the vibration feature for every email you receive. This will not only extend your battery life, but it may help you live a calmer and more peaceful existence. Instead, you can check your BlackBerry for new messages at regular intervals, not as every single message arrives.

- **Use data intensive applications sparingly:** Such applications make a lot of use of the radio transmit/receive feature, which uses up the battery more quickly. For example, mapping programs transmit large amounts of data to show the map moving or related satellite imagery.

- **More email and texting instead of talking on the phone:** The phone is one of the largest consumer of battery life. If possible, send an email or SMS text message instead of making a call. Lots of time, this is less intrusive for the recipient, and you may be able to get an answer when a phone call might not be possible. For example, your contact might be in a meeting and able to respond by text, even though he can't talk to you if you were to call instead.

Chapter 2:
Windows PC Setup

This chapter shows you how to install BlackBerry Desktop Software (formerly called "BlackBerry Desktop Manager") on your Windows computer. Then it shows you how to synchronize your contacts, calendar, tasks and memos, backup and restore, and more.

If you want to transfer files and media, then check out Chapter 3: "Windows PC Media and File Transfer."(You will need some of the instructions in this chapter on how to install Desktop Software if you want to use it as your method to transfer files.)

Have an Apple Mac computer? Please go to Chapter 4: "Apple Mac Setup."

Unless you work at an organization that provides you access to a BES (BlackBerry Enterprise Server), if you are a Windows user, you will need to use BlackBerry Desktop Software to do a number of things:

- Transfer or synchronize your personal information (addresses, calendar, tasks, and notes) between your computer and your BlackBerry.
- Back up and restore your BlackBerry data.
- Install or remove an application. (You can also do this right on your BlackBerry– see Chapter 20: "BlackBerry App World")
- Transfer or sync your media (songs, videos, and pictures) to your BlackBerry (see Chapter 3.)

CAUTION: Do not sync your BlackBerry with several computers .You could corrupt your BlackBerry and/or other databases, and end up with duplicates, or worse yet, deleted items.

Downloading Desktop Software for Windows

Each new version of RIM's Desktop Software program has come with more functionality and more versatility than the previous versions. So, it is always a good idea to keep up to date with the latest version of the Desktop Software.

The Disk from the BlackBerry Box

It is fairly likely that the disk that arrived with your brand new BlackBerry has a version of Desktop Software that is already out of date. This is because many times, the CD was produced more than a month ago, and in the meantime a new version has been released. So we recommend grabbing the latest version from the internet directly from www.blackberry.com, as shown below.

Checking Your Current Version

If you have already installed Desktop Software, you should check which version you currently have. The easiest was to do that is start up your Desktop Software program by clicking on the icon or doing a search from your Windows search bar.

1. Click on the Help (Question mark) icon in the upper right corner as shown.

2. Select **About BlackBerry Desktop Software**.

3. The version number of your particular version will be shown here under the name. If you don't have version **6.0.1** or higher, it is time to upgrade.

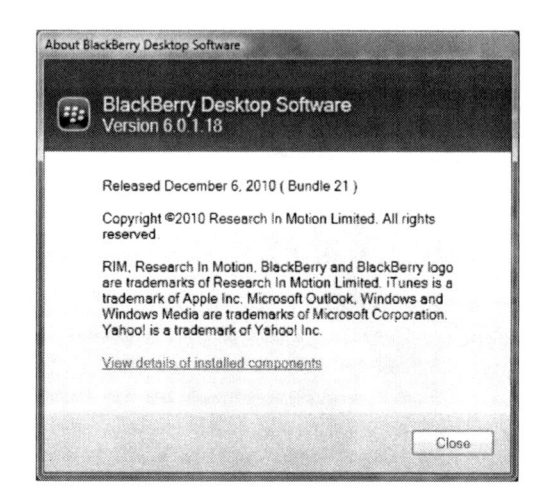

4. If you are not on the latest version, or just want to verify, then from the same **Help** menu (question mark), select **Check for Updates**.

5. Then you will see a pop-up window saying you have the latest version or there is a new one available.

Getting the Latest Version of Desktop Software

If you have Desktop Software installed already use the Check For Updates method above. If you don't have it installed, then go to BlackBerry.com by using this link: http://us.blackberry.com/apps-software/desktop/ (link will be different if you are from a country other than the US). Otherwise, perform a web search for "BlackBerry Desktop Software download," pick the search result that looks correct, and go to the BlackBerry web site. Then follow the links to get to the download screen shown in Figure 2-1. Click the **Download** button at the bottom to get started.

CHAPTER 2: Windows PC Setup

TIP: We offer a free video tutorial showing you how to Update your version of BlackBerry Desktop Software on YouTube: http://www.youtube.com/watch?v=dnJpGugmyVc

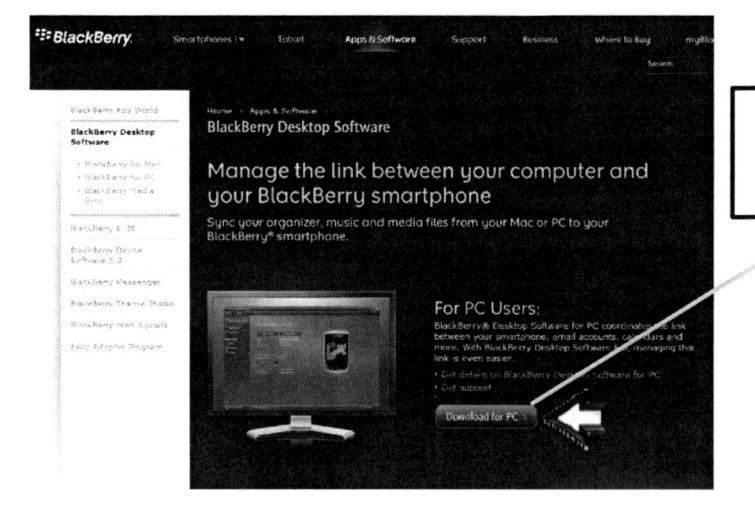

Figure 2-1. BlackBerry Desktop Software download page

Save the file to a place where you will remember it. This is a large file so it may take some time to download.

Installing BlackBerry Desktop Software

Locate and double-click the installation file that you downloaded. It will usually be in your **Downloads** folder unless you changed the default and will probably look something like Figure 2-2.

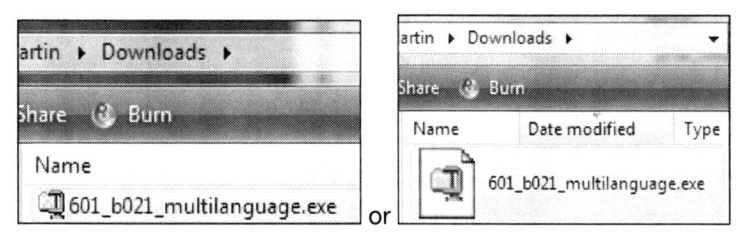

Figure 2-2. Locating the downloaded installation file on your computer

What the file looks like will depend on your view in Windows Explorer (e.g., Small Icons, Large Icons, List, or Details). The first few numbers in the file name correspond to the version of Desktop Software; in Figure 2-2 it shows **601** for version **6.0.1**.

After you double-click the install file, then follow the directions to complete the installation.

Overview of BlackBerry Desktop Software

One of the great things about your BlackBerry is the amount of information, entertainment, and fun that you can carry in your pocket at all times. But what would happen if you lost your BlackBerry or lost some of your information? How would you get it back? What if you wanted to put music from your computer on your BlackBerry? Fortunately, your BlackBerry comes with BlackBerry Desktop Software, which can back up information and load new applications on your BlackBerry, as well as synchronize your computer to your BlackBerry and add media to it.

TIP: We offer a free video tutorial showing you a Tour of BlackBerry Desktop Software v6.0 on YouTube:
http://www.youtube.com/watch?v=uOkiJdllycg

To get started, click the **BlackBerry Desktop Software** icon on your computer, or go to **Start ➤ All Programs ➤ BlackBerry ➤ BlackBerry Desktop Software.** When it starts, you should see a screen similar to Figure 2-3.

Discovered Music Question

You will only see this pop-up window if you have already synced music from your computer library. We recommend selecting **Yes** to avoid having duplicates added to your computer media library (iTunes or Windows Media Player).

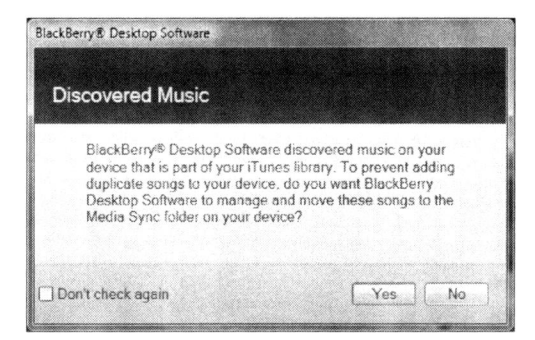

Make sure your BlackBerry is attached to your computer via the USB cable provided.

CHAPTER 2: Windows PC Setup

Figure 2-3. *BlackBerry Desktop Software main screen showing BlackBerry Torch 9800 connected via USB cable*

NOTE: Version 6.0.1 is shown in Figure 2-3; your version may be higher.

Entering Your Device Password

If you have enabled password security on your BlackBerry, you will have to enter your password on your computer right after you connect your BlackBerry to your computer.

View Video Tutorials and Free Tips at www.MadeSimpleLearning.com 99

Switch Devices (Old Phone to New Torch)

TIP: We offer a free video tutorial showing you how to Switch Devices using BlackBerry Desktop Software v6.0 on YouTube: http://www.youtube.com/watch?v=sDUcgGoOPBA

If you are upgrading from another BlackBerry, Palm, or Windows Mobile device, you will want to use the Switch Devices function in Desktop Software.

1. To start the process, select **Device** > **Switch Devices** from the menu.

2. Then connect your old device to your computer with the USB cable and select it from this screen.

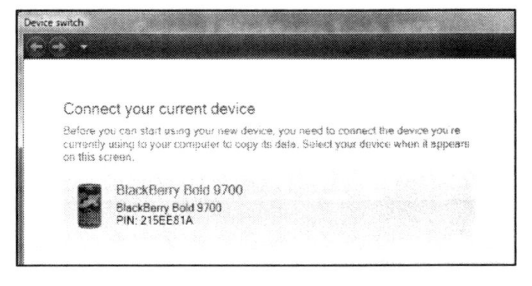

3. Then select whether or now you want both **Device Data** and **Third Party Applications**,(the default) and click **Next**.

NOTE: Some applications from your old BlackBerry may not work on the new Torch.

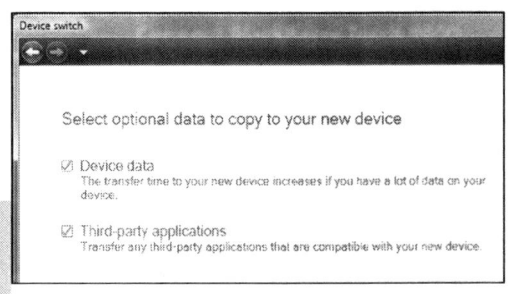

4. Then all your data, settings and applications will be backed up from your old device.

5. Finally, you will be asked to connect your new Torch to your computer to complete the process.

> **NOTE:** You will not be able to copy any software from your Palm/Windows Mobile device to your BlackBerry. If you have a favorite application from Palm/Windows Mobile, then check out Chapter 20: "BlackBerry App World" or the software vendor's web site to see if they have a version compatible with your BlackBerry Torch.

Synchronizing Organizer Data (Contacts, Calendar, Tasks and MemoPad)

You will probably come to rely on your BlackBerry more and more as you get comfortable using it. Think about how much information you have stored on it.

> **TIP:** If you use Google for your contacts and calendar, you should use the Google Sync application instead of BlackBerry Desktop Software – see Chapter: 10 "Your Contact List" and Chapter 11:" Manage Your Calendar" for detailed steps.

Now ask yourself, "Is all my information safely stored in my computer or other safe place?" Then ask, "Is all my BlackBerry information synchronized or shared with the information in my computer software?" Synchronizing your BlackBerry with Desktop Software is very important. Your data will be safe and backed up or shared with the correct program on your computer—making things like your calendar, address book (contacts), and tasks more useful.

> **TIP**: We offer a free video tutorial showing you how to Synchronize your Organizer Data using BlackBerry Desktop Software on YouTube:
> http://www.youtube.com/watch?v=Q_i5eMcbsYE

Setting Up the Sync

Start up Desktop Software on your computer as shown above.

1. Connect your BlackBerry to your computer using the USB cable and make sure you see the green USB logo next to the name of your Blackberry in the main screen.

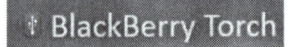

2. Click **Organizer** in the left column.

3. Click the **Configure Settings** button as shown in Figure 2-4.

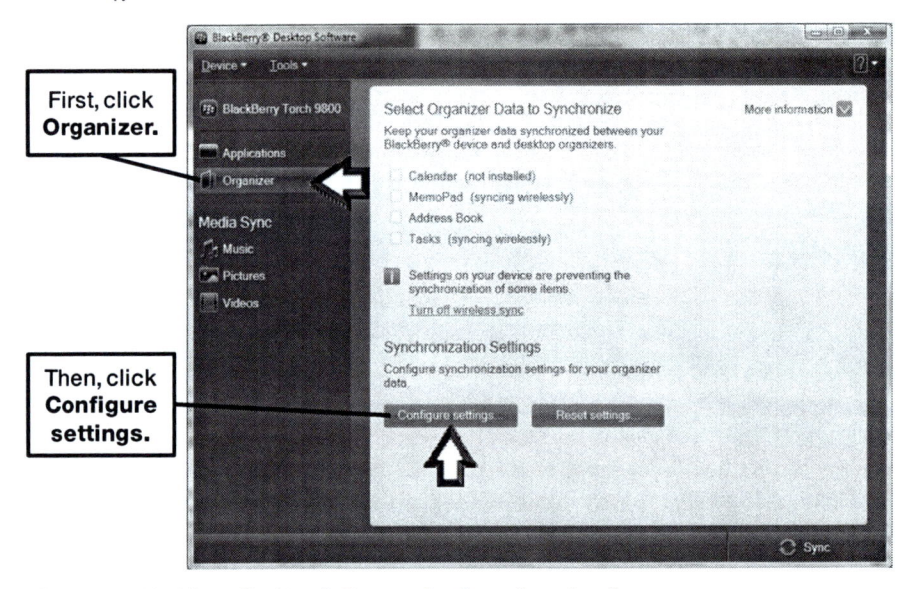

Figure 2-4. BlackBerry Desktop Software – Configure Organizer Sync

4. Now you will see the main IntelliSync program window shown in Figure 2-5.

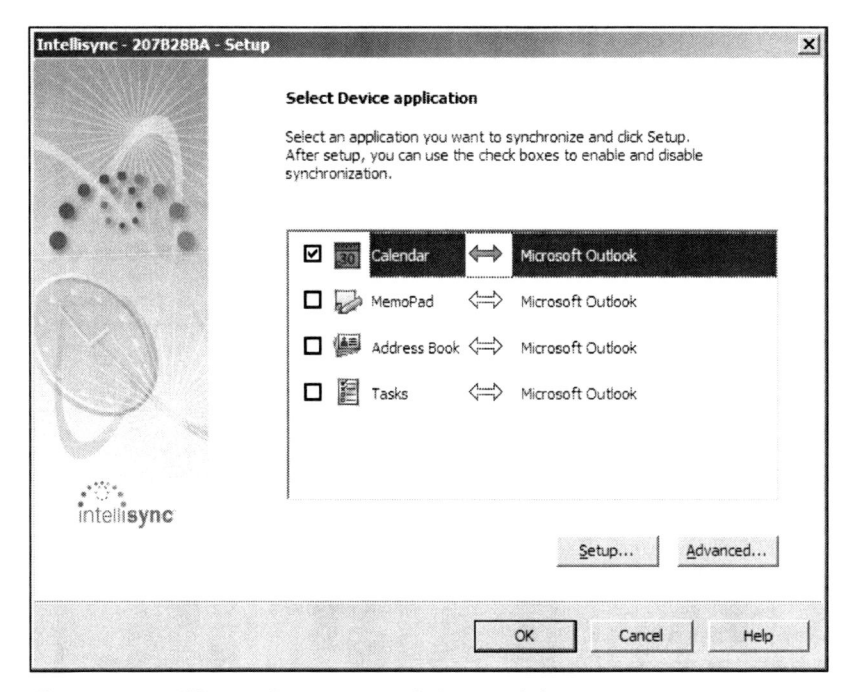

Figure 2-5. IntelliSync main setup screen in Desktop Software

5. To get started, just check the box next to the icon you want to sync (or click the name of the icon, or click the check box and then click the **Setup** button at the bottom). For example, clicking **Calendar** and **Setup** will bring you to screens with details on how to sync your computer's calendar to your BlackBerry.

6. Select your desktop application from the list in Figure 2-6 and click **Next**.

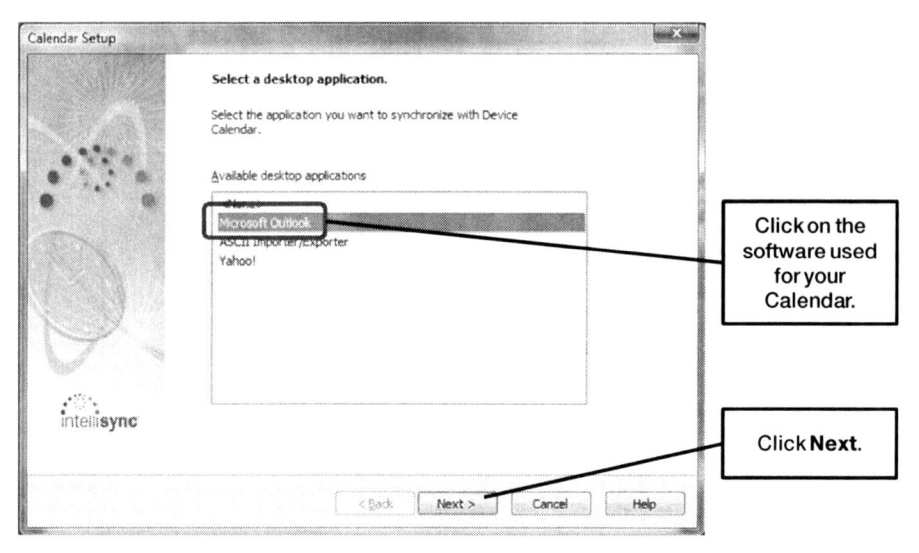

Figure 2-6. Select your desktop application to sync using IntelliSync in Desktop Software.

7. Now you will see options for two-way or one-way sync. Two-way sync means that any changes you make on your computer or BlackBerry will be synchronized to the other device. This is what you usually will want. Under special circumstances, you might require or want one-way sync.

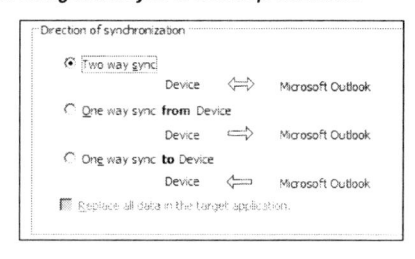

8. Click **Next** to see an advanced screen with more options. This screen shows options for the calendar. The screens for the address book, tasks, and MemoPad may have different options. We recommend the settings shown to help make sure you never miss out on any data you enter on your BlackBerry if you forget to sync every day. These settings will sync calendar events up to 30 days old from your

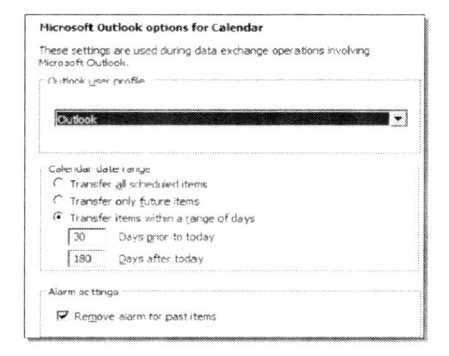

CHAPTER 2: Windows PC Setup

BlackBerry and 180 days into the future.

9. Repeat the procedure for all the applications you want synced. Click **Next** then click **Finish** on the next screen. You'll see similar screens for all four applications; with some minor variations (see Figure 2-7).

Repeat the steps for the **Tasks** application over the next few windows.

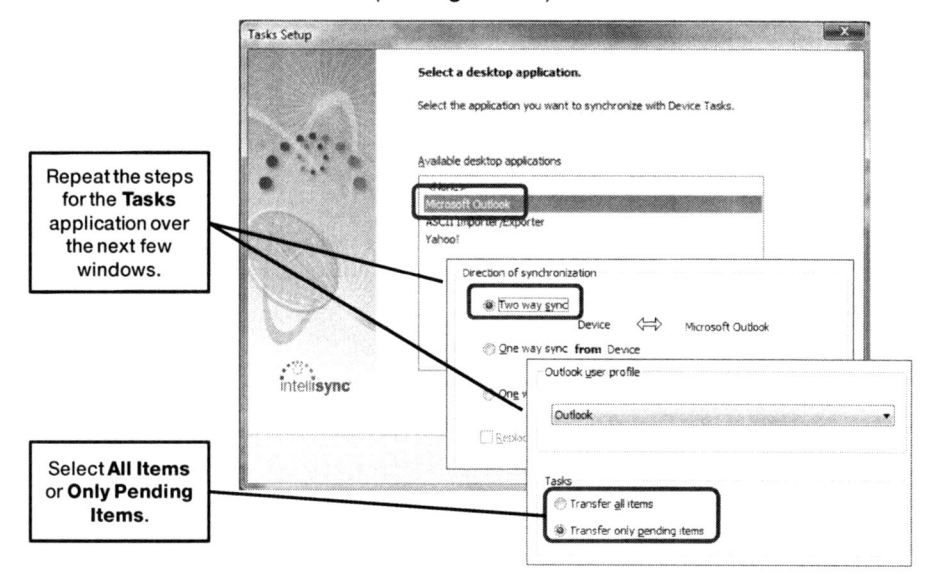

Select **All Items** or **Only Pending Items**.

Figure 2-7. Task sync options in Desktop Software

Once the setup is complete for two-way sync for all four applications, your screen should look similar to the image shown here.

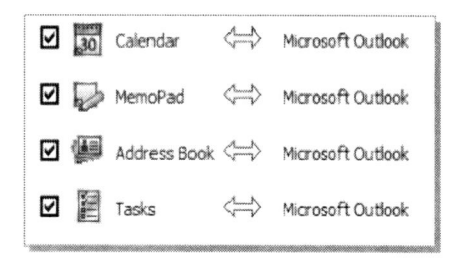

After the configuration is set, go back to the main synchronization screen and check the **Synchronize Automatically** box if you want Desktop Software to automatically synchronize as soon as you connect your BlackBerry to your computer.

Finally, close out all the sync setup windows to save your changes.

Running the Sync

To get the sync started the first time, you have a few options. Desktop Software allows you to Sync All (which includes both the Organizer and Media sync), or sync either type of data separately (see Figure 2-8). Click the **Device** menu and select **Sync All** or **Sync by Type > Organizer**. Or you may click the **Sync All** button in the lower right corner.

Click the **Device** menu then **Sync by Type > Organizer**

or

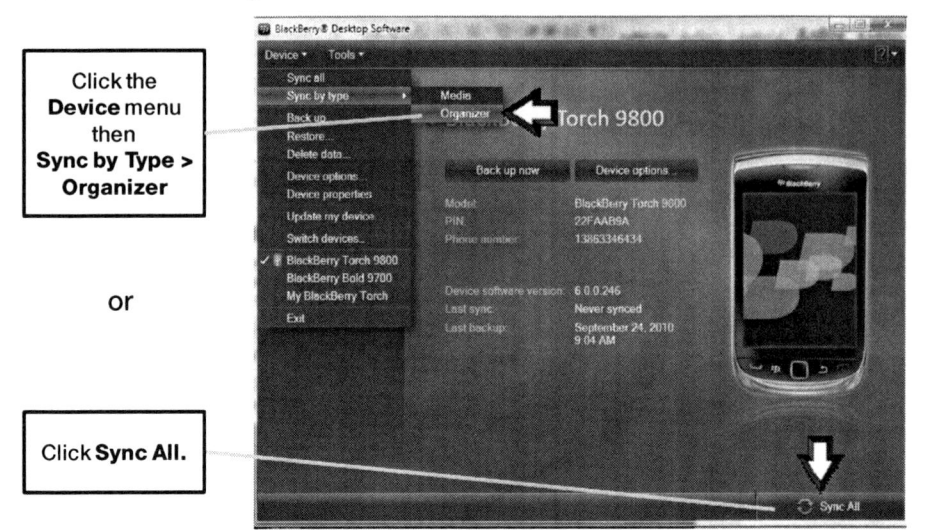

Click **Sync All**.

Figure 2-8. Steps to start the sync in Desktop Software

After starting the sync, you will see a small window pop up showing you status of the current sync. We show you some common messages below.

CHAPTER 2: Windows PC Setup

Automating the Sync

You can automatically sync your BlackBerry every time you connect it to your computer in the Device Options screen.

1. From the menu, select **Device** then **Device Options**.

2. Check the box next to **Organizer data** as shown. Click the box next to **Media files** to automatically sync those as well.

3. Click **OK**.

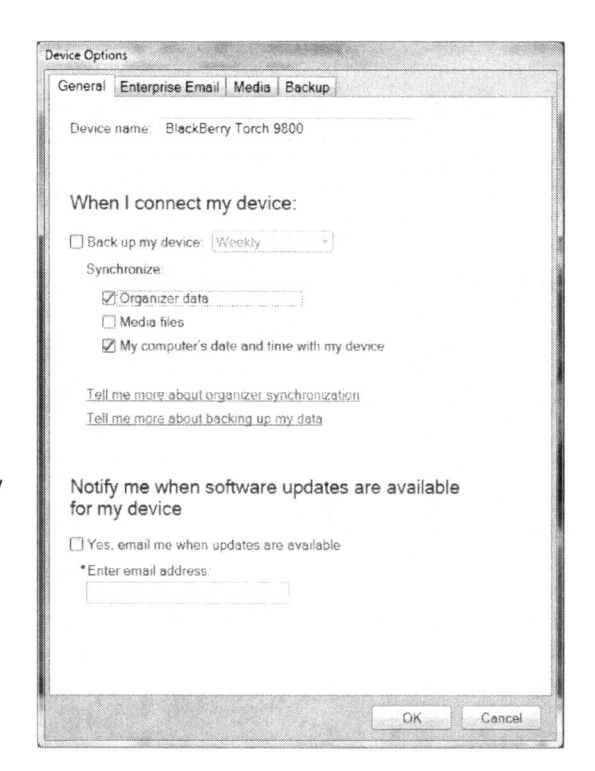

Accepting or Rejecting Sync Changes

During the sync, if there are additions or deletions found in either the BlackBerry or the computer application, a dialog box will appear giving you the option to accept or reject the changes. Click **Details** if you want to see more about the specific changes found.

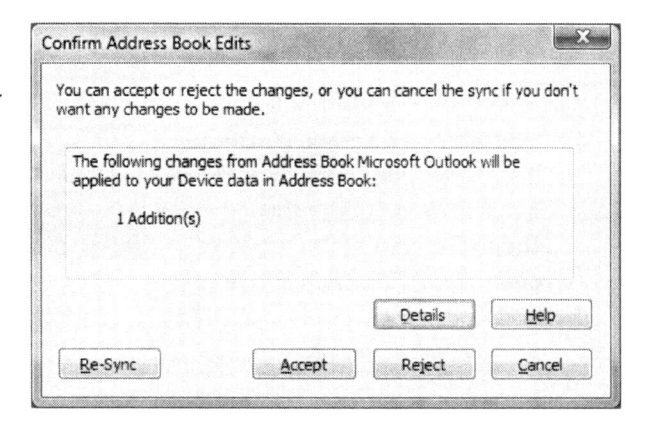

Usually, we recommend accepting the changes unless something looks strange. Finally, the synchronization process will come to an end and your data will be transferred to both your BlackBerry and your computer. If you want more details

View Video Tutorials and Free Tips at www.MadeSimpleLearning.com 107

on what has changed on your BlackBerry and computer, click the **Details** button.

Troubleshooting the Organizer Sync

Sometimes you will encounter errors or warning messages when you try to sync. In this section we try help you through some of the more common issues.

Message That Default Calendar Service Has Changed

Sometimes you may see a message similar to the one shown. This could happen when you add new email accounts to your BlackBerry. Click **Cancel** on the screen shown to the right and follow the steps below to verify everything is **OK** before you sync again.

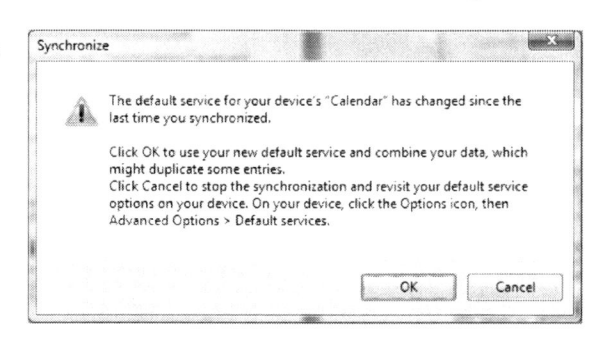

After pressing the **Cancel** button, follow these steps:

1. Tap the Search icon and type **Default**.
2. Tap the **Options** icon.
3. Click **Default Services**.
4. You will see a screen similar to the one shown. Verify that the email address under the **Calendar** item at the top is set correctly. If not, click and adjust it.
5. Press the **Menu** key and select **Save**.
6. Re-sync using Desktop Software, and if you see the same error—ignore it by clicking **OK**.

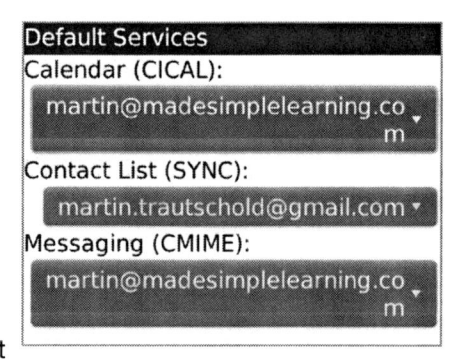

CHAPTER 2: Windows PC Setup

Closing and Restarting Desktop Software

Try closing down Desktop Software and restarting it; sometimes this can help with many issues you may be experiencing.

Removing and Reconnecting your BlackBerry

Sometimes a simple disconnect and reconnect can also help. Give it a try.

Getting More Help for Desktop Software Issues

Some of the Desktop Software sync issues can be particularly tricky. Before you pull out too much hair, you should go try the BlackBerry Technical Solutions Center. You should also try some of the more popular BlackBerry online discussion forums to see if others have experienced and solved similar issues.

To use the **BlackBerry Technical Solution Center** use your computer's web browser and find the site by doing a web search for it, or you can go to this web link: http://www.blackberry.com/btsc

Alternatively, you can pull up your favorite web browser search engine and try a web search for the particular issue you are facing.

The following are some of the BlackBerry forums that could be helpful with a variety of issues:

- www.crackberry.com
- www.blackberryforums.com
- www.berryreview.com
- www.blackberryrocks.com
- www.blackberrycool.com

Backup Data

TIP: We offer a free video tutorial showing you how to Add or Remove Applications using BlackBerry Desktop Software on YouTube: http://www.youtube.com/watch?v=8Fn1T5D_Vw0

In order to protect the information on your BlackBerry from being lost, damaged or stolen, we highly recommend you backup your BlackBerry on a periodic basis. You can even automate the backup in Desktop Software.

1. Start up **Desktop Software** and connect your BlackBerry to your computer.

2. Click the **Back up now** button in the middle of the screen or you click the **Device** menu and select **Backup** as shown.

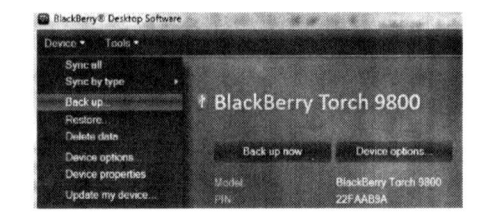

3. You will be shown the **Back Up Options** screen the first time where you can adjust the items related to the backup (see Figure 2-9).

Backup Type:

Full includes all email and settings, **Quick** does not include email and **Custom** allows you to select individual types of data to backup.

Include files saved on built-in media storage – we recommend checking this box to back up any photos, videos or other items that may be stored on your internal memory rather than your media card.

File name – leave this alone.

Save backup files to – we recommend changing this to a location on your computer that is backed up regularly or to an external drive such as a network disk or USB thumb drive for safe keeping.

Encrypt Backup Files – for additional security you can force a password to be required to access the backup file.

Check the box next to **Don't ask for these settings again** if you want to skip seeing this screen.

Click the **Back up** button to start the back up.

CHAPTER 2: Windows PC Setup

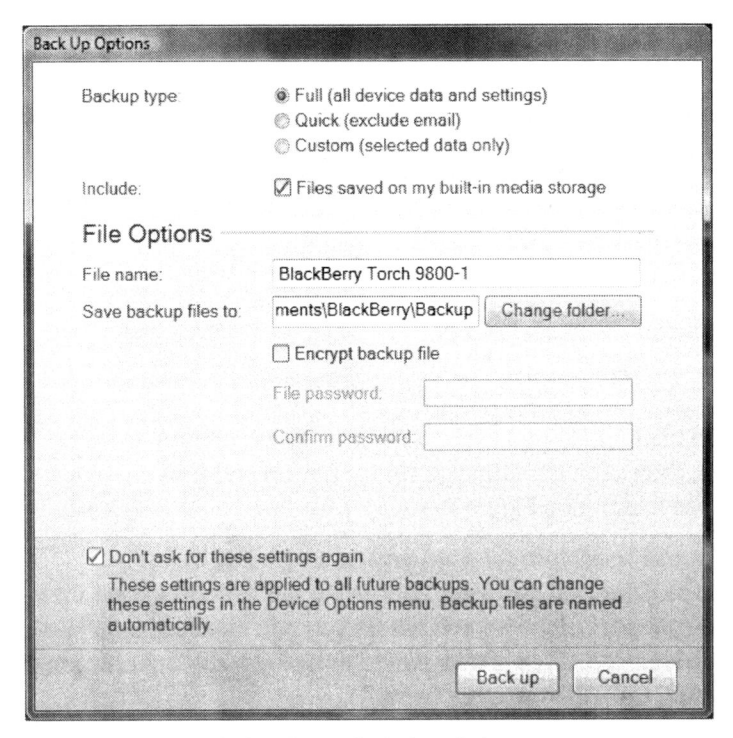

Figure 2-9. Backup Options Screen in Desktop Software

Automating the Backup Process

Backups are like chores you like to forget – that's why there is a way to automate them.

1. Go to the **Device** menu and select **Device Options** to see this screen.

2. Check the box next to **Back up my device** and select **Daily**, **Weekly**, **Bi-weekly** (every 2 weeks), or **Monthly**.

3. Click **OK** to save your settings.

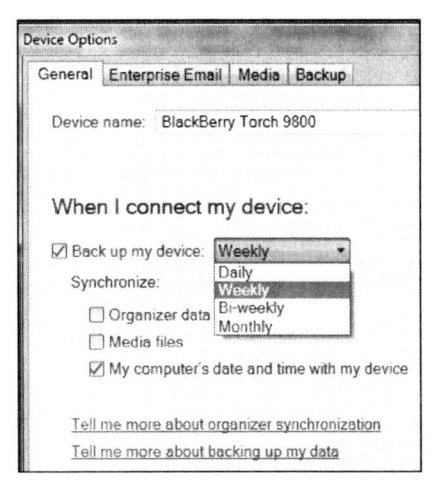

View Video Tutorials and Free Tips at www.MadeSimpleLearning.com   111

Restore Data

Sometimes you may need to restore the data you have backed up to your BlackBerry, we show you how below.

1. Start up **Desktop Software** and connect your BlackBerry to your computer.
2. Click the **Device** menu and select **Restore** as shown. 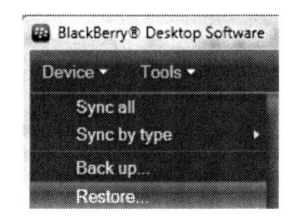
3. Click on the backup file you wish to restore. If you cannot see it listed, then click the **Change folder** button and move to the correct folder (see Figure 2-10).
4. You can choose to restore **All data and settings** (the default) or **Select device data and settings**. Choosing the second option then opens up a new part of the screen where you can check of individual data types to restore. This allows you to selectively restore only your Address Book, Calendar or other items.
5. Click **Restore** to start.

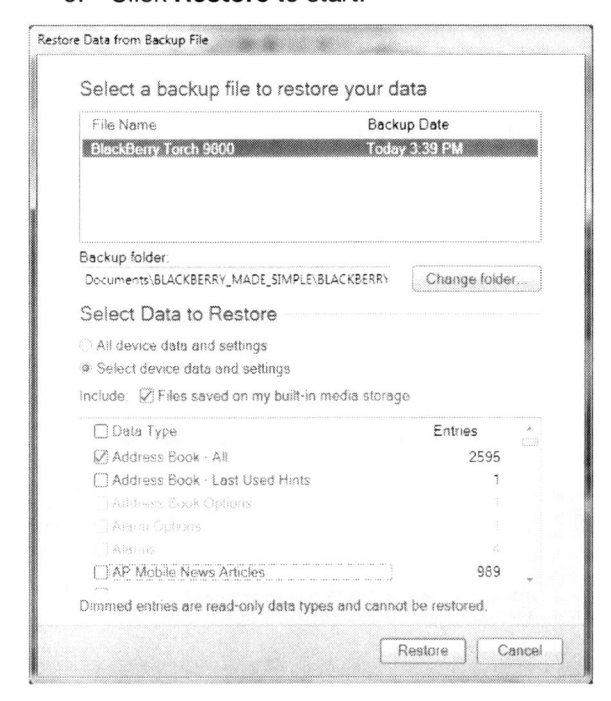

Figure 2-10. Restore Options Screen in Desktop Software

Delete Data

There are times when you may want to delete all the data on your BlackBerry or selectively delete just your address book (contacts), calendar, tasks or other data. In this section we show you how to accomplish this.

1. Start up **Desktop Software** and connect your BlackBerry to your computer.

2. Click the **Device** menu and select **Delete data** as shown.

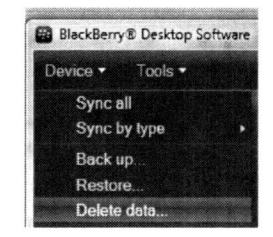

3. You can choose to delete **All data** (the default) or **Selected data**. Choosing the second option then opens up a new part of the screen where you can check of individual data types to delete. This allows you to selectively delete only your Address Book, Calendar or other items (see Figure 2-11).

4. You also have check boxes allowing you to delete files from the internal media storage and to backup the data file before deleting (a good precaution). Finally, you can encrypt the backup if you choose.

5. Click **Delete** to start.

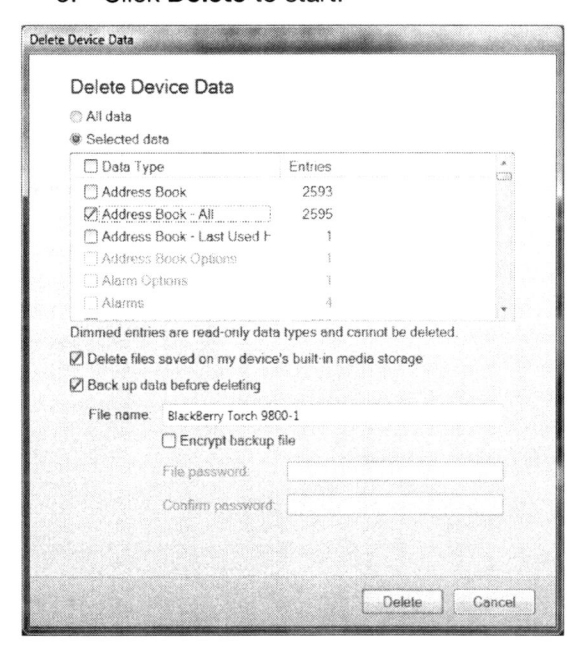

Figure 2-11. Delete Device Data Screen in Desktop Software

Working with Applications

Use **Applications** function in **Desktop Software** to add or remove software from your BlackBerry. This is also used to update your BlackBerry OS or system software. There are easier ways to load or remove software from your device—see Chapter 20: "BlackBerry App World."

TIP: We offer a free video tutorial showing you how to Add or Remove Applications using BlackBerry Desktop Software on YouTube:
http://www.youtube.com/watch?v=8Fn1T5D_Vw0

1. First, make sure your BlackBerry is connected to your computer.
2. Now click **Applications** in the left column to see the screen shown in Figure 2-12.

Click **Applications** first.

Click **Import files** to add files downloaded to this list.

Click the **X** to remove an application.

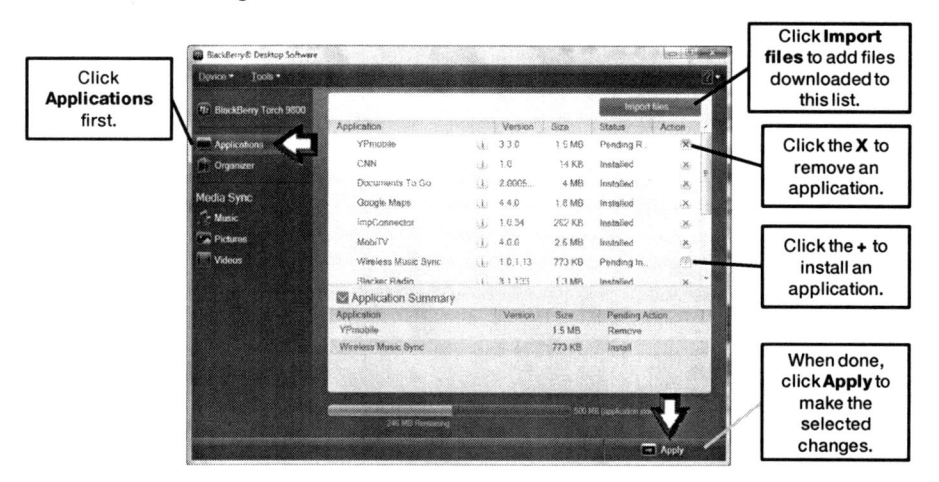

Click the **+** to install an application.

When done, click **Apply** to make the selected changes.

Figure 2-12. Applications main window in Desktop Software

3. To remove applications, click the **X** in the right column.

CHAPTER 2: Windows PC Setup

4. Sometimes after you download applications to your computer from the Internet, they do not automatically appear in this list. To add these files, you will need to click the Import Files button in the upper right corner and look for files that end with .ALX and .COD file extensions. Once you import these files, they will appear in the list.

5. When you're done selecting files, click the **Apply** button in the lower left corner.

Updating the BlackBerry System Software

You can update your the device software, which is the system software running on your BlackBerry smartphone using Desktop Software. You can upgrade and downgrade using this feature.

TIP: We offer a free video tutorial showing you the BlackBerry System Software update process on YouTube:
http://www.youtube.com/watch?v=1clnMFp6lAY

Getting an Email when Software Updates are Available

Select **Device** then **Device Options** from the menu. Near the bottom you will see the section where you can be notified of software updates. Check the box or boxes and enter your email address. Click **OK** to save.

Notify me when software updates are available for my device

☑ Keep me subscribed to email notifications
☑ Use a different email address:
  artin@madesimplelearning.com

Checking the BlackBerry Site for System Software Updates

You can check yourself at any time for the latest software for any BlackBerry wireless carrier from this web page:
http://us.blackberry.com/support/downloads/download_sites.jsp

Click the carrier from your country.

Here is the link for AT&T software downloads: https://www.blackberry.com/Downloads/entry.do?code=577BCC914F9E55D5E4E4F82F9F00E7D4

Once you get to the correct site, you need to select your BlackBerry from the drop down list and scroll down to see the available software.

Compare the version number shown in the main screen of Desktop Software to the version next to Applications shown on the web site. See Figure 2-13.

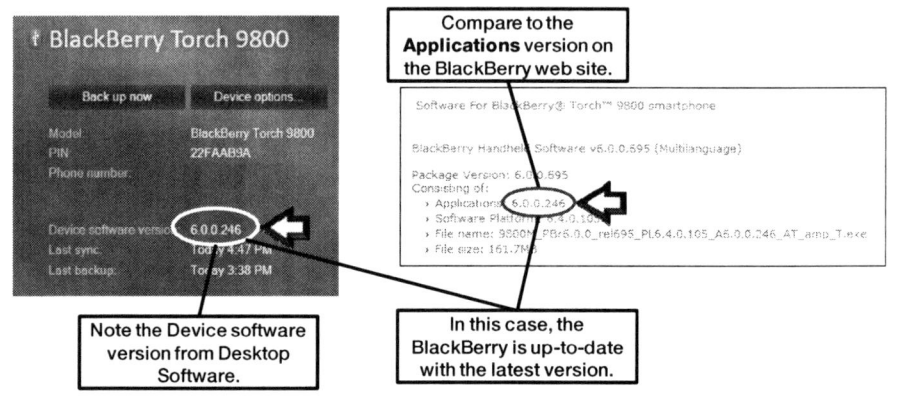

Figure 2-13. Applications main window in Desktop Software

If you see the version available on the BlackBerry web site is newer (higher), then click the download link.

After the software is downloaded, you need to install it on your computer by double-clicking on the downloaded file. You will have to close the Desktop Software in order to complete the installation.

Installing the BlackBerry System Software Update

When an update is available, you will see a new button in the main window in Desktop Software. This button will be called **Update My Device** and show you the new version that is available. Click that button and follow the steps to backup your BlackBerry and install the new system software.

> **NOTE:** This system software update can take 30 minutes or longer, so be prepared to do it when you don't need your BlackBerry for quite a while.

Chapter 3: Windows PC Media and File Transfer

In this chapter we will help you get your important files and media from your Windows computer to your Torch. Your Torch is quite a capable media player on which you can enjoy music, pictures, podcasts, and videos. Like your computer, your BlackBerry can even edit Microsoft Office (Word, Excel and PowerPoint) documents.

You have a variety of choices about how to transfer documents and media and we explore all the most popular ones in this chapter. You will quickly see that some methods such as **USB Drive** transfer work well for pictures and documents (spreadsheets, word processing or presentation files), whereas you will want to use the Media Sync feature in BlackBerry Desktop Software to transfer your music playlists. Here are a few ways to transfer files and media between your BlackBerry and your computer:

- BlackBerry Desktop Software Media Sync
- USB Drive File Transfer
- Email the files to yourself as attachments *(if they are small enough)*

Verify Your Memory Card is Installed

Before you go too wild transferring all your music, videos, pictures and other documents to your Torch, you want to make sure you have a Media Card installed. While your Torch has 4GB (gigabytes) of on-board memory for media files, you may or may not have an additional Media card installed. It's easy to verify that you have the Media Card installed and see how much free space is available. We show you how in the "Verifying the Memory Card Installation and Free Memory" section of Chapter 1: "Getting Started." If you don't have a memory card, we show you how to install the MicroSD format card in your Torch also in Chapter 1.

USB Drive File Transfer Method

The **USB Drive** transfer method allows you to directly copy or drag-and-drop any file types to your BlackBerry when it looks like another disk drive to your computer. We describe how to set up and use this method in the "Transferring Content to your Torch Using USB Drive Mode" section of Chapter 1.

Sync Media with BlackBerry Desktop Software

> **NOTE:** In order to use the Media Sync, you need to install the latest version of BlackBerry Desktop Software. We show you how in Chapter 2: "Windows PC Setup."

Media Sync is a great tool that allows you to sync music, playlists, pictures and videos to your BlackBerry. You can even sync music using the Wireless Music Sync that takes place over your Wi-Fi network.

> **NOTE:** For those of you who have used BlackBerry smartphones for a while you will know that BlackBerry used to provide a separate stand alone Media Sync application but that application has now been absorbed into the Media Sync functionality within Desktop Software.

Confirm Your Media Sync Options

Before you use the Media Sync in Desktop Software, it is a good idea to make sure the Media options are set up correctly.

1. Start Desktop Software and connect your BlackBerry to your computer with the USB cable.
2. Click the **Device options** button in the main screen or select **Device** > **Device Options** from the menu (see Figure 3-1).
3. Click the **Media** tab at the top.
4. If you want to enable **Wi-Fi music sync**, check the top box.

CHAPTER 3: Windows PC Media & File Transfer

5. Select your **Music source**. You can choose between **iTunes** and **Windows Media Player**.

6. We recommend leaving the box checked that says **Start iTunes to acquire album art**, otherwise you may not see any album art on your BlackBerry.

7. The next two buttons allow you to adjust where pictures and videos that you have taken on your BlackBerry will be stored on your computer. The defaults are the **pictures** and **videos** folders on your computer, but you might want to change it to a location that is backed up regularly.

8. You can also adjust where the synced media is stored on your BlackBerry. We recommend the media card.

9. You can use the slider at the bottom to adjust how much free space should be left over after the media sync is complete. This is space for any other files such as Microsoft Office documents, PDF files, or files you save from email attachments. Click **OK** to save your settings.

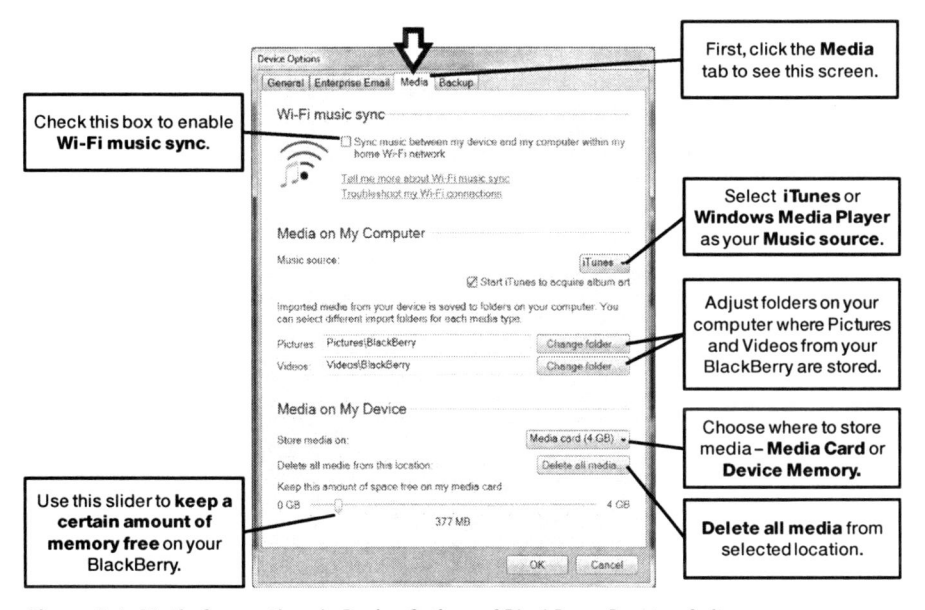

Figure 3-1. Media Sync options in Device Options of BlackBerry Desktop Software

Once you have your media options set, you are ready to sync your media.

View Video Tutorials and Free Tips at www.MadeSimpleLearning.com   119

Syncing Music

In order to set up the music, pictures and videos sync, you need to have BlackBerry Desktop Software running and your BlackBerry connected to your computer with the USB cable.

1. Click **Music** in the left column (see Figure 3-2).

2. If you want to sync all your music check the **All music** box at the top. Unless you have a smaller music library, it is fairly likely all your music will not fit on the BlackBerry. In that case use the other checkboxes to select more specifics.

3. Check boxes in the **Artists**, **Playlists** and **Genres** to sync all music from that particular artist, playlist or genre.

4. To view songs in a particular artist, playlist or genre, click on that item to highlight it and view the songs listed at the bottom of the screen.

5. If you want to find a song, click the **Search** area in the top right of the screen.

6. If you are feeling like trying something new or want to just be surprised by what is synced, click the **Random Music** button in the lower right corner. This will fill your BlackBerry with random songs from your library.

7. Click the **Sync** button in the lower right corner to start the music sync.

CHAPTER 3: Windows PC Media & File Transfer

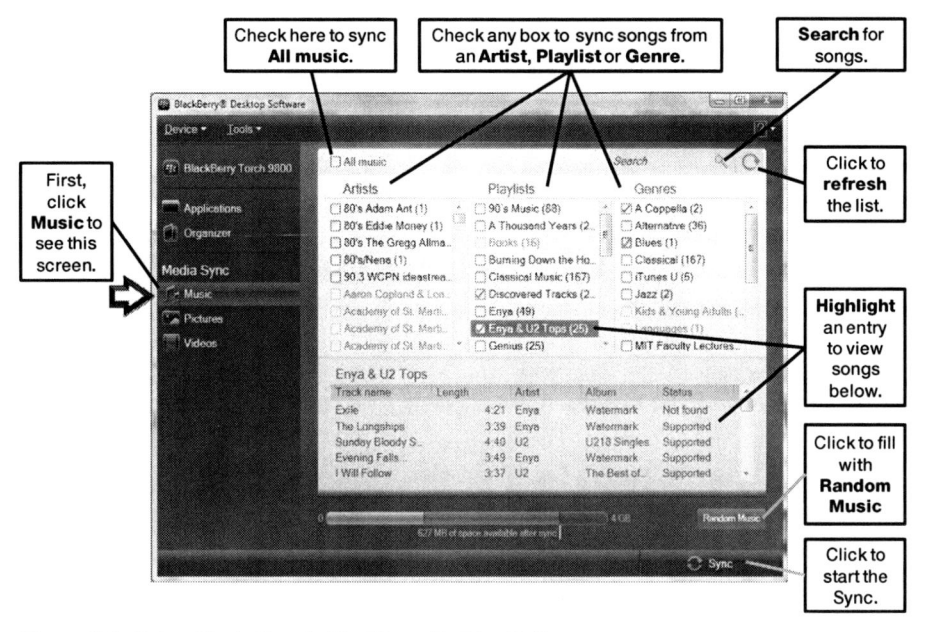

Figure 3-2. Set up Music Sync in BlackBerry Desktop Software

Some Music Cannot Be Synced

After you sync your music, you may see a pop-up message similar to this one saying certain songs could not be synced. Usually, this is because the songs are protected with DRM (Digital Rights Management) and can only be played on certain devices (e.g. iPod.) Sometimes songs are just not found.

Click the link at the bottom to view songs and determine exactly why they were not able to be synced.

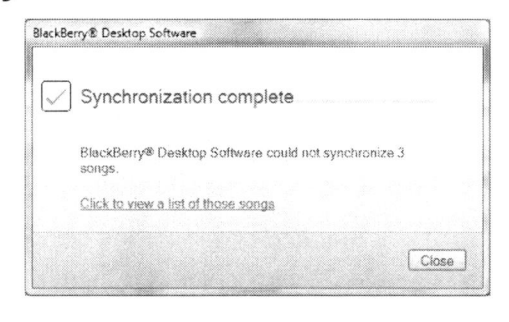

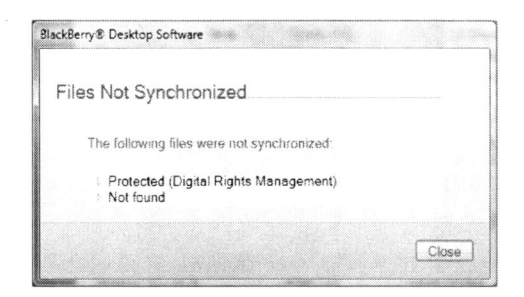

View Video Tutorials and Free Tips at www.MadeSimpleLearning.com   121

Wi-Fi Wireless Music Sync

BlackBerry now offers a new feature called **Wireless** or **Wi-Fi Music Sync**. What this allows you to do is view your computer's entire music library from your BlackBerry.

After you enable the **Wi-Fi Music Sync** as shown below, inside your **Music** app on your BlackBerry you will see that Wi-Fi Sync has been enabled.

Below are the steps to enable **Wi-Fi Music Sync**, then we show you how to use it on your BlackBerry.

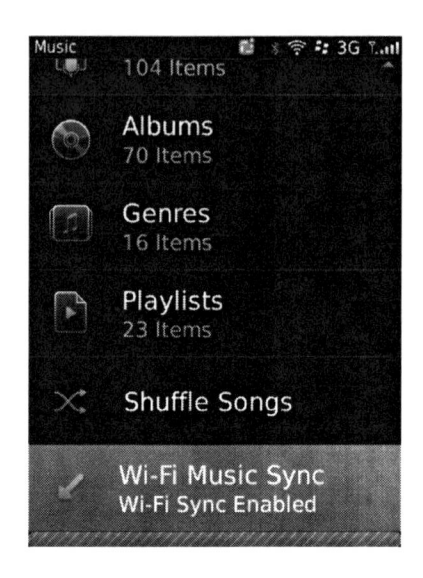

Set Up Step 1: Installing the Music Sync App on Your BlackBerry

Before you can sync music wirelessly, you need to use the Applications function to load up the Wireless Music Sync app on your BlackBerry. Here's how to get that done:

1. Start up **BlackBerry Desktop Software** and connect your BlackBerry with the USB cable to your computer.

2. Click **Applications** in the left column.

3. Click the **+** next to **Wireless Music Sync** as shown in Figure 3-3.

4. Click the **Apply** button in the bottom right corner to load the app on your BlackBerry.

CHAPTER 3: Windows PC Media & File Transfer

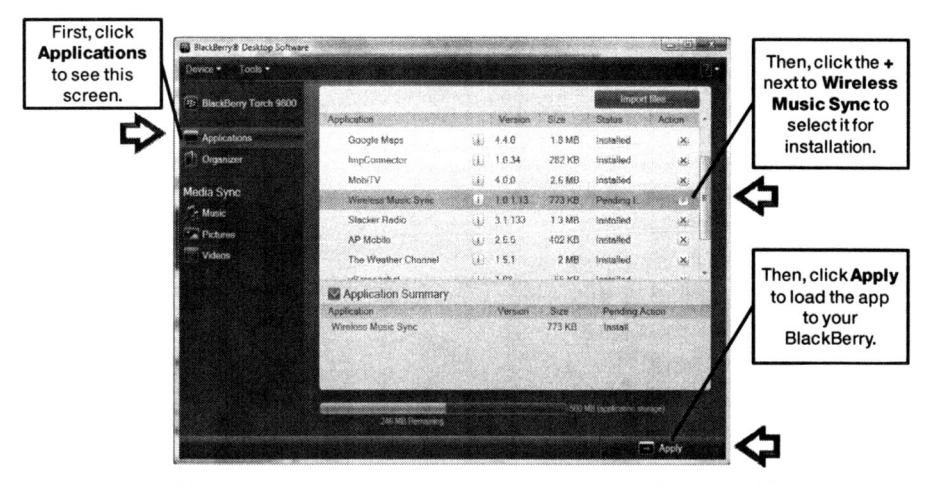

Figure 3-3. Load the Wireless Music Sync App using the Applications funciton in BlackBerry Desktop Software

Set Up Step 2: Starting Wi-Fi Music Sync on your Computer

Next, you need to start up the **Wi-Fi Music Sync** in the **BlackBerry Desktop Software** on your computer.

1. Start BlackBerry Desktop Software and connect your BlackBerry to your computer with the USB cable.

2. Select **Tools > Wi-Fi music sync** as shown.

3. On the next screen, select **Turn wireless sync on**.

TIP: This is the same screen where you will turn off wireless sync if you want to do that later.

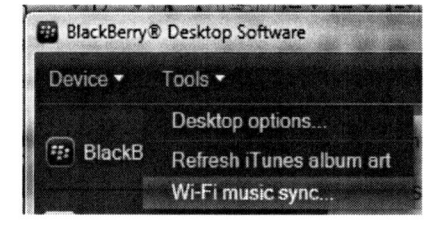

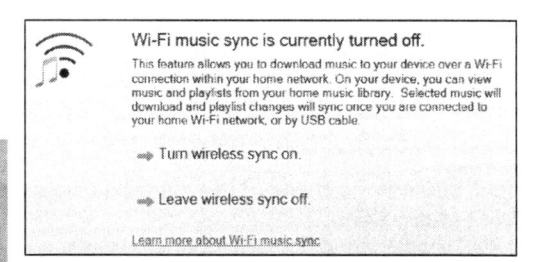

4. Then, the app will make sure you have installed the **Wireless Music Sync** app on your BlackBerry. If it was not installed, the **Wireless Music Sync** will be installed at this time. It will then load up a list of your computer's music library on your BlackBerry including album art.

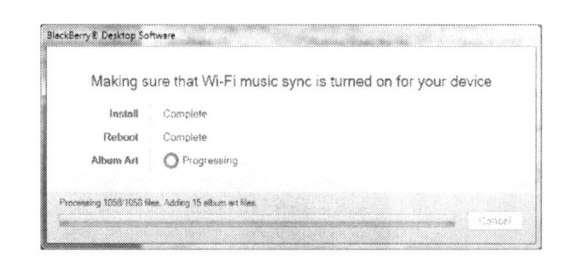

5. Finally, you will see a status message the Desktop Software screen showing that **Wi-Fi Music Sync** is **ON**.

TIP: Click this status message to turn Wi-Fi Sync off.

After this process is done, you will see a screen which gives you the requirements for Wi-Fi Music Sync to operate properly:

- Turn on Wi-Fi on your BlackBerry (See "Connecting with Wi-Fi" section of Chapter 1: "Getting Started.)

- Turn on your Wi-Fi router (This is a little strange -- Has your router ever been turned off?)

- Turn on your computer and have it connected to the Wi-Fi network.

- Click on any dimmed song or playlist on your BlackBerry to add it to your sync queue. When you come in contact with your home network, these songs are synced.

Now, you can grab any of the dimmed songs shown on your BlackBerry using Wi-Fi Music Sync.

NOTE: Some newer dual-band routers have more than one network to which you can connect. Your BlackBerry and computer must be connected to the same network for Wi-Fi Music Sync to work properly.

CHAPTER 3: Windows PC Media & File Transfer

Using Wi-Fi Wireless Music Sync

After you setup the **Wi-Fi Music Sync**, you will notice that some songs, playlists and albums appear grayed out or dimmed in your BlackBerry **Music** app. These songs are in your computer library but not on your BlackBerry. If you click on any of these dimmed items, you can request that they be synced to your BlackBerry (see Figure 3-4).

> **TIP:** You can even request songs, artists or playlists when you are away from your home Wi-Fi network. What happens is that these requests are logged and when you return home the items are synced wirelessly and automatically behind the scenes!

The next time you come in contact with the Wi-Fi network where your computer is located (and your computer is turned on), you will receive the requested songs wirelessly via your Wi-Fi network.

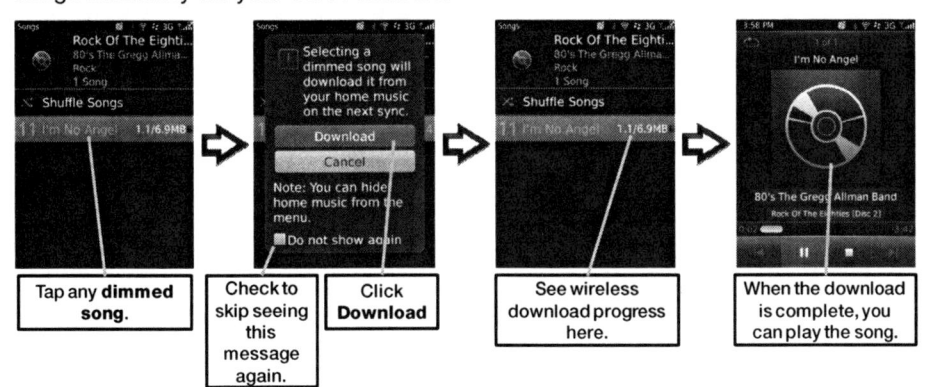

| Tap any **dimmed song**. | Check to skip seeing this message again. | Click **Download** | See wireless download progress here. | When the download is complete, you can play the song. |

Figure 3-4. How Wireless Music Sync works in the Music app on your BlackBerry.

Some Songs Cannot Be Wirelessly Synced

You will notice that when you try to click on a dimmed song from your **Music** app, you will see a couple of different error messages.

Unsupported Format – this means the media is not playable on your BlackBerry or that it has been protected with Digital Rights Management (DRM).

Cannot be found – this means the song cannot be located on your main music library on your computer.

In both cases, you cannot sync these songs to your BlackBerry. Try checking to see if you can play these songs on your computer. Some of these songs may have been purchased in iTunes and are protected to only play on your Apple device.

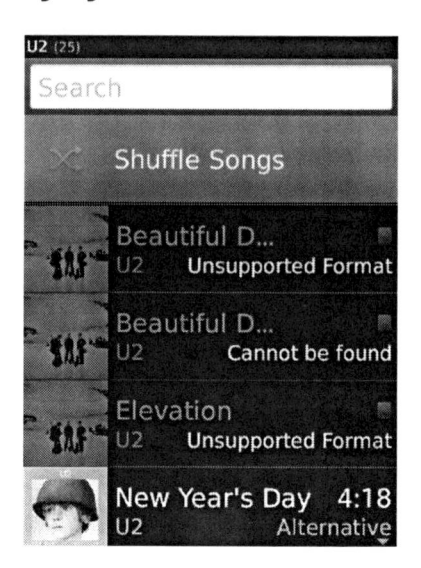

Working with Pictures

You can transfer the pictures you snap on your BlackBerry back to your computer and sync selected pictures to your BlackBerry using media sync. Start up BlackBerry Desktop Software and connect your BlackBerry to your computer with the USB cable.

Click **Pictures** in the left column.

Working with Pictures on your Device (BlackBerry)

Click the **Device Pictures** tab at the top of the screen (see Figure 3-5).

If you want to change the default folder (Pictures/BlackBerry) location where your pictures will be imported, then click the **Change Import Folder** button.

If all the settings look fine, then click the **Import** button in the lower right corner to copy all pictures from your BlackBerry to the selected folder.

CHAPTER 3: Windows PC Media & File Transfer

> **NOTE:** You will see all the pictures disappear from view in the main window after they are copied, however they are not deleted from your BlackBerry. They are no longer marked as new pictures. In order to view the pictures on your BlackBerry, you need to use the pull-down menu under the **Device Pictures** tab and select **All Pictures**.

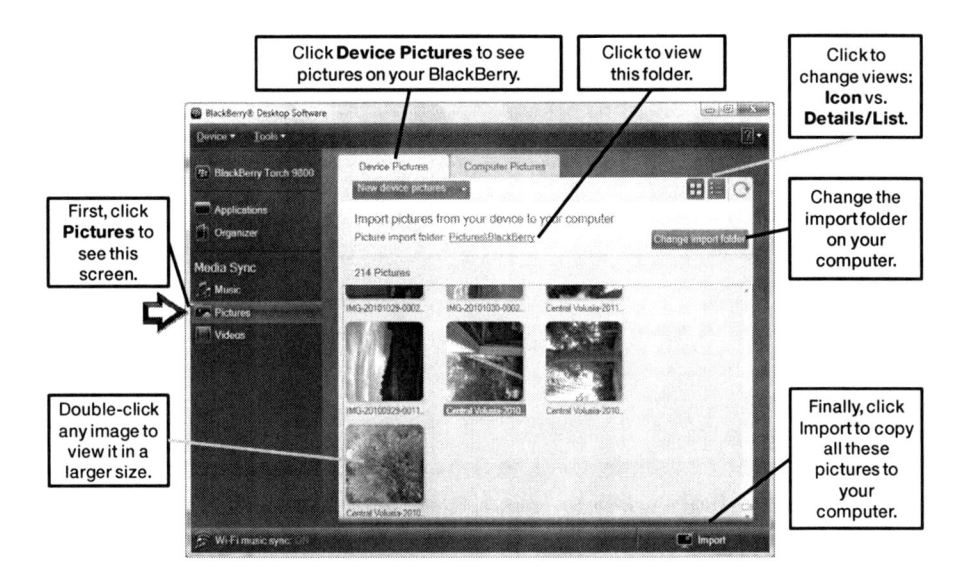

Figure 3-5. Importing Pictures taken on your Device (BlackBerry) to your Computer.

How to Delete Pictures from Your BlackBerry

You cannot delete pictures directly from within the view shown in BlackBerry Desktop Software, so you need to take an extra step.

1. You need to be able to click on one of the pictures shown on your device. If you no longer see any pictures, use the pull-down menu and select **All Device Pictures** near the top of

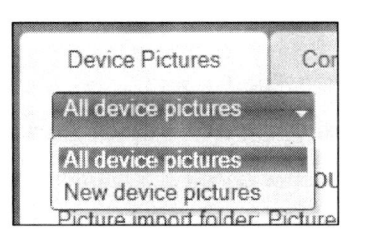

View Video Tutorials and Free Tips at www.MadeSimpleLearning.com 127

the screen.

2. **Right-click** on any image from the Device Pictures view and select **Show in Explorer**.

3. Now you will see all the pictures in the Microsoft Explorer program. From here, you can delete, move, copy or do whatever you want to one, several or all of the pictures.

4. Select one or more pictures using these methods:

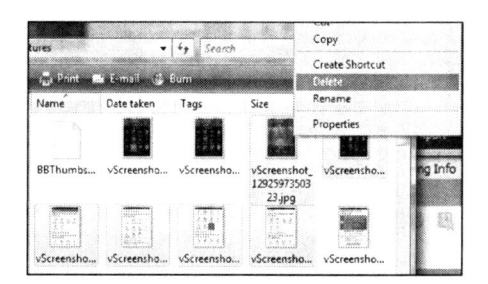

    a. Use **Shift + Right-Click** to select several pictures in a row, or **Ctrl + Right-Click** to select individual pictures. **Ctrl + A** will select all pictures.

    b. Right click any highlighted image to **Delete**, Copy, or Cut them.

Working with Pictures on your Computer

Click the **Computer Pictures** tab at the top of the screen (see Figure 3-6).

You can do the following on this screen:

- If you want to add new folders to sync to your BlackBerry click the **Add Folder** button.
- Check or uncheck entire folders to sync.
- Check or uncheck individual pictures to sync.

When you're done selecting items, click the **Sync** button in the lower right corner to start the sync.

CHAPTER 3: Windows PC Media & File Transfer

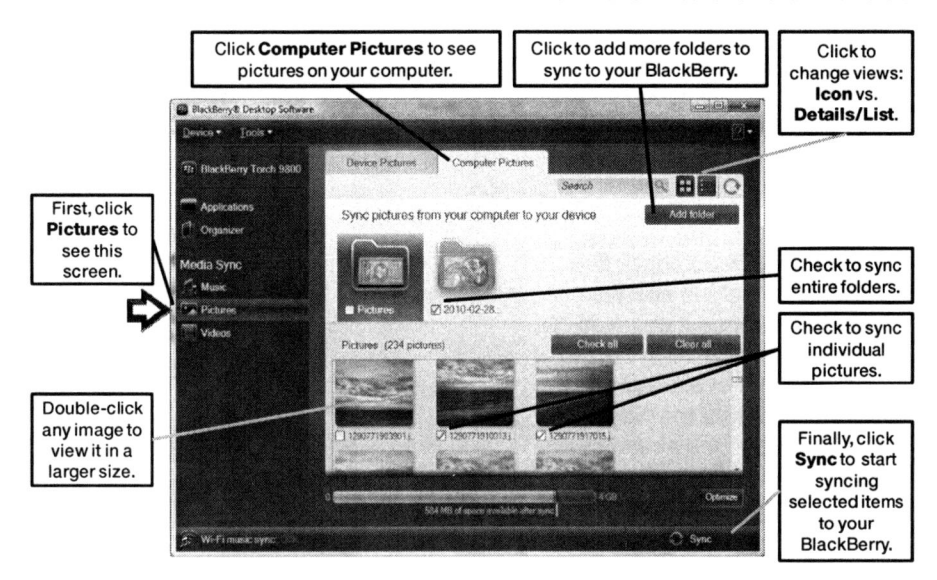

Figure 3-6. Syncing Pictures and Folders of Pictures from your Computer to your BlackBerry.

Working with Videos

You can transfer the videos you take on your BlackBerry back to your computer and sync selected videos to your BlackBerry using media sync. Start up BlackBerry Desktop Software and connect your BlackBerry to your computer with the USB cable.

Click **Videos** in the left column. The functionality for working with Videos is identical to the Pictures shown above, so once you get the feeling for working with pictures, you can use exactly the same steps to work with your videos. You can import, sync and delete in exactly the same way as with pictures.

Video Conversion and Quality

The only thing that is different with videos compared to pictures is that many of the videos need to be converted in order to play on your BlackBerry. This will happen by default if you don't change anything. However, you may want to make some adjustments to the conversion settings or turn it off altogether.

View Video Tutorials and Free Tips at www.MadeSimpleLearning.com   **129**

You will notice in the lower right corner of the screen when you click on the **Computer Videos** tab at the top of the screen that there is a check box and pull down menu.

The **Convert videos** check box is checked by default and we recommend leaving it checked unless you have a strong reason to uncheck it. (For example if you are certain that the videos you want to sync are already formatted and encoded to play perfectly on your BlackBerry.)

If you want to save space and are OK with a slightly grainier look and feel on your videos, then use the pull-down menu and select **Medium quality** or **Low quality**.

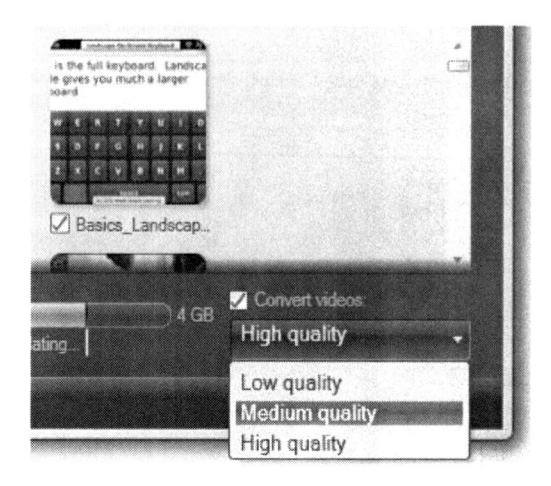

Troubleshooting Desktop Software

Sometimes when you are previewing a file or performing some other function, Desktop Software Manager may crash and stop responding. If this is the case, you can stop the program by following these steps:

1. On your computer, press **Ctrl + Alt + Delete**.
2. If you are given a choice, then select **Start Task Manager**.
3. From Window Task Manager (Figure 3-7), click the **Processes** tab.
4. Then scroll down and highlight the image name of **Rim.Desktop.exe**, as shown.
5. Click the **End Process** button at the bottom.

CHAPTER 3: Windows PC Media & File Transfer

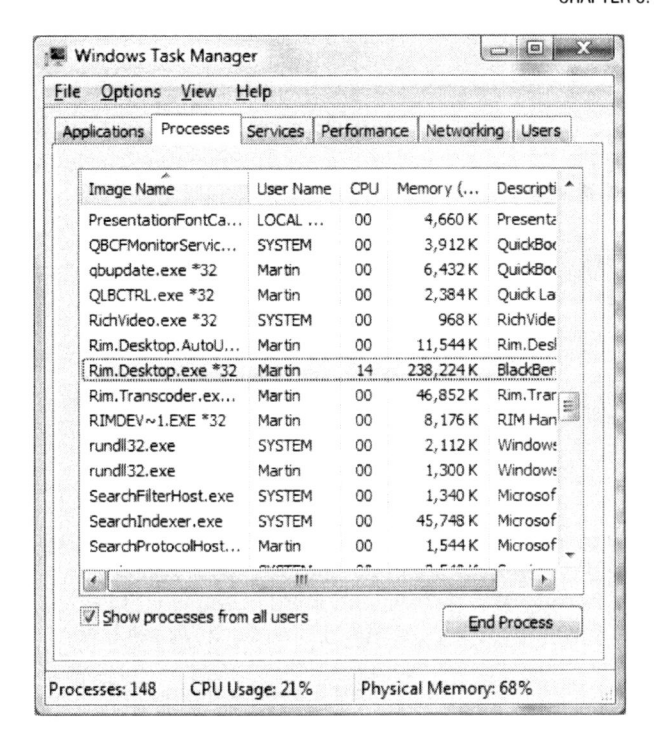

Figure 3-7. Windows Task Manager

6. On the screen that says **Do you want to end this process?** click **End process** to stop the program.

7. Now you can restart the program and try again.

Chapter 4: Apple Mac Setup

This chapter shows you how to install the new BlackBerry Desktop Software on your Apple Mac computer and do the basics of synchronizing your contacts, calendar, tasks and memos, backup and restore, and more. If you want to transfer files and media with your Mac, then check out Chapter 5 "Apple Mac Media and File Transfer." (You may need some of the instructions in this chapter on how to install Desktop Software if you want to use it as your method to transfer files.)

Do you have a Microsoft Windows computer? If so please go to Chapter 3.

> **CAUTION:** If you plan on syncing with more than one computer (e.g., work, home, etc.)? Be sure to check the **with other computers (safer sync)** option on the **Device Options** screen (discussed later) if you plan on syncing your BlackBerry with multiple computers using Desktop Software. Otherwise, you could end up corrupting your BlackBerry and/or computer databases!

If you are a Google Contacts and Calendar User

Do you want a wireless, two-way automated sync and you use Google Contacts and Google Calendar? Try using **Google Sync** for your BlackBerry. This is a free application and give you a full two-way wireless sync.

BlackBerry Desktop Software for Mac

For years, Windows users have enjoyed seamless synchronization of their contacts, calendar, notes, and tasks with their PC via the BlackBerry Desktop Software. Now, the peace of mind that comes with knowing your data is fully backed up is available to the Mac user.

If you have never used BlackBerry Desktop Software, you will now be able to not only synchronize your data, but you will be able to back up, restore, sync your iTunes playlists, and more.

The Disk from the BlackBerry Box

It is fairly likely that the disk that arrived with your brand new BlackBerry has a version of Desktop Software for Mac that is already out of date. This is because many times, the CD was produced more than a month ago, and in the meantime a new version has been released. So we recommend grabbing the latest version from the internet directly from www.blackberry.com, as shown below.

Downloading and Installing Desktop Software for Mac

Desktop Software for Mac software is available for free from BlackBerry.com. The following exercise shows you how to install it:

1. Open up your web browser and go to the download page (Figure 4-1):
   http://na.blackberry.com/eng/services/desktop/mac.jsp.

2. Fill out the required information on the download page and then click **Download Now** to download the software.

Figure 4-1. Locating the download file on the BlackBerry web site

3. Once the file is downloaded, follow the steps you normally would perform to install any software on your Mac. If you have trouble or need assistance, please do a web search for "installing BlackBerry Desktop Software for Mac" to find all the latest online help.

> **NOTE:** If you have been using another application to sync your BlackBerry to your Mac (e.g. Pocket Mac or the Missing Sync), you will receive another warning note letting you know that in order to proceed, the connection between your BlackBerry and the third-party synchronization software will need to be discontinued.

The installation process will begin. Follow the onscreen prompts as your Mac installs the new **Desktop Software** starting with the screen shown in Figure 4-2.

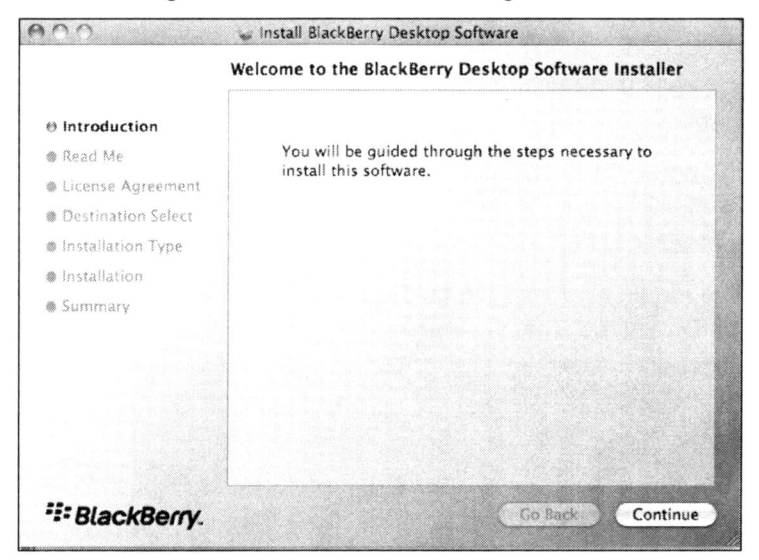

Figure 4-2. First BlackBerry Desktop Software installation screen

Starting Desktop Software – Device Options Screen

To locate the Desktop Software app, click the **Finder** icon and then click your **Applications** icon. The **BlackBerry Desktop Software** icon will be in your **Applications** directory.

Double-click the **BlackBerry Desktop Software** app and the Welcome/Device Options screen will appear, showing you information about your particular BlackBerry (Figure 4-3). On this screen you can adjust your device options.

CHAPTER 4: Apple Mac Setup

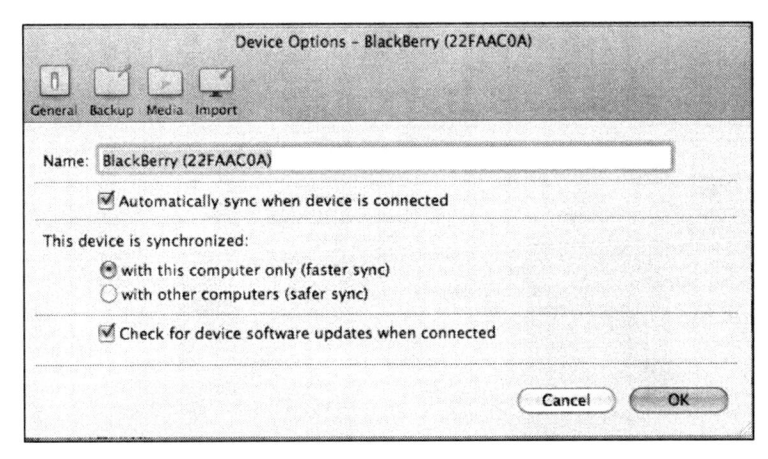

Figure 4-3. BlackBerry Desktop Software for Mac Device Options screen

1. You can adjust the **Name** of your BlackBerry if you choose.

2. We recommend leaving the check box to Automatically sync when device is connected checked.

3. Next you see the **This device is synchronized** option.

    a. If you synchronize your BlackBerry with other computers, a network server, Google Sync for Calendar, or Google Contacts, select **with other computers (safer sync)**.

    b. If you are only planning on syncing your BlackBerry with this one Mac, you can choose **with this computer only (faster sync)**.

4. Finally we recommend leaving the last box checked as well that has the software check for any device software updates (that is the operating system software for your BlackBerry).

5. Click **OK** when done.

Main View in Desktop Software

After you connect your Torch the first time, Desktop Software will show you a picture of your BlackBerry, a clean interface displaying information along the left-hand bar, and commands along to top bar (Figure 4-4).

NOTE: This figure shows a Torch 9800; your BlackBerry may look slightly different.

View Video Tutorials and Free Tips at www.MadeSimpleLearning.com   135

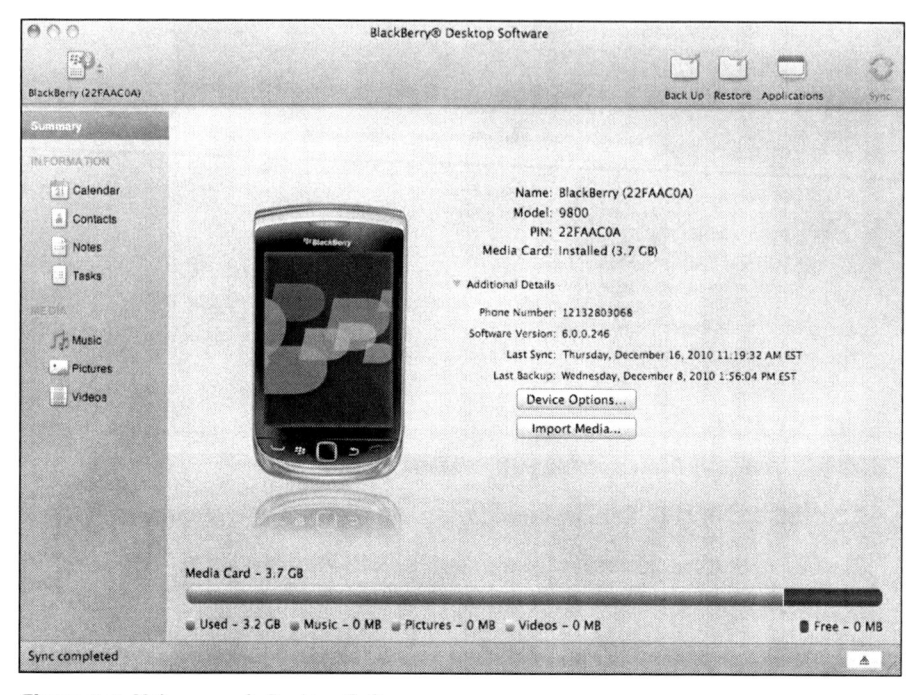

Figure 4-4. Main screen in Desktop Software

Using Desktop Software for Mac

One of the first things you will notice is the **Device Options** button next to the picture of your BlackBerry (Figure 4-4).

Clicking this brings you to the same **Device Options** screen you saw when you first started Desktop Software; see Figure 4-3).

Backup Options

From the **Device Options** screen, click **Backup** icon along the top.

CHAPTER 4: Apple Mac Setup

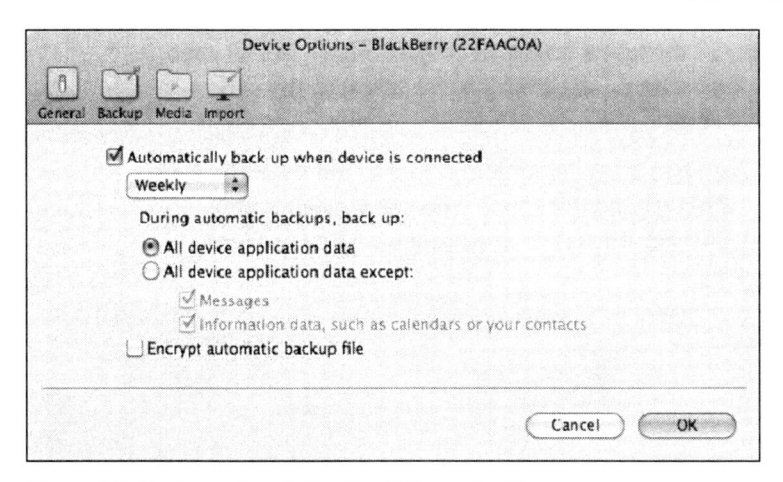

Figure 4-5. Backup options in Desktop Software for Mac

If you want to create a backup each time you connect your device, just check the **Automatically back up when device is connected** box. You can then specify exactly what you wish to be backed up. (We cover the specific backup options in greater detail later in the chapter.)

Setting Up Your Sync Options

Click **OK** or **Cancel** to return to the main screen of Desktop Software (Figure 4-4), and look at the left column under where it says **INFORMATION** (Figure 4-6).

Set your **Sync** options here.

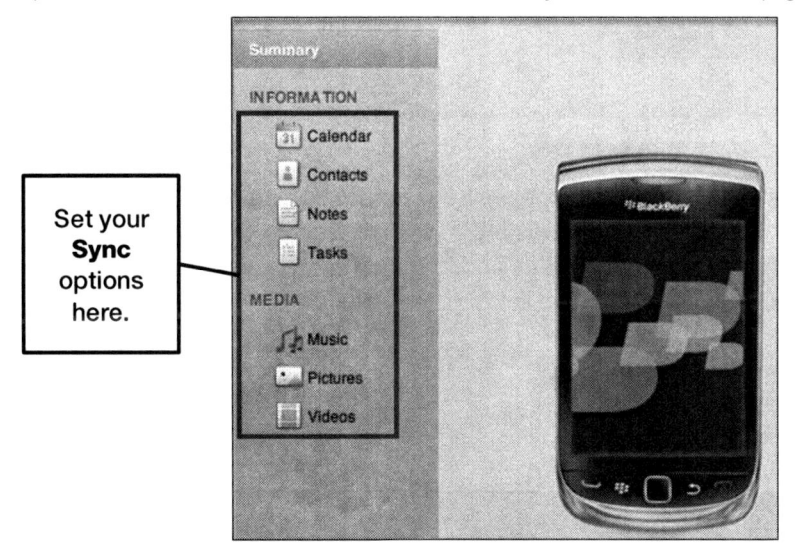

Figure 4-6. Sync options in Desktop Software for Mac

This is where you set the sync options for your calendar, contacts, notes, tasks as well as Media (Music, Pictures and Videos which we cover in Chapter 5).

View Video Tutorials and Free Tips at www.MadeSimpleLearning.com   137

1. Click any of the items below **INFORMATION**. In this case we will start with the **calendar** to be taken to the sync setup screen.

2. This screen has a similar look and feel to the calendar sync screen within iTunes, for those who are familiar with syncing an iPad, iPhone or iPod touch and a Mac (see Figure 4-7).

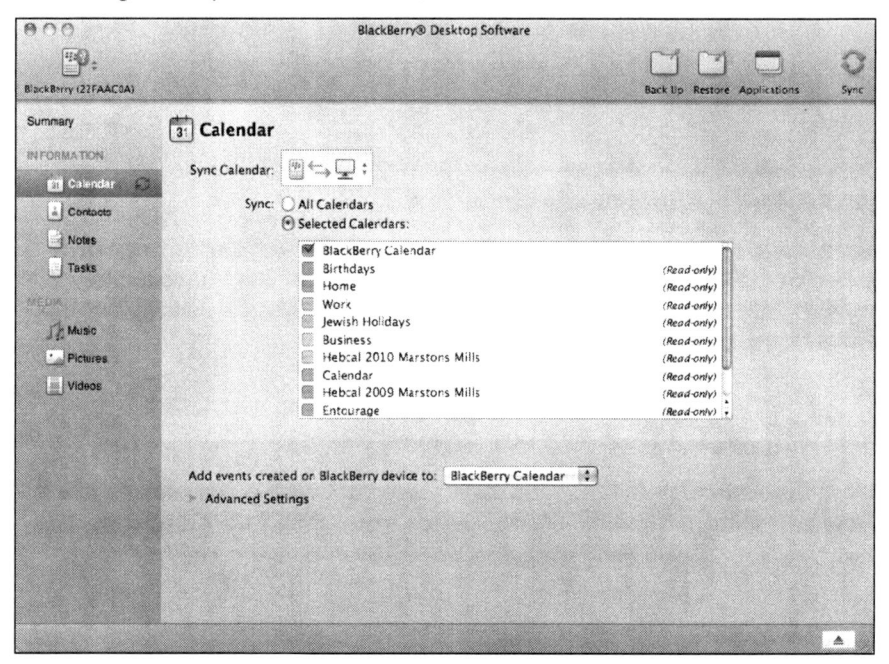

Figure 4-7. Calendar sync options screen

3. Desktop Software will notice all the calendars you have on your BlackBerry. In this example, we use Google Calendar, and have many different calendars, each set to a unique color. You see this in the list of calendars.

4. Click the box with the **Sync Calendar** option. The red X on this icon shows that the BlackBerry calendar will not be synced with the Mac calendars.

CHAPTER 4: Apple Mac Setup

5. If you click the box and select the next item in the drop-down, the picture will change to show that now you desire a two-way sync between the Mac calendar and the BlackBerry calendar.

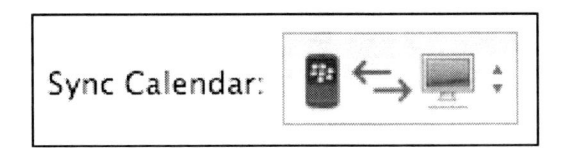

Sync Calendar:

6. You can also select into which calendar you want events to go that you create on your BlackBerry. The default is the Business calendar, but that can be changed to any calendar that you have set up on the device.

Add events created on BlackBerry device to: BlackBerry Calendar

Advanced Settings

Click the **Advanced Settings** tab at the bottom, and the options shown in Figure 4-8 will be revealed.

Add events created on BlackBerry device to: BlackBerry Calendar

Advanced Settings

Sync: ● All events
○ Only future events
○ Only events 14 days prior and 90 days after

☐ Replace all calendar events on this BlackBerry device

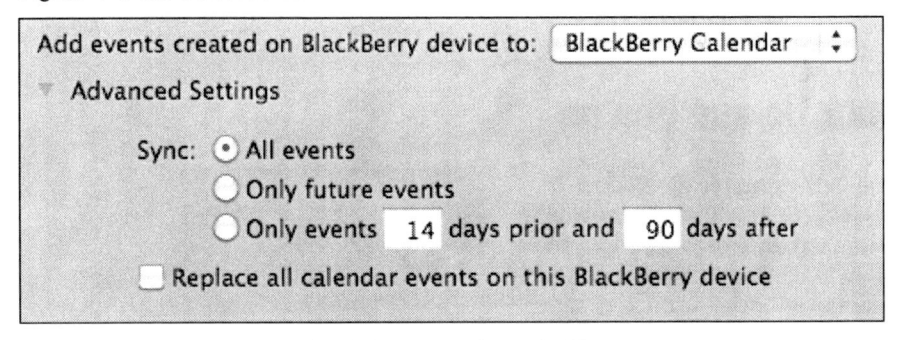

Figure 4-8. Advanced sync settings in Desktop Software for Mac

Like Desktop Software for the PC, you can specify as to whether you want to sync all events or future events only, or you can set individual parameters for synchronization.

> **TIP:** We do not recommend selecting **Only future events** unless you have a strong reason to do so. For example, say you made some notes on an event that was held yesterday in the BlackBerry calendar notes field. If you selected **Only future events**, these important notes would not be transferred to your Mac.

View Video Tutorials and Free Tips at www.MadeSimpleLearning.com   139

To replace all calendar events on the BlackBerry with events from your Mac's calendar, just click the check box at the bottom of the screen.

Syncing Contacts, Calendar, Notes, and Tasks

The procedure for setting the sync options for contacts, tasks, and notes is identical to what was just shown. The only things that change are the groups or events to choose within each category (see Figure 4-9).

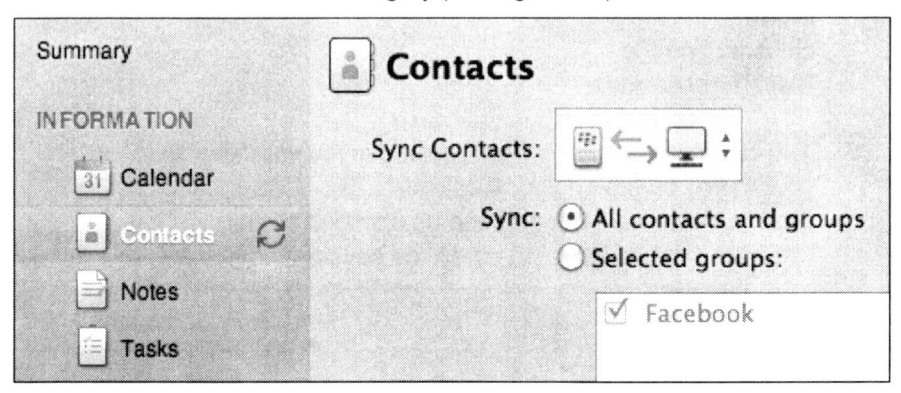

Figure 4-9. Sync Contacts setup screen

On this screen, you can click next to **Sync Contacts**, as discussed previously, and choose either to not sync with the Mac or to perform a two-way sync. You can then choose to sync either all contacts or groups, or only selected groups from your address book.

Back Up and Restore

One of the great features now available to Mac users is the ability to backup and restore either you entire BlackBerry contents or just selected information on your Mac. **Back Up** and **Restore** begins with the two icons at the top of the main screen in Desktop Software.

Using Backup

The following explains how to use back up:

CHAPTER 4: Apple Mac Setup

1. Click the **Back Up** icon and you will be taken to the next screen where you specify exactly which information you wish to be backed up on your Mac (Figure 4-10).

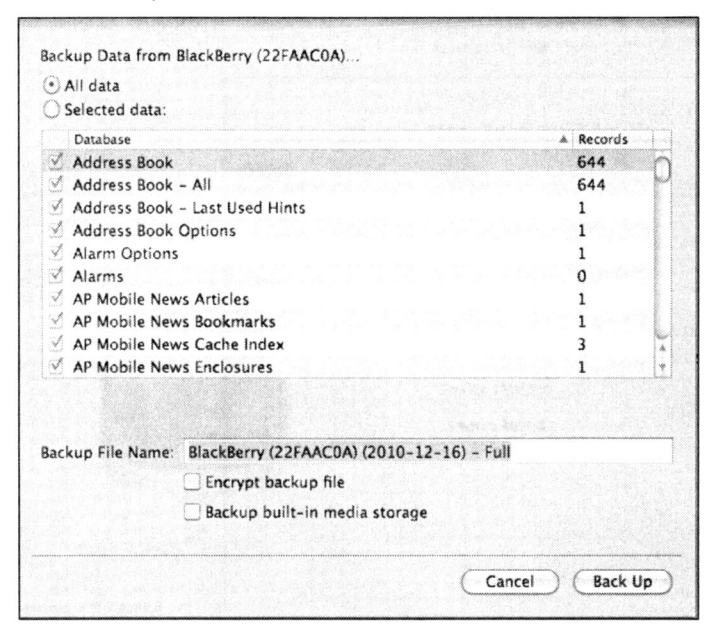

Figure 4-10. Backup screen in Desktop Software for Mac

2. Select either **All data** or **Selected data**, and then choose exactly which items you wish to back up.

3. Let's say that you are only really concerned with backing up your contacts, calendar, and notes—just check off each of those boxes and your backup will complete much faster.

4. You can specify the name of your backup for easy retrieval in the future.

5. Once you have made all your backup selections, click the **Back Up** button and the progress of the backup will be displayed in a dialog box.

Restoring from Backup

We all know that sometimes-unexplained things happen and we lose information in our BlackBerry. Maybe we try to update the OS and make a mistake, or maybe we sync with other computers and the information gets corrupted. Now Mac users have a reliable and safe way to restore data on their devices.

1. Click the **Restore** icon ![Restore] along the top row of the main screen in Desktop Software. You will then be taken to the restore options screen (Figure 4-11).

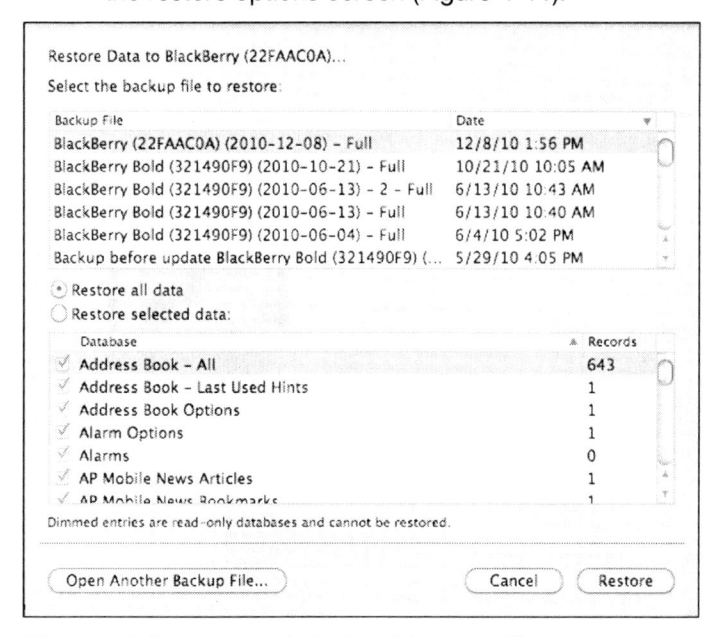

Figure 4-11. Restore screen in Desktop Software for Mac

2. If you have made a backup file on your Mac (which is required so that you'll have a file you can restore), it will be shown in the top box under **Backup File**. If you have multiple backup files, they will all be listed here.

3. Select the file from which you wish to restore information (or, if you did a selective backup, selected data will be displayed in the second screen below.)

4. Click the **Restore** button and your BlackBerry will be restored just as it was when you made the backup file.

Adding and Removing Applications

For the first time, Mac users are now able to add or remove applications on their BlackBerry using the Desktop Software.

CHAPTER 4: Apple Mac Setup

> **TIP:** Most of the time, you will be using the BlackBerry App World app on your BlackBerry to add or remove applications. See Chapter 20: "BlackBerry App World" for more info.

1. Click the **Applications** icon ![Applications] (in the upper-right corner of the main screen) and you will be taken to the Install/Remove Applications screen in Desktop Software (see Figure 4-12).

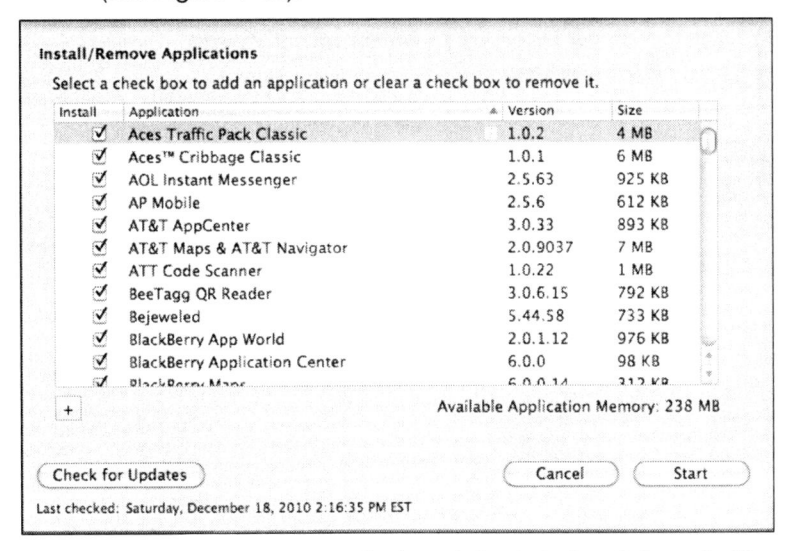

```
Install/Remove Applications
Select a check box to add an application or clear a check box to remove it.

Install   Application                          Version    Size
  ☑       Aces Traffic Pack Classic            1.0.2      4 MB
  ☑       Aces™ Cribbage Classic               1.0.1      6 MB
  ☑       AOL Instant Messenger                2.5.63     925 KB
  ☑       AP Mobile                            2.5.6      612 KB
  ☑       AT&T AppCenter                       3.0.33     893 KB
  ☑       AT&T Maps & AT&T Navigator           2.0.9037   7 MB
  ☑       ATT Code Scanner                     1.0.22     1 MB
  ☑       BeeTagg QR Reader                    3.0.6.15   792 KB
  ☑       Bejeweled                            5.44.58    733 KB
  ☑       BlackBerry App World                 2.0.1.12   976 KB
  ☑       BlackBerry Application Center        6.0.0      98 KB
  ☑       BlackBerry Maps                      6.0.0.14   312 KB

                          Available Application Memory: 238 MB

( Check for Updates )                        ( Cancel )  ( Start )
Last checked: Saturday, December 18, 2010 2:16:35 PM EST
```

Figure 4-12. The Install/Remove Applications window in Desktop Software for Mac

2. To check for updates to any of the applications installed including your BlackBerry operating system software, click the **Check for Updates** button in the lower-left corner.

3. Place a check mark in the box for any application that isn't checked already, and it will be installed on your device. Conversely, uncheck any box, and that application will be removed from your BlackBerry.

4. Click the **Start** button, and the selected or deselected applications will be either installed or uninstalled, depending on your selection.

View Video Tutorials and Free Tips at www.MadeSimpleLearning.com   143

> **TIP:** For peace of mind, it is always a good idea to perform a backup both before and after you add or delete applications from the device.

Automating the Sync

In order to have your BlackBerry sync every time you connect it to your Mac, you will need change a setting in the Device Options screen.

1. Click the **Device Options** button below the picture of your BlackBerry on the main screen (Figure 4-13).
2. Then check the check box next to **Automatically sync when device is connected**.

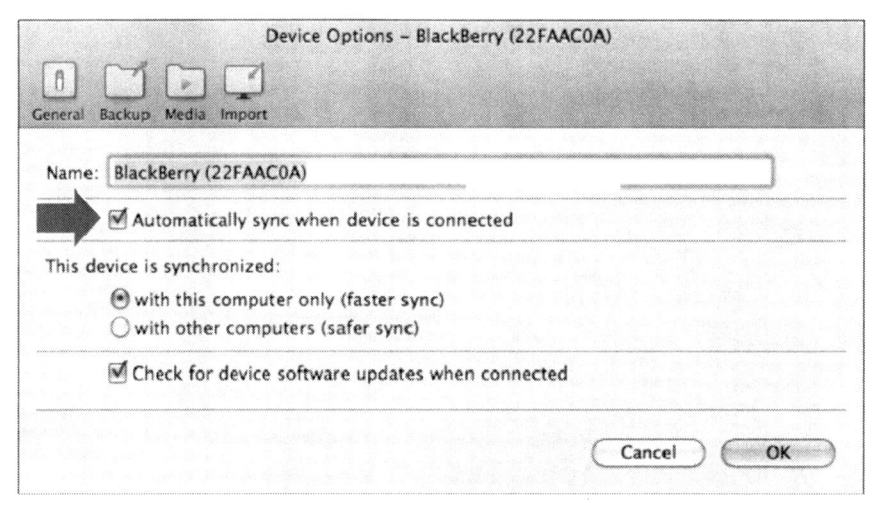

Figure 4-13. Automate the sync in Desktop Software for Mac.

Chapter 5: Apple Mac Media and File Transfer

Your BlackBerry can be a great media player. In order to get all of your songs, videos, and other media from your Mac onto your BlackBerry, you'll need to learn some of the information in this chapter.

There are a few ways to load up media (music, videos, and pictures) and Microsoft Office documents (for use with **Documents to Go**) onto your BlackBerry:

- Use BlackBerry Desktop Software for Mac (described in this chapter).
- Use USB Drive Mode transfer (See the "USB Drive Mode Transfer" section of Chapter 1: "Getting Started" for help with this method).
- Attach a file or two to an email message you send yourself from your computer and open them on your BlackBerry (if they are under about 2 MB file size). This method will work for smaller files such as low-resolution pictures, small Microsoft Office documents, and smaller PDF files. This email attachment method will not work for music or videos because they are usually too large to email.

Setting Your Media Sync Options

Before you get set up to sync your Media, you will want to verify the options screen.

1. Start up Desktop Software for Mac as shown in the previous chapter.

2. Connect your BlackBerry to your Mac with your USB cable.

3. Click the **Device Options** button.

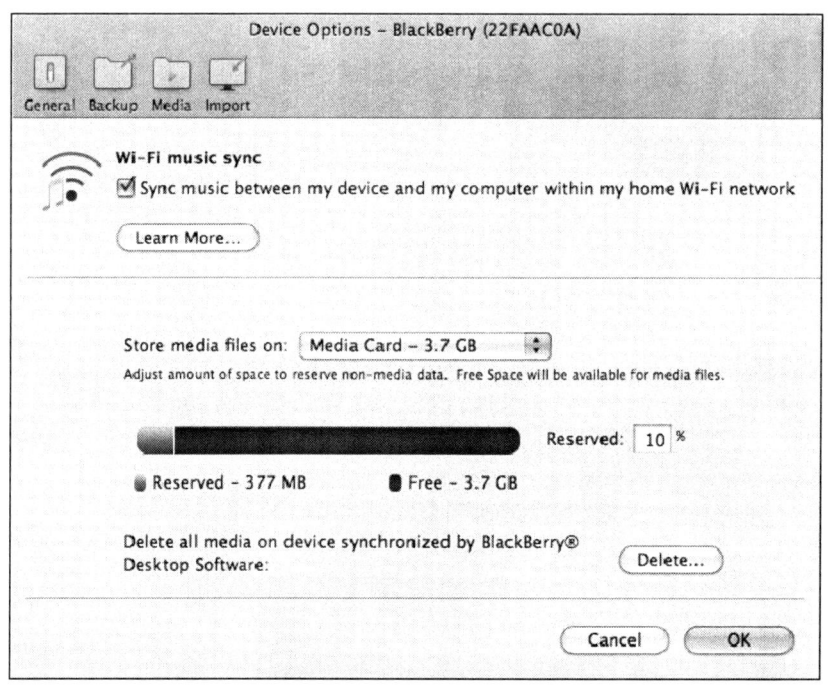

4. Click **Media** icon along the top of the screen (see Figure 5-1).

Figure 5-1. Media Sync in Desktop Software for Mac

5. The top check box for Wi-Fi music sync will allow you to sync your music wirelessly. We cover this in detail later in this chapter.

6. Set the **Store media files on** either your **Media Card** or **Device Memory**. We recommend using the **Media Card** as your **Device Memory** is usually much more limited and you want to leave that free for emergencies.

7. We recommend increasing the **Reserved** space beyond the default 10 percent to something like 20-25 percent.

> **TIP:** The reason you might want more **Reserved** space is to allow you to open or save items that you receive as email attachments on your BlackBerry or might decide to download while you are web browsing on your BlackBerry.

   8. Click the **Delete** button next to the **Delete all music on device** statement (see Figure 5-1), and you can remove any or all the music synced to your device using Desktop Software.

Why would you want to do that? Let's say you had previously synced a number of music playlists, videos and pictures you no longer want on your BlackBerry. Deleting all media with this button will be faster than changing your sync options.

You now have the option of syncing your iTunes playlists complete with album art, so you might want to start fresh and get rid of the other music on your BlackBerry.

   9. Click **OK** to save your changes.

Syncing Music

BlackBerry Desktop Software allows you to sync your iTunes playlists right onto the media card or device memory of your BlackBerry.

   1. Click the **Music** icon under the **Media** line along the left-hand column of the main screen.

   2. You will then be taken to the **Music Sync** screen. It provides you with some very nice options, as shown in Figure 5-2.

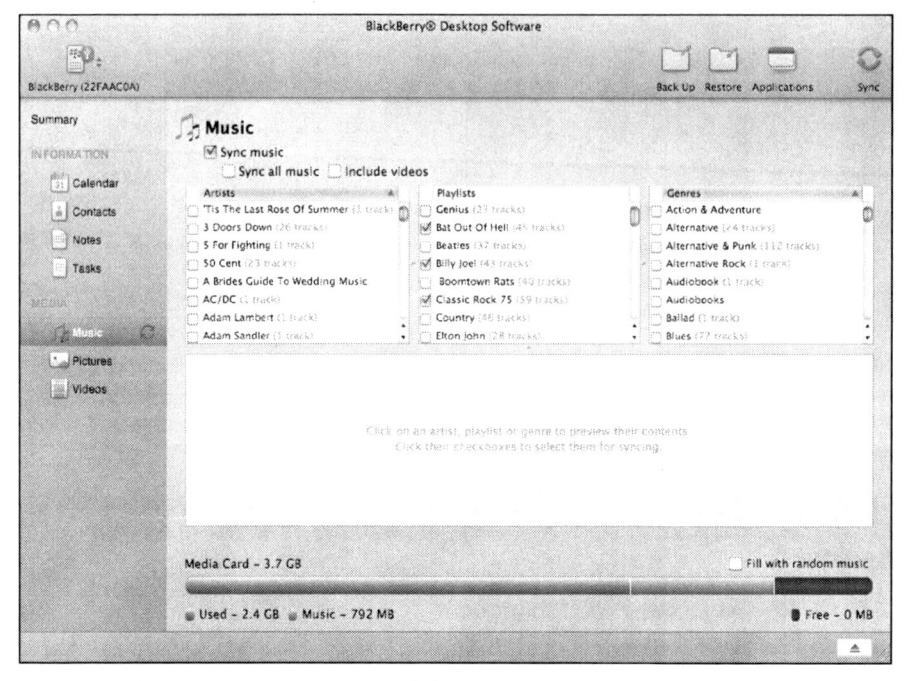

Figure 5-2. Sync Music screen in Desktop Software for Mac

As with iTunes, you can choose to sync all music and playlists or just selected playlists. New to the Mac Desktop Software is the ability to also sync specified **Artists** and/or **Genres**.

3. Just place a check mark in the **Sync Music** box at the top of the screen, and you can then select which playlists you wish to sync between your Mac and your BlackBerry.

4. Place a check mark in the **Fill with random music** box to have additional songs randomly placed on to the BlackBerry until the **Reserved** free space is met. (You set the **Reserved** space in the Media Options screen above).

5. In this example, we selected three playlists that we wanted on our BlackBerry—so we placed check marks in the appropriate boxes (see Figure 5-3).

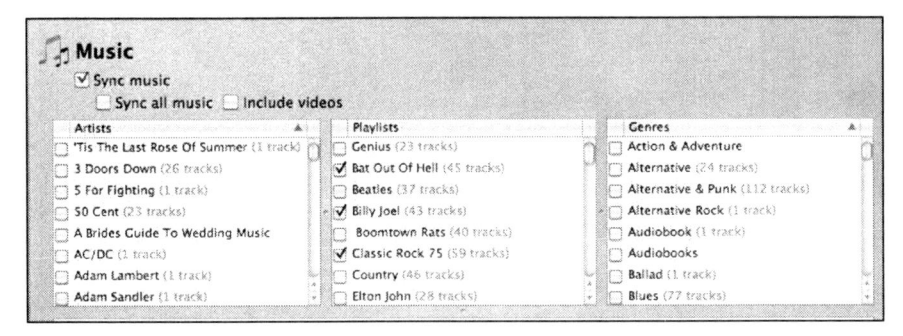

Figure 5-3. Syncing specific playlists, artists or genres in Desktop Software for Mac

Click the **Sync** icon at the top right of the screen to perform the music sync to your BlackBerry.

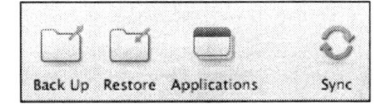

Syncing Pictures and Videos

Also new to Desktop Software for Mac is the ability to sync your pictures and videos in the same way you sync your music.

Syncing Pictures

Under the **Media** heading in Desktop Software, you will see **Music**, **Pictures** and **Videos**.

1. Click on **Pictures** in the left column to see the **Picture** sync screen.

2. The Mac Desktop Software will attempt to determine which program(s) you have available for pictures and where your picture albums are stored.

3. In Figure 5-4, our Mac has the photos stored in iPhoto, so that is what is shown.

4. Place a check mark next to any picture, album or event listed and it will be synced to your BlackBerry.

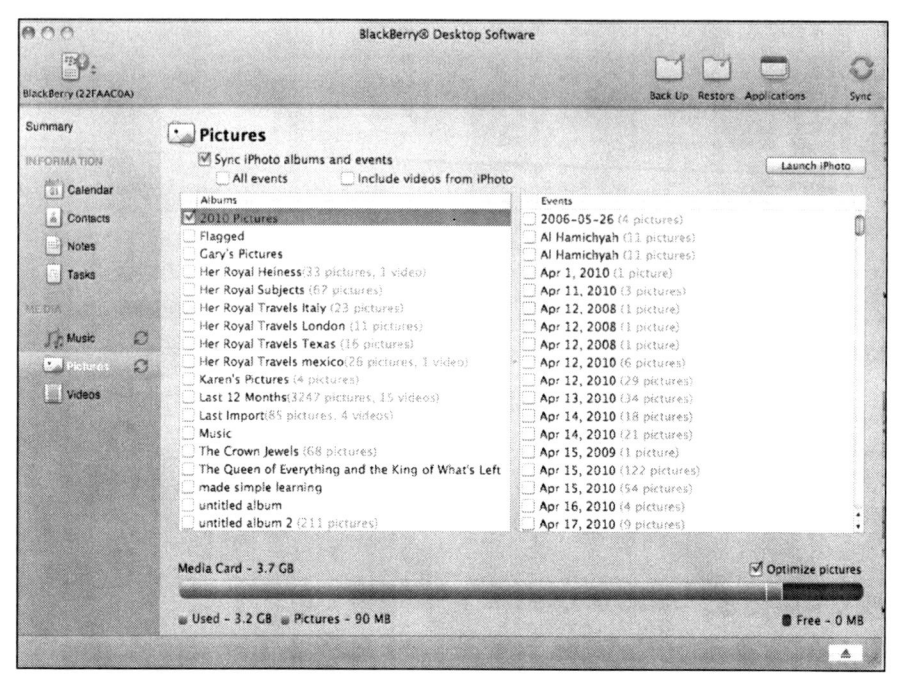

Figure 5-4. Syncing specific pictures, albums or events in Desktop Software for Mac

Syncing Videos

You can sync video content from your Mac in a very similar fashion to pictures.

1. Under the **Media** heading, click on **Videos**. You will see a very similar screen for selecting video content from your Mac to be synced to the BlackBerry.

2. Assuming that iTunes is your primary method for handling video, the screen will ask you to simply select which videos from iTunes you wish to be transferred to your BlackBerry.

3. Click any of the videos shown (See Figure 5-5.) You cannot check dimmed items that have DRM next to them as these are protected to only run on your Apple devices.

CHAPTER 5: Apple Mac Media and File Transfer

Figure 5-5. Syncing specific videos, albums or events in Desktop Software for Mac

Why can some videos and songs not be synced? (DRM)

In Figure 5-5 you see a red **DRM** next to many of the videos. These videos are purchased or rented in iTunes and are **DRM** protected (Digital Rights Management), so they can only be synced to and played on Apple devices (iPhone, iPod or iPad) registered with your iTunes.

> **NOTE**: If your videos are protected with **DRM** (digital rights management) you will not be able to copy them to the BlackBerry. Also, if you choose too many videos for your memory card, you will receive a **Memory Full** error message in the lower right hand corner as shown above.

View Video Tutorials and Free Tips at www.MadeSimpleLearning.com   151

Importing Pictures and Videos from Your BlackBerry

On the main screen of the BlackBerry Desktop Software, under the **Device Options** button is an **Import Media** option.

Import Media will show you all the new media on your BlackBerry – from songs purchased to pictures and videos taken and give you the option of importing this media to your Mac.

Click the **Import Media** button and select which media items you want imported to your BlackBerry (see Figure 5-6).

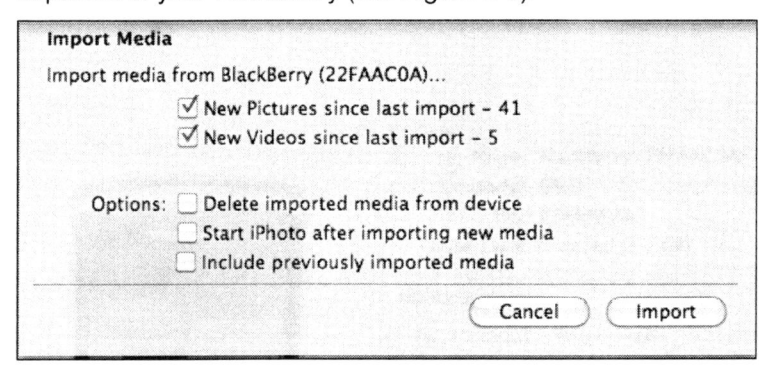

Figure 5-6. Import Media from your BlackBerry to your Mac

Using Wireless Music Sync

New to Desktop Software for Mac is Wireless Music Sync. Wireless Music Sync lets you keep your synced playlists up to date on your BlackBerry using your home Wi-Fi network.

Setting up Wi-Fi Music Sync

Click on **Device Options** and then on the Media tab. You will see a box next to **Wi-Fi Music sync**.

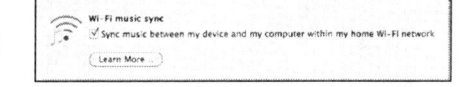

Check the box and you will be prompted that **Wi-Fi Music Sync** needs to be installed on the BlackBerry.

CHAPTER 5: Apple Mac Media and File Transfer

Wireless Music Sync will be installed on the BlackBerry. Once installation is complete, the BlackBerry will need to be restarted and then an information screen will appear. You will also be reminded to make sure your router is working and that your computer is turned on and connected to the same Wi-Fi network as your BlackBerry.

You will be reminded to turn on Wi-Fi on your BlackBerry (see the "Wi-Fi Connections" section of Chapter 1 for help).

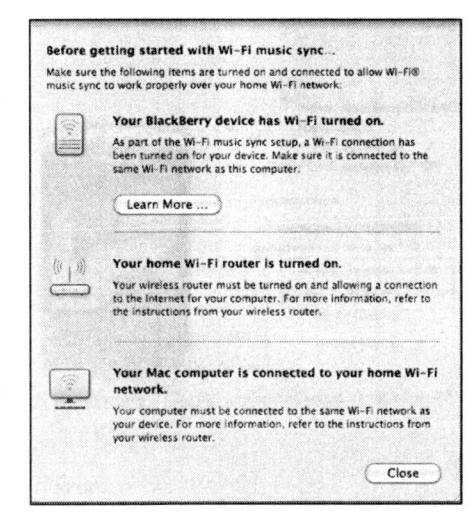

Using Wi-Fi Wireless Music Sync on Your BlackBerry

After you setup the **Wi-Fi Music Sync**, you will notice that some songs, playlists and albums appear grayed out or dimmed in your BlackBerry **Music** app.

> **NOTE**: If you are not connected to the Wi-Fi network, you will see a "waiting for connection to desktop..." message. If you see that, make sure you connect to the Wi-Fi network. You may be prompted to connect to the Desktop Software once again with the USB cable the first time you enable Wi-Fi.

These songs are in your computer library but not on your BlackBerry. If you click on any of these dimmed items, you can request that they be synced to your BlackBerry (see Figure 5-7).

> **TIP:** You can even request songs, artists or playlists when you are away from your home Wi-Fi network. What happens is that these requests are logged and when you return home the items are synced wirelessly and automatically behind the scenes!

The next time you come in contact with the Wi-Fi network where your computer is located (and your computer is turned on), you will receive the requested songs wirelessly via your Wi-Fi network.

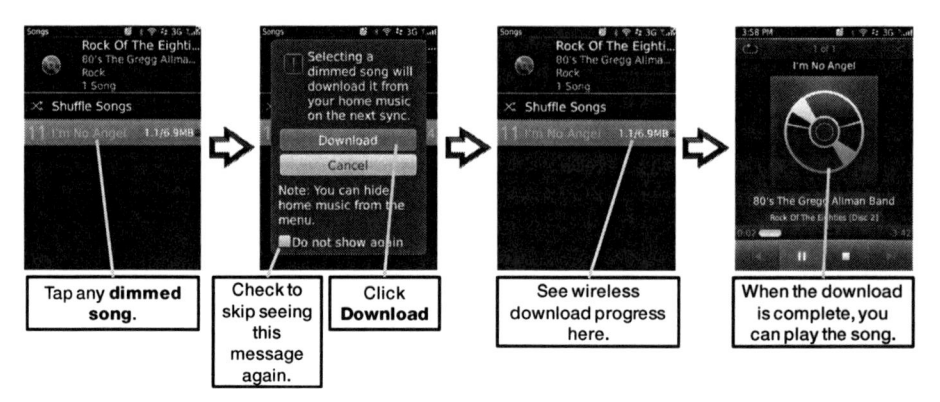

| Tap any **dimmed** song. | Check to skip seeing this message again. | Click **Download** | See wireless download progress here. | When the download is complete, you can play the song. |

Figure 5-7. How Wireless Music Sync works in the Music app on your BlackBerry.

> **NOTE**: If the Wi-Fi Music Sync does not work, the wireless connection was most likely not made. Try the following:
>
> On your device, in the browser, type the following URL: **http://*type your computer IP address here*:4481/mediasync/music**. If you are prompted, accept the certificate warning. If the webpage displays the text "Wireless music sync", then your device is connected to the same network as your computer. If you receive a display error, your device is not connected properly to the same network as your computer or there might be a firewall issue or routing issue.
>
> If your computer, router, or wireless network has a firewall, verify that ports 4481 and 4482 are open for TCP and UDP, and that the BlackBerry Desktop Software is allowed by the firewall. For more information and instructions, see the documentation for your firewall software and/or router.
>
> If your router has UDP broadcast, verify that UDP broadcast is turned on. Most routers support this feature and do not mention it, but your router might need special configuration. For instructions, see the documentation that came with your router.
>
> Verify that your gateway IP address is the same on both your device and your computer, and that both your device and computer IP addresses are on the same subnet.
>
> Tip Courtesy of www.blackberry.com - Copyright 2010 Research In Motion Limited. All Rights Reserved.

Some Songs Cannot Be Wirelessly Synced

You will notice that when you try to click on a dimmed song from your **Music** app, you will see a couple of different error messages.

Unsupported Format – this means the media is not playable on your BlackBerry or that it has been protected with Digital Rights Management (DRM).

Cannot be found – this means the song cannot be located on your main music library on your computer.

In both cases, you cannot sync these songs to your BlackBerry. Try checking to see if you can play these songs on your computer. Some of these songs may have been purchased in iTunes and are protected to only play on your Apple device.

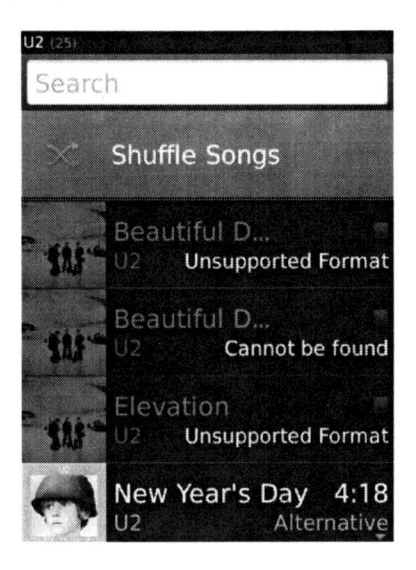

USB Drive Mode Transfer for Your Media Card

This works whether you have a Windows or a Mac computer. We will show images for the Mac computer process, but it will be fairly similar for a Windows PC. This transfer method assumes you have stored your media on a MicroSD media card in your BlackBerry.

1. To get to this screen, click the **Search** icon from your **Home** screen and type "storage."

2. Click the **Options** icon and tap **Storage**.

3. Make sure **Media Card Support** for your media card is checked (turned on), and other settings are as shown.

View Video Tutorials and Free Tips at www.MadeSimpleLearning.com

4. Now connect your BlackBerry to your computer with the USB cable. Choose **USB Drive** from the three options shown on the screen.

5. Your media card will look just like another hard disk to your computer (similar to a USB flash drive).

> **TIP:** If you choose the **Sync Media** option then the Desktop Software will open and attempt to sync media using the options described earlier in this chapter.

Using Your BlackBerry in USB Drive Mode

> **NOTE:** You will need to install Desktop Software for Mac in order to be able to use this mass storage option. This is because Desktop Software contains drivers required to connect your BlackBerry to your Mac.

Once connected, your Mac will see your BlackBerry as a mass storage device and mount it as an external drive.

The BlackBerry will also be visible if you click the **Finder** icon in the dock. It will be listed under **Devices**.

> **NOTE**: In the image to the right, there are two devices listed – one is the internal memory for the BlackBerry called "BlackBerry 1," and "No Name" is the Media Card.

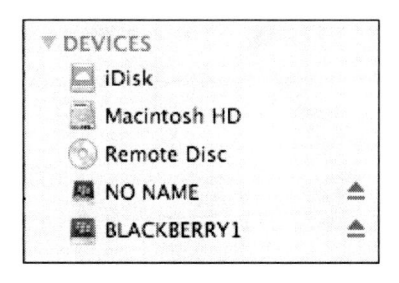

Exploring the Drive

Right-click the icon for the BlackBerry or Media Card and choose **Open**—or double-click on the **Desktop** icon and open the drive (see Figure 5-8).

Now you can explore your BlackBerry as you would any drive. You can copy pictures, music, and video files by just dragging and dropping to the correct folder, or you can delete files from your BlackBerry by clicking the appropriate folder, selecting files, and dragging them to the trash.

> **NOTE:** Your music, video, ring tone, and picture files are located in the folder called **BlackBerry**.

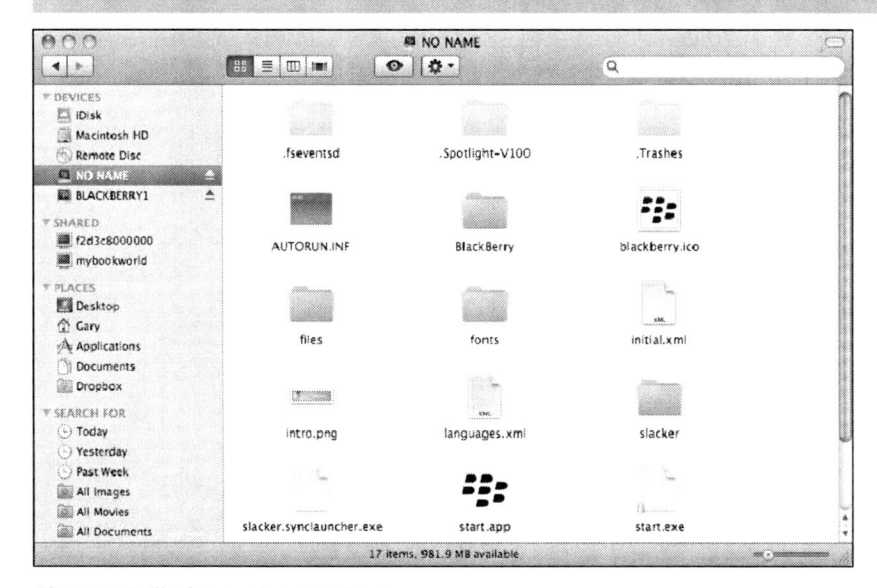

Figure 5-8. Finding media on the Torch

Copying Files Using USB Drive Mode

After your BlackBerry is connected and in USB Drive Mode, just open up your computer's file management software. On your computer, start your Finder. Look for another hard disk or BlackBerry model number that has been added.

> **NOTE:** You will see your own BlackBerry model number (e.g., TORCH_9800) or another way of identifying your BlackBerry like BlackBerry 1.

When you plug your BlackBerry into your Mac, it will identify the main memory and the contents of the MicroSD card as two separate drives, and place them right on your desktop for easy navigation.

On your computer, click the **Finder** icon in the lower-left corner of the dock. On Windows, click the disk drive letter that is your BlackBerry media card. You will see your devices (including both BlackBerry drives) on the top and your places (where you can copy and paste media) on the bottom.

To copy pictures (or other items) from your BlackBerry, follow these steps:

1. Select the pictures from the **BlackBerry/pictures** folder using one of the following methods:
    - Draw a box around some pictures to select them.
    - Click one picture to select it.
    - Press **Command+A** to select them all.
    - Press the **Cmd** key and click to select individual pictures.
2. Once selected—Ctrl-click or right-click one of the selected pictures and select **Cut** (to move) or **Copy** (to copy) (see Figure 5-9).

CHAPTER 5: Apple Mac Media and File Transfer

Figure 5-9. Status of copying files to your BlackBerry

3. Click any other disk/folder on your computer (e.g., **Documents**), and navigate to where you want to move/copy the files.

4. Once there, right-click again in the right window where all the files are listed, and select **Paste**.

You can also delete all the pictures/media/songs from your BlackBerry in a similar manner. Navigate to a BlackBerry/(media type) folder such as **BlackBerry/videos**. Press **Command+A** on your computer keyboard to select all the files, and then press the **Delete** key to delete all the files.

You can also copy files from your computer to your BlackBerry using a similar method. Just go to the files you want to copy and select (highlight them). Then right- click **Copy** and paste them into the correct **BlackBerry/[media type]** folder.

> **NOTE:** Not all videos, images, or songs will be playable or viewable on your BlackBerry. Use Desktop Software for Mac to transfer the files; most files will be automatically converted for you.

Chapter 6: Typing, Spelling, Search, and Help

In this chapter, we help you get typing as fast and accurately as possible on your Torch, show you the useful **Word Substitution** (known as "AutoText" in previous BlackBerry devices) feature, look at the built-in **spell checker**, dig into the **Universal Search** and also show you how to find **Help** right on your Torch.

> **TIP**: Check out the "Touch Screen Gestures" and "Keyboard" sections of the Quick Start Guide at the beginning of this book to learn some good touch screen, keyboard, and typing basics as well as tips and tricks.

When to Use Touch Screen or the Trackpad

Use the **touch screen** for many functions: tap, zoom, scroll, etc.

Use the **Trackpad** for carefully moving the cursor (typing, etc.)

The touch screen works very well for browsing, zooming, and selecting icons or menu items. You will find that there are times when you like the accuracy and control of the **Trackpad** better than the touch screen. For times when you are trying to carefully move the cursor when you are editing text or scrolling up/down, the **Trackpad** may actually be easier to use than trying to exactly position your finger on the screen.

CHAPTER 6: Typing, Spelling, Search & Help

Clicking versus Tapping

Clicking the **Trackpad** is the same as tapping a selected item with your finger on the screen.

Clicking and holding the **Trackpad** is the same as touching and holding (long-pressing) the screen.

You'll also find some great **Trackpad** features you can adjust to suit your own preferences.

Trackpad Sound and Sensitivity

The nice thing is you can adjust the speed of the Trackpad as well as whether or not it makes a soft clicking sound when you move it.

1. Tap the **Search** icon and type "trackpad."
2. Tap the **Options** icon (wrench).
3. Tap **Trackpad Sensitivity**.
4. Click on the **Horizontal Sensitivity** or **Vertical Sensitivity** number. The default is 70; higher is more sensitive, lower is less.
5. Check the box next to **Audible Roll** to hear a soft clicking sound as you move or glide the **Trackpad**.
6. Press the **Menu** key and select **Save**.

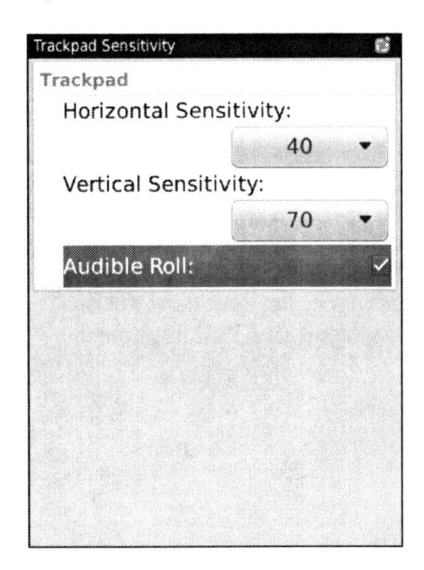

TIP: You may find that your Trackpad moves too quickly up or down lines. If so, then reduce the **Vertical Sensitivity** to 50 or lower.

Learning the Three On-Screen Keyboards

In addition to your physical slide-out keyboard, your BlackBerry comes with three on-screen keyboards that will allow you to type just the way you like. Please see the "Keyboards" section in the **Quick Start Guide**.

Keyboard:	When to use:
SureType keyboard (Portrait Mode)	Short, quick typing tasks, like a quick email or SMS text message.
Multitap keyboard (Portrait Mode)	If you are used to quickly typing on another type of phone, this will be familiar for you.
Full keyboard (Landscape or Portrait Mode)	If you prefer a single letter per key for ease of entry and accuracy, this is the best keyboard for you.

Landscape or Portrait - Full Keyboard

Landscape - Full keyboard Mode

Close your physical keyboard and tilt your BlackBerry Torch sideways into Landscape mode from any program in which you can type, the keyboard will be displayed as a **Full keyboard**.

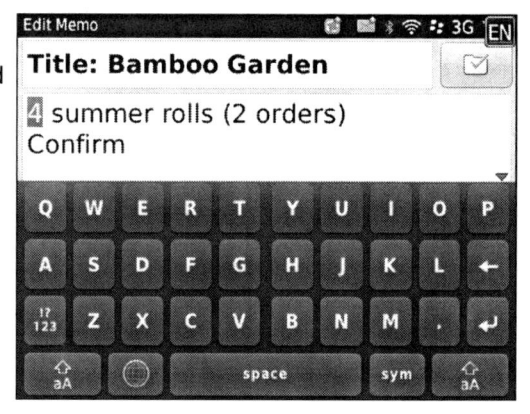

Portrait - Full keyboard Mode

You can also choose **Enable Full keyboard** after pressing the **Menu** key in Portrait Mode.

CHAPTER 6: Typing, Spelling, Search & Help

Portrait – Multitap Keyboard

You must select **Enable Reduced Keyboard** from the menu to see the **Enable Multitap** keyboard option.

This is the more standard cell phone typing technology, where you press the key once for the first letter on the key and twice for the second letter. For example, with the 3/DEF key, you would press it once for "D" and twice to get the "E" and three times to get the "F" in Multitap mode.

TIP: You will see the symbol in the upper right corner whenever you are viewing the **Multitap** keyboard.

Portrait – SureType Keyboard

You must select **Enable Reduced Keyboard** from the menu to see the **Enable SureType** keyboard option.

This is an innovative technology from BlackBerry that predicts what you are typing from the keys pressed, even though most keys have two letters on them. You only press each key once and the BlackBerry guesses which letter you meant to type based on the context (what you have typed before it) and/or what is in your Address Book. It even learns from you!

View Video Tutorials and Free Tips at www.MadeSimpleLearning.com   163

When typing with **SureType**, you will see the pop-up window below (or above) what you are typing. The highlighted word is the one currently being guessed.

If the highlighted word is correct, then press the **space** bar to instantly select it.

If you need to correct it, scroll, tap to highlight, or select a different word or group of letters.

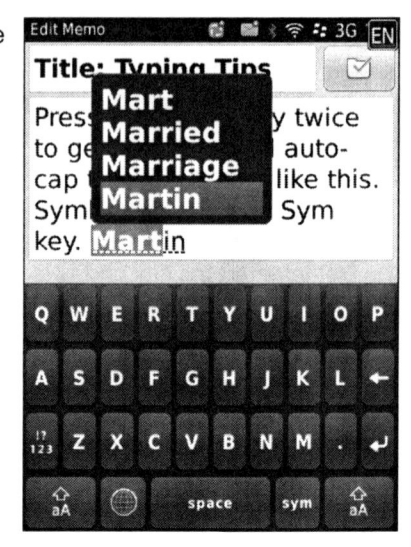

Wait to Select Corrections in SureType Mode

An important tip when using **SureType**: wait until the end of the word to select a correction. Many times the **SureType** system will show you the correct word at the second to-last or last letter of the word. If you keep adjusting what it guesses after each letter, it will take you all day to type.

Scrolling up/down the list of SureType options can be quite time consuming.

So we recommend that you continue typing letters until you see your word on the list shown on the screen.

Typing Tips for Your Torch

As we showed you in the Quick Start Guide, you have three on-screen virtual keyboards in addition to your slide-out physical keyboard. In this section we get you started with a lot of good typing tips to help you save time and increase accuracy.

Press and Hold for Automatic Capitalization

Here's an easy way to get uppercase letters as you're typing:

On the physical keyboard, simply press and hold the letter to capitalize it.

On the on-screen keyboards, you can press and hold a letter then select the uppercase letter from the pop-up window. Otherwise, it may be faster to simply press the shift key on the screen then tap the letter.

CHAPTER 6: Typing, Spelling, Search & Help

Caps Lock and Num/Alt Lock

Using the physical keyboard, to lock the **Caps** key to TYPE ALL UPPERCASE, press the **Alt** key, then press the right **Shift** key. Tap the either **Shift** key to turn off **Caps** lock.

Using the on-screen keyboard, press and hold the Caps key until you see a small lock appear on it. Tap the same key to turn off Caps lock.

Using the physical keyboard, if you want to type only numbers or the symbols shown on the top of the keys, turn on **Num/Alt** lock by pressing the **Alt** key and then the left **Shift** key. Tap the either **Shift** key to turn off **Num/Alt** lock.

Automatic Period and Cap at End of Sentence

At the end of a sentence, just press the **Space** key twice to get a "." (period). The next letter you type will be automatically capitalized.

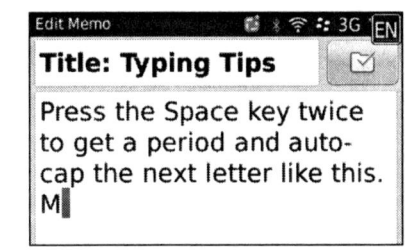

Typing Symbols

There will be times when you want to type the symbols shown on the physical keyboard as well as typing symbols while you are on the on-screen virtual keyboards, in this section we show you how.

On the Physical Keyboard

Using the physical keyboard, you can type the symbols and numbers shown on the top of the keys by pressing the **Alt** key then the key on the keyboard. For example, to type an @, you would press **Alt** then press the **P/@** key.

To type symbols not shown on the keyboard, you need to press the **Sym** key just to the right of the **Space** key. Then you will see the first of two extra symbols screens. Roll the **Trackpad** to the left or right to switch between the two screens (see Figure 6-1).

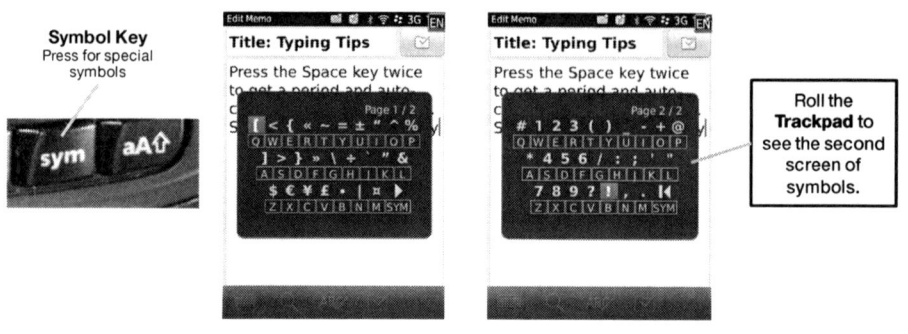

Figure 6-1. Accessing symbols not shown on the keyboard

Select the symbol by pressing the associated letter or gliding and clicking the **Trackpad**. In the image above, if you press the letter **C** on your keyboard while viewing Page 1/2, you'd get the symbol for the Japanese Yen currency. Notice that Page 2/2 are simply the same symbols that you can easily access by pressing the **Alt** key.

On the Virtual Keyboards

When typing using the on-screen keyboards, to get a symbol you have to press the **?!123** or **Sym** keys. On the SureType and Multitap keyboards you will need to press the **Sym** key more than once to see the **sym1**, **sym2** and **sym3** keyboards. Depending on which particular keyboard you are using, the layout and availability of various symbols changes (see Figure 6-2, 6-3 and 6-4).

TIP: To switch between the various on-screen keyboards, press the **Menu** key and select **Enable Full Keyboard**, **Enable Reduced Keyboard**, **Enable SureType** or **Enable Multitap**.

Figure 6-2. Accessing symbols with the Full Screen on-screen virtual keyboard

CHAPTER 6: Typing, Spelling, Search & Help

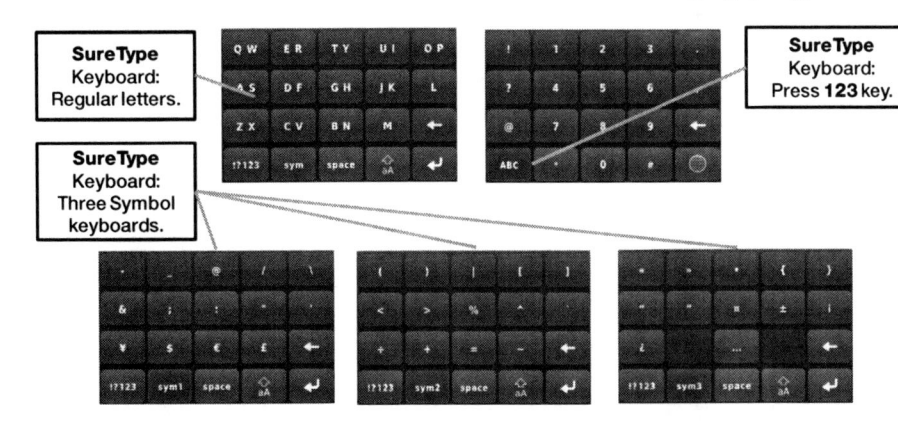

Figure 6-3. Accessing symbols with the SureType on-screen virtual keyboard

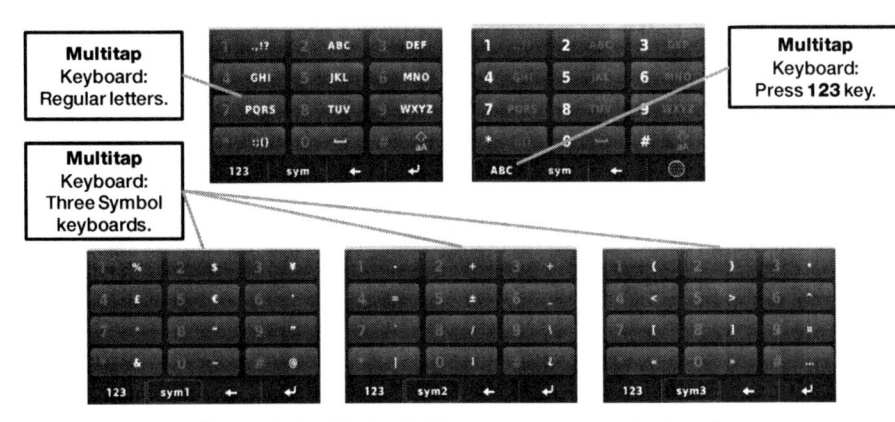

Figure 6-4. Accessing symbols with the Multitap on-screen virtual keyboard

Quickly Typing Accented Letters and Other Symbols

If you want to type common accented letters, you can do so quickly on both the physical and on-screen keyboards. The process is a little different as described below.

On the Physical Keyboard

You can easily type standard accented characters such as Á by pressing and holding a letter on the physical keyboard and gliding the **Trackpad**.

Try this: Press and hold the letter **A** to scroll through these accented characters related to A: **À Á Â Ã Ä Å Æ** (both upper and lower case).

> Letters this trick works on are: E, R, T, Y, U, I, O, P, A, S, D, K, C, V, B, N, M.

View Video Tutorials and Free Tips at www.MadeSimpleLearning.com   167

Letters that give you some other common characters: V key for ¿, T key for trademark symbol, C key for copyright symbol, R key for registered trademark symbol.

On the Virtual Keyboards

When typing using the on-screen keyboards, press and hold any letter to see a pop-up window of associated accented or other symbol characters (see Figure 6-5).

Figure 6-5. Accessing accented letters and other symbols by pressing and holding letters

Editing Text

Making changes to your text is easy with the BlackBerry.

Tap anywhere on the screen or move your finger over the **Trackpad** to position the cursor.

Edit text using **Del** or **Alt+Del**.

Use the **Del** key to erase characters to the left of the cursor.

Hold the **Alt** Key and press **Del** to delete characters under the cursor.

The Mighty Space key

Like many of the keys on the BlackBerry, the space key can do some very handy things for you as you type.

Using the Space key While Typing an Email Address

On most handhelds, when you want to enter the @ or a period for an email address, you need a complicated series of commands—usually a **SHIFT** or **ALT** plus another key.

On the BlackBerry, you don't need to take those extra steps. When you're typing the email address, just press the **Space** key once after the user name (martin, for instance) and the BlackBerry will automatically insert the @— "martin@". Skip typing the dot as well. Another tap of the **Space** key will type the dot before **.com**.

Quickly Navigating Drop-Down Lists

The **Space** key is also great at moving you down to the next item in a list. In a minute field, for example, pressing **Space** jumps to the next 15 minutes. In an hour field, you jump to the next hour, and in a month field, you jump to the next month. In any other type of field, pressing the space key jumps you to the next entry.

Give it a try.

1. Click on your **Calendar** icon.
2. Open up a new calendar event by clicking the **Trackpad** anywhere in **Day** view.
3. Glide down to the month and press **Space.** Notice you advance one month.
4. Glide over to the hour and press **Space.** You advance one hour.
5. Finally, glide to the minutes and press **Space**. Notice that you move 15 minutes forward.

These great tricks let you quickly reschedule calendar events.

Jump to First Letter Trick

You can use the letter keys on your physical keyboard to instantly jump to the first item matching either letter on the key (if there are two letters), or to jump to a matching menu item, or to a matching item in a list (like the long list that appears when you press the Options icon).

Using Number Keys to Type Dates and Times

You can use the number keys on your physical keyboard to instantly type a new date or time or to select an entry in a drop-down list with that number.

Examples include:

> Typing 40 in the minute field to set the minutes to 40;

> Typing 9 in the hour field to get to 9 AM or PM.

This also works in fields where drop-down list items start with numbers, as in the Reminder field in calendar or tasks. Typing a 9 immediately jumps you to the **9 Hours** setting.

Fine-Tuning Your Keyboards

You can customize your on-screen keyboards in the **Options** app on your BlackBerry. You might want to hear a small key-click when you press each key, you may prefer to have the reduced keyboards (SureType or Multitap) automatically appear when you close the sliding keyboard. You can set all these options and more by following these steps.

1. From your **Home** screen, click the **Search** icon and type "**keyboard**."

2. Tap the **Options** icon.

3. Tap **Keyboard**.

From here you can set the following:

- Place a check next to **Key Tone** to hear a click every time you press a key.

- Uncheck the **Open Virtual Keyboard on Slider Close** if you do not want to see the on-screen keyboard automatically appear.

- Adjust the **Key Rate** between **Slow** to **Fast**.

- Change the **Keyboard layout** between **QWERTY** (US standard), **AZERTY**, **QWERTZ** (European standards).

- Change the **Portrait Keyboard Type** between **Full** and **Reduced** (**Multitap** and **SureType**).

- Adjust **Show Key Indicator** between **Always**, **Never** and **On Touch and Hold**.

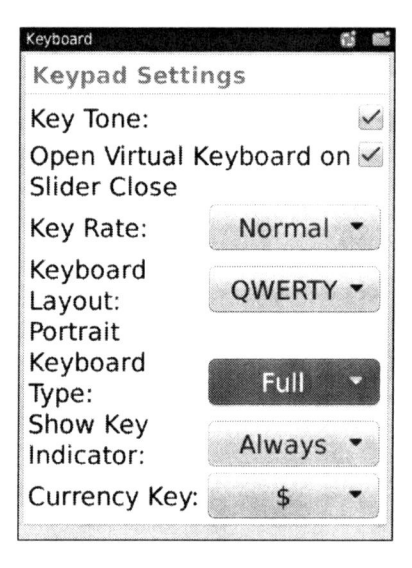

- Finally, change **Currency Key** between **Dollars**, **Pounds**, **Euros**, and **Yen**. (You may see different options depending on your selected languages and locale.)

4. Press the **Menu** key and **Save** your changes.

Fine-Tuning Your Typing, Spelling and More

You can also customize how your BlackBerry handles your typing, spelling and related items.

1. From your **Home** screen, click the **Search** icon and type "**typing**."
2. Tap the **Options** icon.
3. Tap **Typing**.

You can select the following for your Portrait virtual and Landscape virtual and Physical keyboards:

Select the style of typing –

Predictive – this is where the BlackBerry tries to guess the word you are typing and shows the guesses in a drop-down list for you to select.

Direct – this shows only the letters you type without any guessed words.

Corrective – this automatically selects corrections from the predictions.

For the **Predictive** and **Corrective** options, you can tap the **Advanced Style Options** button to see additional screens of options (see Figure 6-6).

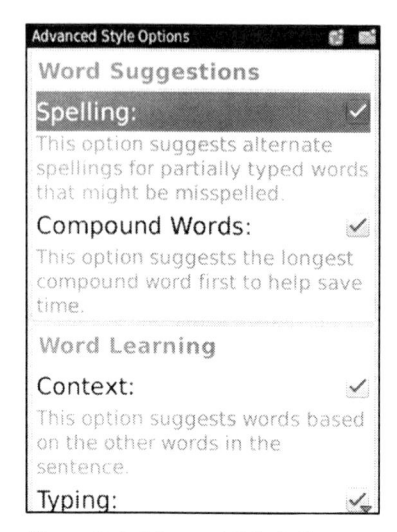

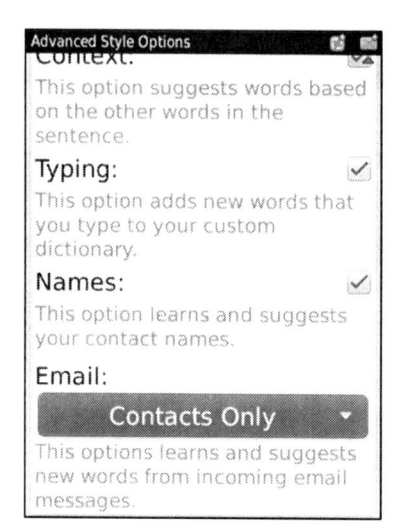

Figure 6-6. Advanced Syle Options screens for Predictive and Corrective typing.

Selecting Text to Copy, Cut or Delete

The fastest way to select text to copy, cut or delete is to simultaneously touch the screen with two fingers to highlight the text, then drag the handles that appear.

Once you have the text selected, press the **Menu** key and select **Copy** or **Cut**. You can also use the soft keys that appear at the bottom of the screen to do the same.

> **TIP:** Check out the "Copy and Paste" section of the Quick Start Guide for more on copying and pasting text between apps.

> **TIP:** If using the on-screen keyboards is not working well for you, simply slide out the physical keyboard and use that.

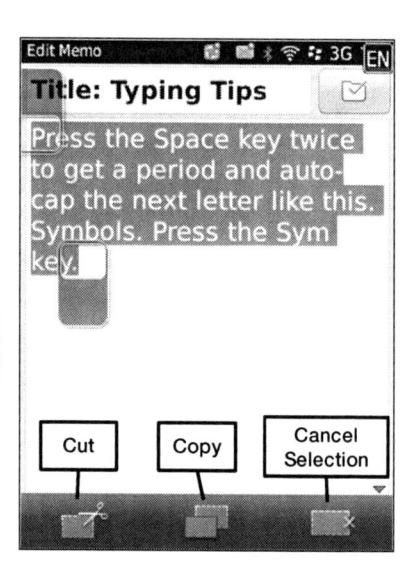

Save Time with Word Substitution (AutoText)

In this section, we show you some great tips and tricks to save time and increase accuracy for words and phrases that you may have to type many times. If you have used BlackBerry smartphones before, this feature was called **AutoText** on older devices. **Word Substitution** can help you by allowing you to skip typing the apostrophe in most contractions and automatically common spelling mistakes.

We will show you how to come up with useful **Word Substitutions** for just about anything you can imagine:

- An email you send frequently.
- Legal disclaimer text.
- Language describing your products or services.
- A phrase, paragraph or other text you commonly type.
- A set of directions to your home or office.
- Various different email signatures.
- Your home or office address.
- A special time stamp that automatically types today's date and time.

Just about anything that you might want to type frequently.

Standard Word Substitutions

Sometimes, typing on the little BlackBerry keyboard produces less than desirable results. Fortunately, for the more common misspellings, there are hundreds of standard or pre-loaded Word Substitutions. Below is a list of some of the more common and useful ones.

acn > can	Arent > Aren't	Cant > Can't	Wont > Won't
Didnt > Didn't	Dont > Don't	Teh > The	Hel > He'll
Wel > We'll	Wer > We're	Yr > year	Mn > Minutes
Hr > hours	Lt > current time	Ld > current date	Shel > She'll

TIP: Knowing **Word Subsituion** is there to help you get things right will allow you to type with greater abandon on your BlackBerry. Take a few minutes to browse the **Word Subsituion** pre-loaded entries, especially the contractions, so you can learn to type them without ever using the apostrophe.

In order to view the standard **Word Substitution** pre-loaded entries, follow these steps:

1. From your **Home** screen, click the **Search** icon and type "**word**."
2. Tap the **Options** icon.
3. Tap **Word Substitution.**
4. Scroll up or down or type a few letters to find a particular entry.
5. Press the **Escape** key or **Red Phone** key to back out to your **Home** screen.

Word Substitution	EN
Search	
eyt (yet)	SmartCase
feild (field)	SmartCase
ftp (ftp)	Specified Ca...
hadnt (hadn't)	SmartCase
hasnt (hasn't)	SmartCase
havent (haven't)	SmartCase
hel (he'll)	SmartCase
heres (here's)	SmartCase
hes (he's)	SmartCase
homeaddr (Mart...	SmartCase

You can also create custom Word Substitutions for more advanced things like automatically typing an email signature, driving directions, a canned email, routine text describing your products or services, legal disclaimer text, anything!

Creating Custom Word Substitutions

To create your own custom **Word Substitutions**, navigate to the list of existing Word Substitutions as we described above.

To create a new entry, follow these steps:

CHAPTER 6: Typing, Spelling, Search & Help

1. From the list of **Word Substitutions**, press the **Menu** key and select **New**.
2. Under **Replace**, type the text you want to be replaced. In this example, we typed **homeaddr**.
3. Under **With:** type the text you want to appear when you type your new word. In this case we type our name contact information and home address.

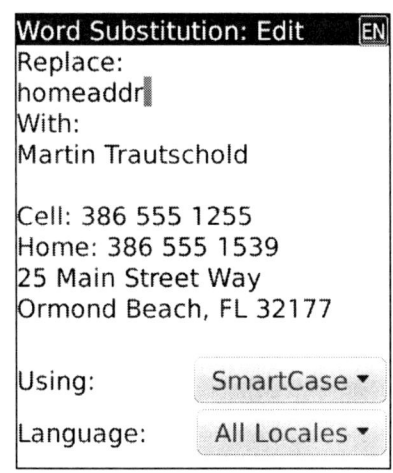

```
Word Substitution: Edit          [EN]
Replace:
homeaddr
With:
Martin Trautschold

Cell: 386 555 1255
Home: 386 555 1539
25 Main Street Way
Ormond Beach, FL 32177

Using:         SmartCase ▼
Language:      All Locales ▼
```

4. For **Using,** you can select **SmartCase** or **Specified Case**. **SmartCase** will replace the text whether or not the case matches exactly, for example **Homeaddr** would work as would **homeaddr**.
5. For **Language** you can select **All Locales** (default) or only specific languages such as **English** or **Spanish**.
6. Press the **Menu** key and select **Save**.

Now, you are ready to test out your new Custom Word Substitution. Open up a MemoPad or compose a new email message. Type your new word (**homeaddr**) and press the **Space** key to see the substitution (See Figure 6-7).

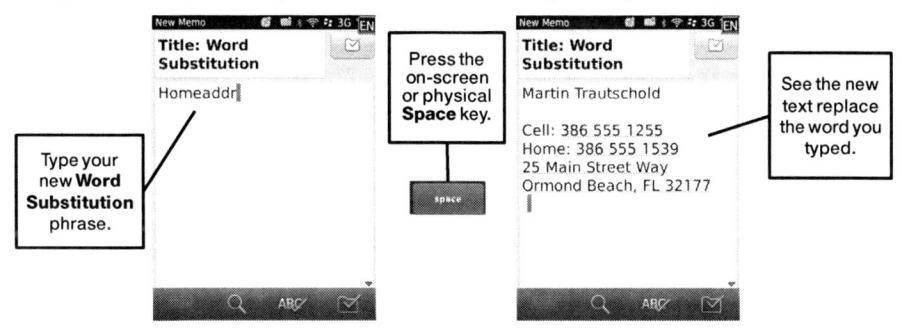

Figure 6-7. Using your new Custom Word Substitution entry.

Edit or Delete a Word Substitution Entry

You may need to edit or remove a **Word Substitution** entry. The steps to get this done are very similar to creating a new one:

View Video Tutorials and Free Tips at www.MadeSimpleLearning.com   175

1. Return to the **Word Substitution** list by repeating the steps shown above.
2. Type a few letters to find the entry you wish to change or delete.
3. Press the **Menu** key, and select **Edit** or **Delete**.

Advanced Features – Macros – Time Stamp

A very useful feature of Word Substitution is that you can actually insert macros or shortcuts for other functions such as the current time and date, your PIN, or owner information. You can even simulate pressing the **Backspace** or **Delete** keys.

Let's create a new entry called **ts** (**time stamp**) that will instantly show the current time and date.

1. Start by creating a new entry as you did above.
2. Type the **ts** under **Replace**.
3. Move the cursor down to under **With**:
4. Press the **Menu** key and select **Insert Macro.**
5. Now, scroll up or down and select the macro you want. In this case, we want a short date (%d) which is **mm/dd/yy** format.
6. Press **Space**, type a hyphen (-), and then press **Space**.
7. Press the **Menu** key and insert the short time (%t) macro. The entry should now look like this image to the right.

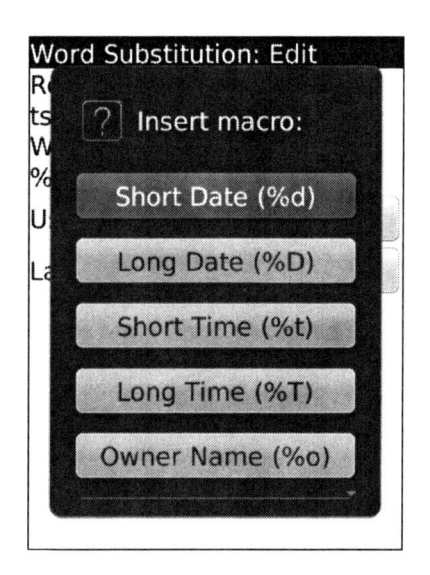

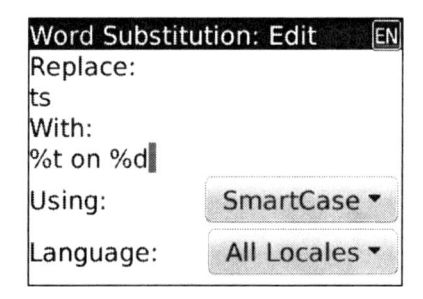

CHAPTER 6: Typing, Spelling, Search & Help

TIP: You could also simply type the **%t** to save a few steps. Read below for a list of all the two-character macro shortcuts.

Now, whenever you want to insert the current date and time, just type your new entry **ts**.

> Title: Meeting ts

Press the **space** key to see the date/time.

> Title: Meeting 1/31/2009 - 9:25a

Here's a list of the standard **Word Substitution Macros:**

%d	Short Date
%D	Long Date
%t	Short Time
%T	Long Time
%o	Owner Name
%O	Owner Information
%p	Your Phone Number
%P	Your PIN
%b	Backspace
%B	Delete
%%	Percent

Title: Macros List

Short Date: 9/22/2008
Long Date: Mon, Sep 22, 2008
Short Time: 8:12p
Long Time: 8:12:29 PM
Owner Name: Martin Trautschold
Owner Info: If found, please contact Martin Trautschold office: 1-386-506-8224.
123 Main Street
Anytown, STATE 38928

Using Your Spell Checker

Your BlackBerry comes with a built-in Spell Checker.

Normally, your Spell Checker is turned on to check everything you type. Words that seem to be misspelled are underlined and suggestions are shown.

1. If you see the correct word just tap it or click on it with the **Trackpad**.

2. If you want to ignore the word, cancel the spell check or add this word to your custom dictionary, then press the **Menu** key.

3. From the menu select one of these

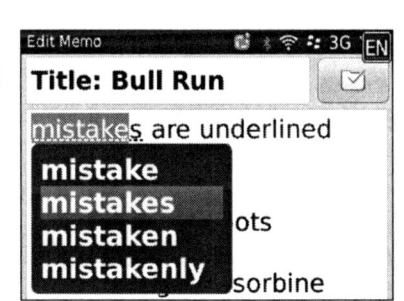

View Video Tutorials and Free Tips at www.MadeSimpleLearning.com

options: **Ignore Once, Ignore All, Add to Dictionary** or **Cancel Spell Check**.

Using the Spelling Custom Dictionary

Sometimes, you may use unique words (such as the names of local places) in your emails that are not found in the standard dictionary. In this case, you may add these words to your own custom dictionary. The advantages of this are (1) that you will never again be asked to replace that word with something suggested and (2) if you misspell this custom word, you will be suggested the correct spelling.

Adding a Word to the Custom Dictionary

Let the Spell Check program notice the word that it believes is misspelled. In this example, we are using "Funfetti" (a brand of cake that Martin Trautschold's family enjoys) that is not in the standard dictionary.

The spell check program will suggest options for replacing the word.

1. Press the **Menu** key.
2. You will see options to **Ignore Once, Ignore All** or **Cancel Spell Check**.
3. Click the **Add to Dictionary** to add this new word to your own custom dictionary.

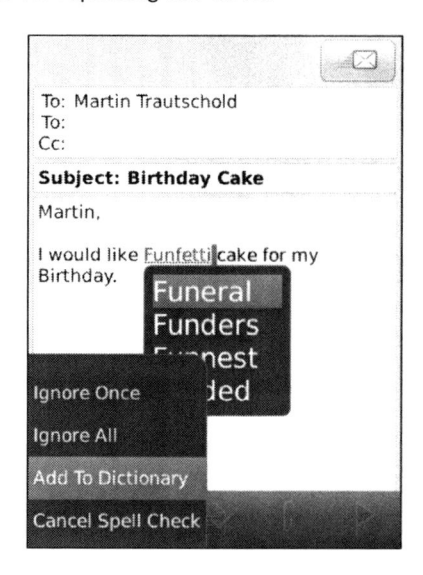

Next time we spell "Funfetti," it will not be shown as misspelled. Even better, the next time we misspell Funfetti (e.g. "Funfeti"); the spell checker will find it and give us the correct spelling.

Edit or Delete Words in the Custom Dictionary

Mistakes happen. Sometimes, you press the wrong menu item and inadvertently add incorrectly spelled words to the Custom Dictionary. The authors have done this plenty of times.

1. From the **Home** screen, tap the **Search** icon and type "Custom."

2. Tap the **Options** icon then **Custom Dictionary**.

3. Once you are viewing the **Custom Dictionary** words, tap any word to simply edit it.

4. To add, delete words or even clear out the entire dictionary, highlight the word by using the Trackpad to roll up or down, then press the **Menu** key and select **New**, **Edit**, **Delete** or **Clear Custom Dictionary**.

5. When done, press the **Menu** key and **Save**.

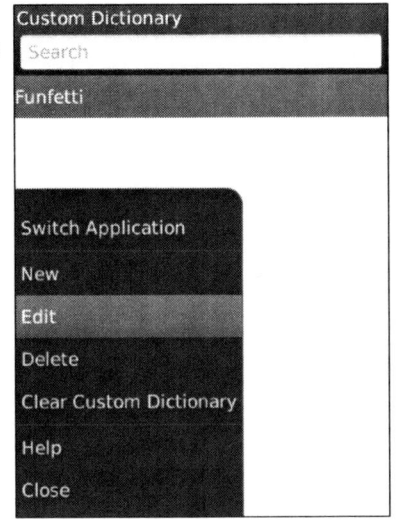

Universal Search

Your BlackBerry can hold so much information that it can sometimes be difficult to keep track of where everything is on the device. Fortunately, the BlackBerry comes with a comprehensive **Search** program (also known as "Universal Search") to help you find exactly what you might be looking for.

In this chapter, we will show you how use the **Search** program across your email, music, contacts, application names, options functions, and even across the web, as well as how to filter your search information to show only what you want.

> **TIP:** You can even use the **Search** program to search through your calendar events and answer questions such as the following:
>
> When is my next meeting with Sarah?
>
> When did I last meet with Sarah?

For this feature to work, you need to type people's names into your calendar events as you enter them. For you example you might enter the following: **Meet with Sarah** or **Lunch with Tom Wallis**.

Understanding How Search Works

As you get used to your BlackBerry, you will begin to rely on it more and more. The more you use it, the more information you will store within it. It is truly amazing how much information you can place in this little device.

At some point, you will want to retrieve something – a name or a word or a phrase – but you won't remember exactly where you stored that particular piece of information. This is where the **Search** program can be invaluable.

TIP: Check out the "Universal Search" section of the Quick Start Guide at the beginning of this book to learn how to enable automatic **Search** when you start typing letters from the **Home** screen.

By default, if you start typing letters from any home screen as shown to the right, you will see those letters appear in the Search field at the top.

If you don't see your letters appear automatically, then you have turned on your Application Shortcuts and need to tap the Search icon (magnifying glass) first before you start typing.

Notice that icons start appearing and changing at each additional character you type.

In the image to the right, three contacts appeared, email, text messages, calendar items, the Help app, and a number of other search apps appeared. If you tap any search app such as Search BlackBerry **Podcasts**, the search phrase that you typed (in this case "help") will be searched in the **Podcasts** app. Keep scrolling down and you will see that you can search the **App World**, **Slacker**, **Facebook**, **YouTube** and even the web in general.

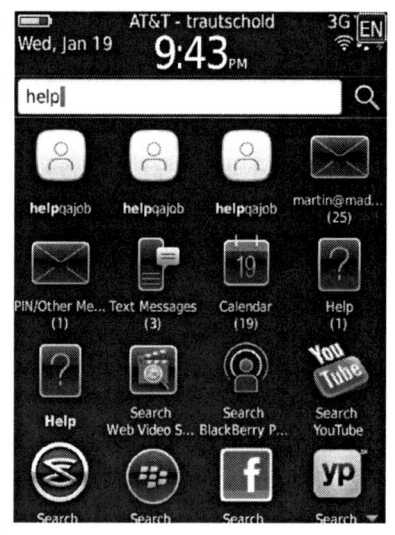

Examples of Searches

You can search for just about anything on your BlackBerry or on the web using the **Universal Search** app. Below are a few types of searches to get you thinking about what is possible.

- Find all videos on **YouTube** that have the word "baby" in the title or description.
- Find every contact in your **Contacts** app (address book) that live on a certain street or in a certain town.
- Find every appointment in your **Calendar** that has the word "dentist" in it – to help you find when your next dentist appointment is scheduled.
- Find all songs or artists in your **Pandora** app that have the word "love" in their song title or artist name.
- Find that **Text Message** where you or someone you were chatting with mentioned the phrase "best book."
- Do a **web search** for pizza in your town or postal code.
- Do a **web search** for the flight number of the person you are picking up at the airport to check to see if their flight is on time.
- Do a **web search** for 303 to understand where that area code is located.

Improving Your Search Quality

So far we've covered some of the things you might search for, as well as how to go about implementing a search. Next, we will cover some valuable tips and tricks that help you create more effective searches.

The more information you enter on your BlackBerry (or enter on your desktop computer and sync to your BlackBerry), the more useful your device becomes. Combining a great deal of useful information with this **Search** tool means you truly have a very powerful and useful handheld computer. As your BlackBerry fills up, however, the possible places where your information is stored increases. Also, a search might not turn up the exact information you are looking for due to inconsistencies in the way you store the information.

Try to be consistent in the way you type someone's name – for example, always use *Martin* instead of *Marty* or *M* or any other variation. This way, the **Search** program will always find what you need.

Occasionally, you should check your address book for contacts with multiple entries. It is easy to wind up with two or three entries for one contact if you add an email one time, a phone number at another time, and an address on still another occasion. Restricting your contacts to one entry each helps you stay organized – and helps you find what information you're looking for more easily.

> **TIP:** It is usually easier to do this cleanup work on your computer, and then sync the changes or deletions back to your BlackBerry.

If you're not sure whether you are looking for *Mark* or *Martin*, just type in *Mar* and then perform your search. This way, you will find both names.

Do your best to put consistent information into calendar events. For example, if you want to find the next dentist appointment for Gary, you could search for *Gary Dentist* in your calendar and find it. But this only works if you made sure to enter the full words *Gary* and *dentist* in your calendar entry. In this case, it would be better to search only for the word, *dentist*.

Fine-Tuning Your Search

You can adjust the breadth of apps included in your **Search**. If you want to avoid searching all your email accounts, or only search specific accounts, you can. If you only wanted to search **Contacts** and **Calendar** items, you could do that as well. After you start a search, press the Menu key and select **Search Options**. Then you will be able to scroll up and down to check or uncheck specific items to include in your search (see Figure 6-8).

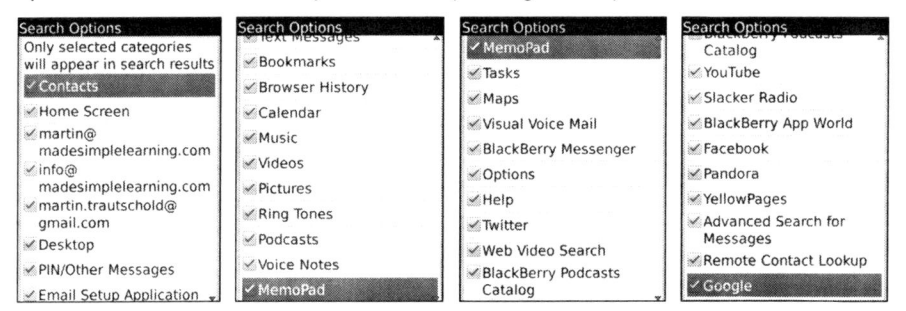

Figure 6-8. Customize your Search with the Search Options screens.

CHAPTER 6: Typing, Spelling, Search & Help

Using the Torch's Built-in Help

There might be times when you don't have this book handy and you need to find out how to do something right away on your Torch.

Two Ways to Get to Help

You can get into the BlackBerry built-in **Help** menus in two ways (see Figure 6-9). One is to press the **Menu** key from almost any app and select **Help** (it's usually near the bottom). The second way is to tap the **Help** icon (which is usually at the bottom of your **All** page of your **Home** screen). The built-in help can be pretty useful when you are in a bind and need to try and find something out in a hurry, but you will notice that the help is 99% text based with sparse directions.

Tap the **Help** icon to see the main help menus.

Tap the **Help** menu item in most apps to see help about that app.

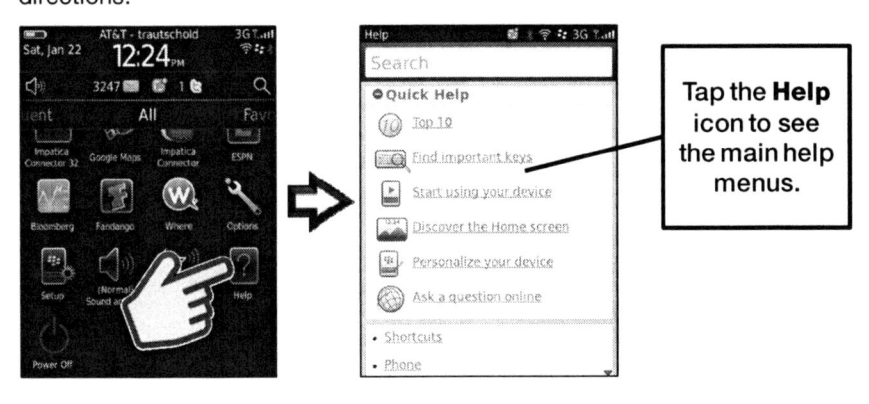

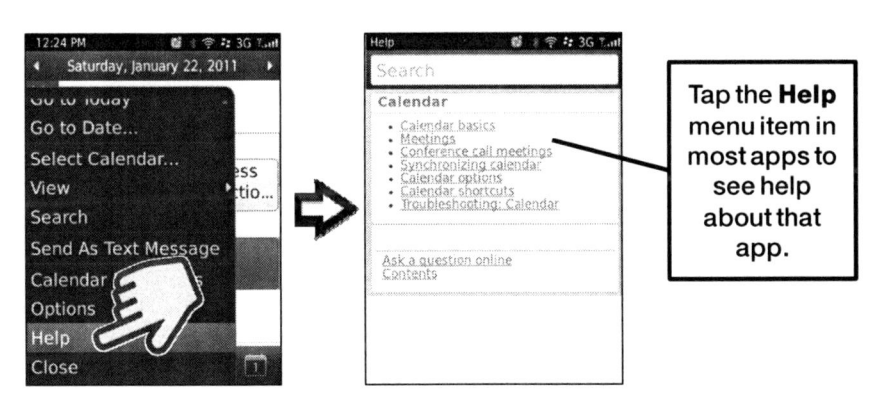

Figure 6-9. Two ways to get to the Help on your BlackBerry.

View Video Tutorials and Free Tips at www.MadeSimpleLearning.com   183

Using the Help Menus

As we showed above, you can get to **Help** from the menu in most apps to see help about that application. For our purposes, we will take a look at the **Help** icon itself.

1. From your **Home** screen, scroll to the bottom and tap the **Help** icon.

2. The main help menu will look similar to this image. From here you can do a number of things:

   - To search for a specific topic, start typing the word or phrase on your keyboard. You will instantly see results appear on the screen. Keep typing more words to narrow the search.

   - To view the ten most frequently asked questions, tap **Top 10**.

   - To learn about important keys on your device, tap **Find Important Keys.**

   - To learn about common setup items such as Wi-Fi, email and more, tap **Start using your device**.

   - Tap any other item in the Quick Help area to learn more.

   - Swipe your finger down to view additional help topics (see Figure 6-10).

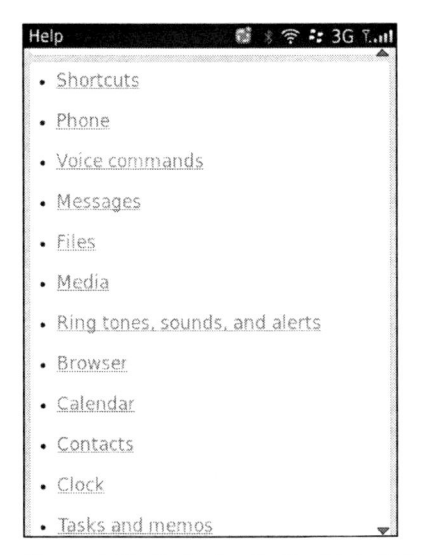

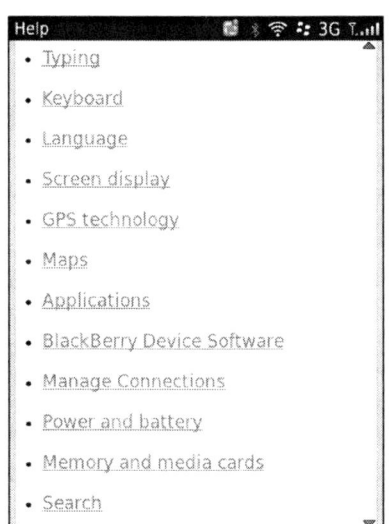

Figure 6-10. Navigating the Torch's built-in help function.

3. Continue to click on topics you would like to learn about.

4. Press the **Escape** key to back up one level in the help menus.

5. As you delve deeper into help you will notice that there are usually links at the bottom for **Related Information**. These can be useful links to help you get a more complete picture of what you are trying to learn.

Help Tips and Tricks

Like pretty much every other feature on the BlackBerry, there are some tips and tricks when using **Help**.

1. **Use your Trackpad.** Sometimes the items on the screen can be fairly small and difficult to tap with your finger. Use your **Trackpad** instead to exactly click the item you want.

2. **Use the Search.** The search built into the Help app is quite powerful and will quickly help you pinpoint the topic you are trying to find. For example, type "con call" to search for all conference call related topics.

3. **Use Related Information links**. At the bottom of most detailed help pages, you will see links to related topics. It can be quite helpful to use these links.

4. **Use the Contents link**. To get back to the main menu of the Help app, tap the **Contents** link you see at the bottom of most every screen.

5. **Use the Top 10**. From the main Help menu, these are the top 10 most frequently asked questions. Chances are, you will or may have already, asked these questions.

6. **Use other Quick Help items**. These can be quite useful to give you a basic overview of how to setup, customize and use your device when you don't have our book handy.

7. **Ask a Question Online**. At the bottom of most screens you will see the link **Ask a Question Online**. Tap this link to be brought to the online help section of BlackBerry.com. Here you can browser and learn more about help and even ask a question.

TIP: When asking an online question, be very brief and only include the most important words. For example, instead of typing "How do I make a conference call?" type just "conference call."

Chapter 7: Personalize Your Torch

In this chapter, you will learn some great ways to personalize your Torch, like changing your Home screen wallpaper, moving and hiding icons, organizing with folders, setting your convenience key, changing your BlackBerry theme (look and feel), and adjusting font sizes and types.

You will also learn how to use the various **Home** screens available to you; **All, Favorites, Media, Downloads** and **Frequent**.

Setting Your Home Screen Preferences

The easiest thing to do to personalize your Torch is to change your **Home** screen preferences which include the background wallpaper, rows of icons, and whether or not you see icons or upcoming calendar events, recent emails, and phone call logs.

Using the Navigation Bar

New to the Torch is the unique **Navigation Bar** that sits atop the rows of icons. This can be touched and dragged up or down to reveal more or less icons on the **Home** screen (see Figure 7-1).

Drag the **Navigation bar** up or down to show more or fewer rows of icons.

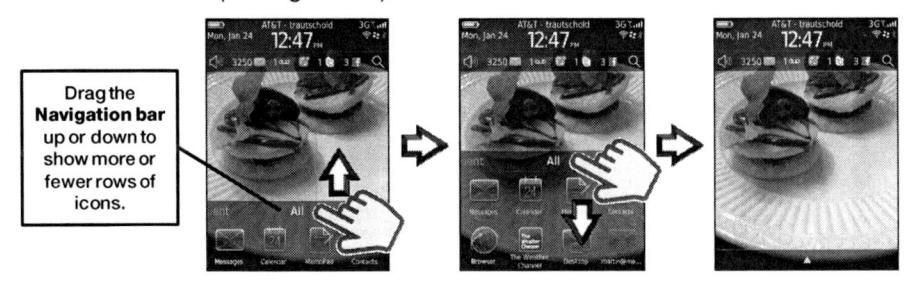

Figure 7-1. Drag the Navigation Bar up or down to reveal more or fewer icons.

Drag the **Navigation Bar** all the way down to just see your wallpaper with no icons on the screen. You can move the **Navigation Bar** up incrementally to reveal one to four rows of icons.

> **NOTE**: You can also press the **Menu** key and then select **Open Tray** to reveal your icons.

Changing Your Wallpaper

You might want to change the background picture on your Torch's Home screen from time to time. To do so, follow these steps:

1. From your **Home** screen, press the **Menu** key and tap **Options**.

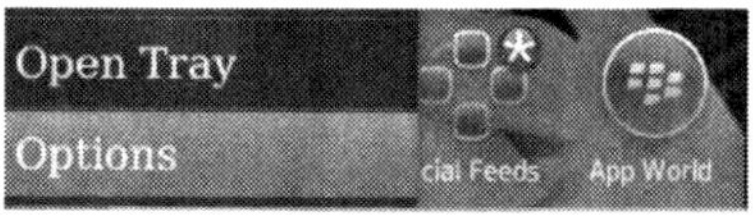

2. Click the **Change Wallpaper** button.

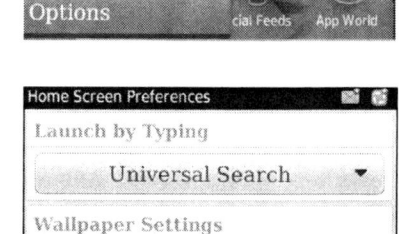

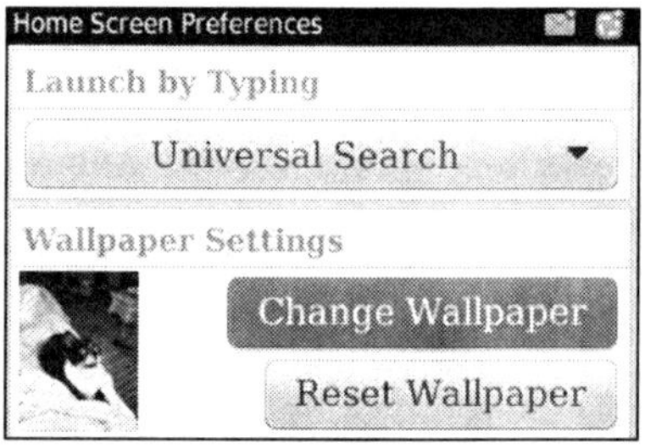

3. If you want to take a picture right now to use as wallpaper, then click on the **Camera** at the top of the page.

    a. Snap a picture.

    b. After taking the picture, press the **Menu** key and select **Set As Wallpaper**.

    c. You can also touch and hold the image and choose **Set as Wallpaper** from the pop-up menu.

4. If you want to use a picture or pre-loaded wallpaper, navigate to a folder; either **Camera Pictures**, **Picture Library**, **Wallpaper**, **BlackBerry Messenger** or **other folder listed** and search for an item to use as your wallpaper.

5. Once you get to the image you want to use, press the **Menu** key and select **Set As Wallpaper**.

Changing the Location of Your Downloads

The default setting on your BlackBerry Torch is to have the your newly installed application icons (your "downloads") go to the **Home** screen. You can customize where you want the **Download** to go by following these steps:

1. From your **Home** screen, press the **Menu** key an select **Options**.
2. Click the drop-down item under the **Downloads Folder** heading.
3. Choose **Home, Media, Instant Messaging, Applications** or **Games**.
4. Press the **Escape** key and Select **Save** on the next screen to save your changes.

Resetting Your Home Screen Preferences

If you find yourself wishing you had not made so many changes, or you just want to revert your Home screen preferences back to the defaults, follow these steps.

1. From your Home screen, press the **Menu** Touch or click on **Options**.
2. Tap on **Reset Settings.**
3. Select the items you want to reset by pressing and clicking on them. You can reset any of the following: **Wallpaper, Layout, Download Folder,** or **Icon Arrangement**.
4. Tap **Apply** when done.

Organizing Your Icons

You may not need to see every single icon on your **Home** screen, or you may want to move your most popular icons to the top row for easy access. The way you move and hide icons may vary a little depending on which Theme you have on your BlackBerry. Below we show you the standard way to move and hide icons.

Picking Favorites

New with BlackBerry OS 6, you can select your favorite icons and place them in the Favorites section of the Home screen.

1. Press and hold the icon you wish to make a favorite until you see the pop-up menu. In this example, we want to make **The Weather Channel** app a favorite.

2. Select **Mark as Favorite** as shown.

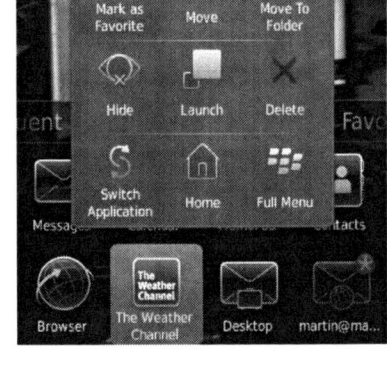

3. Now the icon appears in the **Favorites** section of your **Home** screen for easy access.

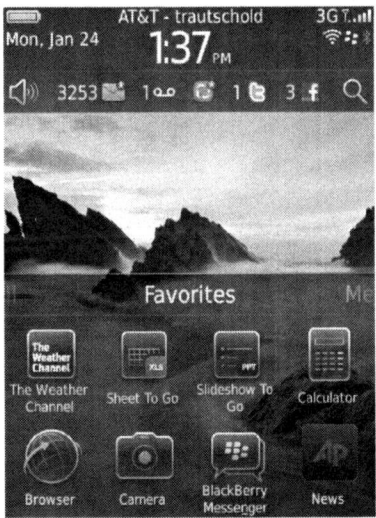

    TIP: The icons you mark as **Favorites** appear at the bottom of the list of icons. You may want to touch and hold the icon to Move it up to the top of the Favorites page.

4. If you want to remove a **Favorite** icon, touch and hold it and select **Unmark as Favorite** from the pop-up menu.

Moving Your Icons within a Folder

Press the **Menu** key to see an array of all your icons. If the icon you want to move is inside a particular folder, like **Favorites, Media, Downloads** or **Frequent**, scroll to that **Home** screen.

1. Press and hold the icon you wish to move until you see the pop-up menu.

2. Select **Move** as shown.

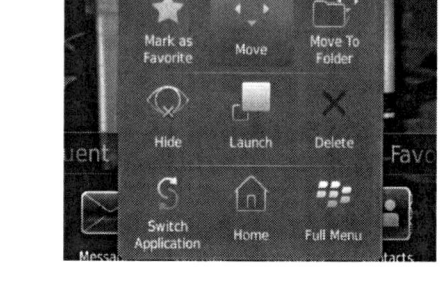

3. Once you select **Move**, you will see arrows around the icon (as shown).

4. Gently touch the screen in the spot where you would like to move the icon. You can also use the **Trackpad** to move the icon which is a bit more precise.

5. Finally, touch or click to set the moved icon at the new location.

> **TIP:** You can also press the **Menu** key and select **Mark as Favorite** to have the icon show up in the **Favorites Home** screen.

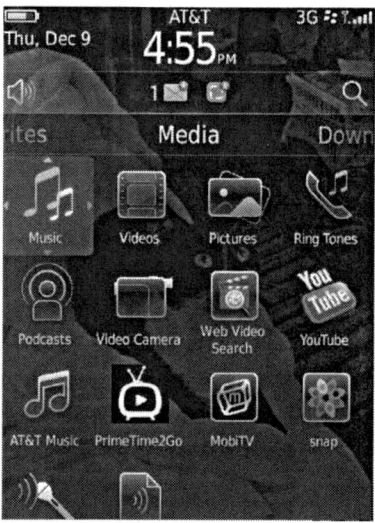

Hiding and Showing Icons

Hiding icons is fairly straightforward. Getting them back takes a few more steps.

How to Hide Icons

To hide an icon, follow these steps.

1. Touch and hold the icon you wish to hide until you see a menu pop-up in the middle of the screen.
2. Select **Hide** from the pop-up menu.

You're done – the icon is hidden.

How to Show Hidden Icons

To show a hidden icon, follow these steps.

1. Press the **Menu** key and select **Show All**.

2. The icons that are hidden appear dimmer or grayed out like the **AT&T Maps** icon to the right.

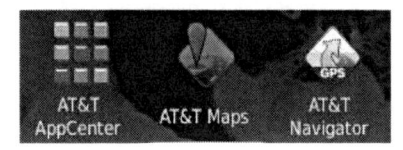

3. Highlight the icon you wish to restore from being hidden.

4. Press the **Menu** key and select **Hide**. Notice that there is a checkmark next to the menu item; this shows the icon is currently hidden. Clicking **Hide** again will un-hide it.

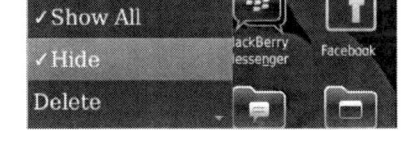

5. Finally, if you want to get rid of all the other hidden icons, you need to turn off the **Show All**. Press the **Menu** key and select **Show All**.

How do I know when I'm in a Folder?

When you are in a folder, you see a little tabbed folder icon at the top of your screen and the name of the folder. In this image, we are in the **Applications** folder.

Moving Your Icons between Folders

Sometimes you want to move icons to your Home folder to make them more easily accessible. Conversely, you might want to move some of the icons you seldom use from your Home folder into another folder to clean up your Home screen.

Let's say we wanted to move our **Text Messages** icon from the **Home** screen to the **Applications** folder.

1. Press and hold the **Text Messages** icon until you see the pop-up menu.
2. Select **Move to Folder** as shown.

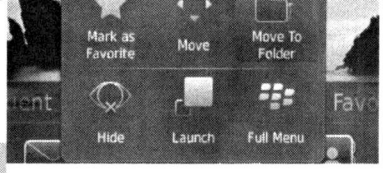

NOTE: You can also press the **Menu** key after highlighting the icon and select **Move to Folder** from the menu.

3. We want to move this icon out of our **Home** folder into the **Applications** folder, so we tap on **Applications**.

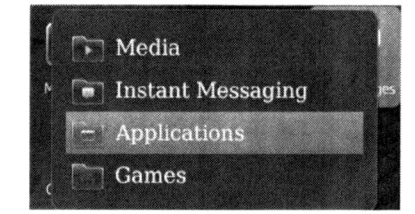

CHAPTER 7: Personalize Your Torch

4. Notice that right after you move an icon, it appears at the bottom of the folder. **Text Messages** is at the bottom of the list of icons in the **Applications** folder.

Moving Important Icons in the Top Row of the Main "All" Folder

Sometimes, you want to get your most used icons into the very top row of the All Folder on the Home screen.

1. If the icon is not in the **Home** folder, then use the **Move to Folder** function to move it to **Home**.

2. Then use the **Moving Icons** function to move it to the very top row of icons. As shown here, the **Calculator** is in the Home (All) folder at the top row for very easy access.

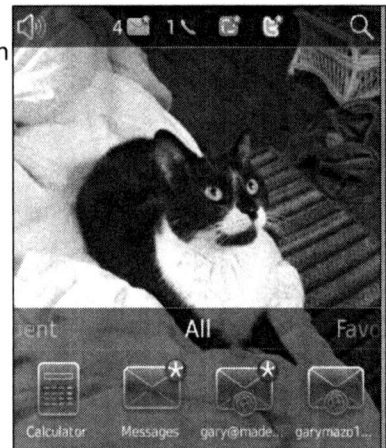

Working with Folders

On your BlackBerry, you can create or delete folders to better organize your icons. There are several folders created by default—typically, the **Media**, **Instant Messaging**, **Applications**, and **Games** folders. You can add your own folders and then move icons into your new folders to better organize them.

View Video Tutorials and Free Tips at www.MadeSimpleLearning.com   195

Creating a New Folder

> **NOTE:** At the time of publication, you could only create folders one level deep. In other words, you can only create new folders when you are in the Home folder, not when you are already inside another folder. This may change with new software versions.

1. To create a new folder, first highlight any icon on the Home screen and then press the **Menu** key and select **Add Folder**.

   NOTE: If you don't see the **Add Folder** menu item, you most likely have the **Navigation Bar** highlighted instead of an **App** icon.

2. After you select the **Add Folder** menu item, you will see this screen. Type your folder name, then touch or click on the folder icon, and scroll left or right to check out all the different possible folder colors and styles.

3. Once you're done selecting the folder icon style, Tap on it, and scroll down to Tap on the **Add** button to finish creating your folder. Now you will see your new folder.

Editing a Folder

You can edit a folder by highlighting it, pressing the **Menu** key, and selecting **Edit Folder**. Then you can change the name and folder icon, and save your changes.

Deleting a Folder

Whenever you want to get rid of a folder, just highlight it and select **Delete,** as shown.

Setting the Date, Time, and Time Zone

You can adjust the Date, Time, and Time Zone in the Setup Wizard, but there are times when you want to adjust it manually.

1. From your home screen, touch or click on the time at the top to open your clock app.

2. Scroll down to **Options** and touch or click.

3. Select **Display** to get to the **Date and Time** setting screen.

4. Click next to **Time Zone** to see a list of all the time zones and select the correct one.

5. If you leave the **Auto Update Time Zone** to the default **On**, then every time you move between time zones, your BlackBerry will automatically detect the new time zone from the cell towers and update your BlackBerry.

6. The only way you can manually adjust the date or time is if you change **Update Time** to **Manual**.

7. You will then see **Set Time** and **Set Date** buttons appear.

8. Then, click on the time or date field to bring up the setting pop-up window. You can swipe up/down to change a value. Or, in number fields, you can click on the field to bring up the keyboard and type a number like "23." Typing digits can be faster and more accurate than scrolling.

9. If you prefer **12 hour** format (7:30 AM/ 4:30 PM) to **24 hour** format (07:30/16:30), you set that in **Time Format**. Tap the **space** bar to toggle between the two options.

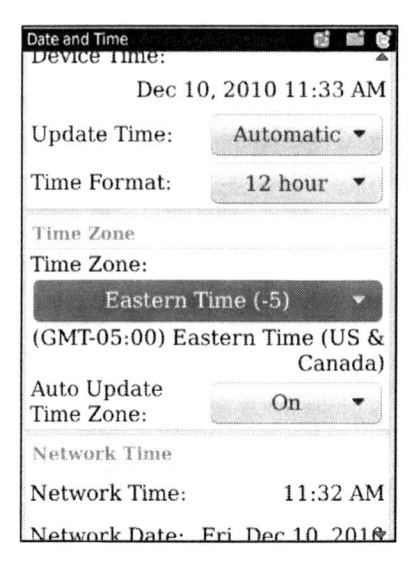

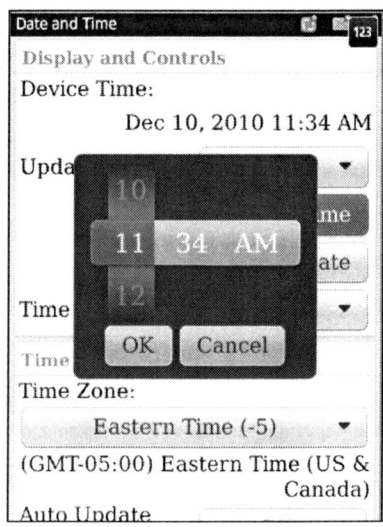

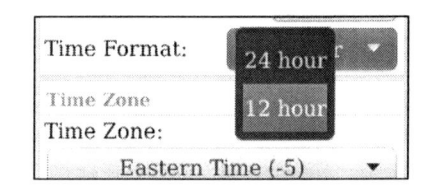

CHAPTER 7: Personalize Your Torch

TIP: You can force the BlackBerry to get updates to the time zone by pressing the **Menu** key and select **Get Time Zone Updates**.

10. Finally, to save your changes, press the **Menu** key and select **Save**.

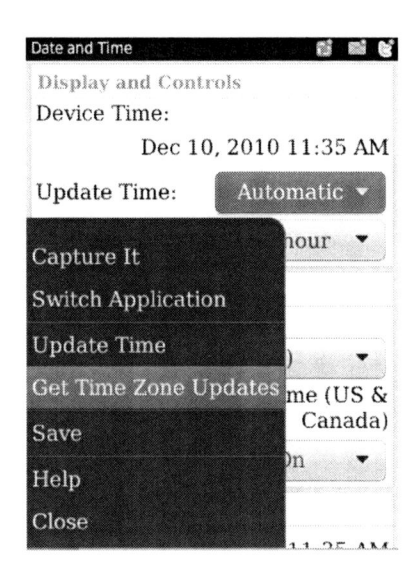

Changing Your Font Size and Type

You can fine-tune the font size and type on your Torch to fit your individual preferences.

Do you need to see more on the screen and don't mind small fonts? Then go all the way down to a micro-size 7-point font.

Do you need to see **bigger fonts** for easy readability? Adjust the fonts to a large **14-point font**.

To adjust your font, follow these steps:

1. Touch or click on the **Options** icon. You may need to press the **Menu** key and **Open Tray** to find it.

2. Scroll down to Display and touch or click and then touch or click on **Screen Display** (Figure 7-2).

3. You can change the **Font Family**, **Font Size**, and **Font Style** on this screen. Touch or click to change any item. Notice the preview of your currently selected style and size to make sure it will fit your needs.

4. When done, press the **Menu** key and select **Save**.

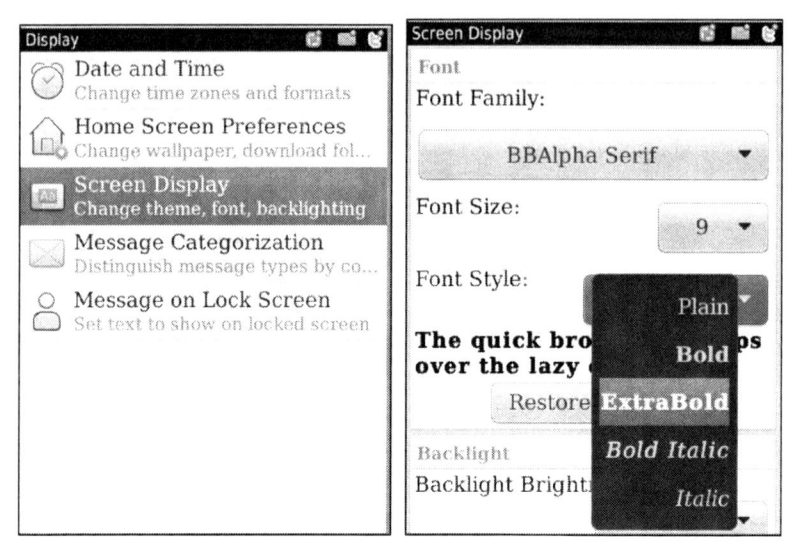

Figure 7-2. Changing your Torch font family, size, and style

Changing Your Theme: The Look and Feel

You can customize your Torch to make it look truly unique. One way to do this is to change the Theme of your Torch. Changing Themes usually changes the layout and appearance of your icons by and the font type and size you see inside each icon. There may be only one or a few themes pre-installed on your Torch. The good news is that you can find hundreds of Themes available for download from various web sites.

CARRIER-SPECIFIC THEMES: Depending on your BlackBerry wireless carrier (phone company), you may see various customized Themes that are not shown in this book.

MORE STANDARD/GENERIC BLACKBERRY THEMES: Most of the Standard Themes shown below are on every BlackBerry (or can be downloaded from http://mobile.blackberry.com).

CHAPTER 7: Personalize Your Torch

1. Scroll and touch or click on the **Options** icon on your BlackBerry. You may have to press the **Menu** key **to Open Tray** see all your icons and then locate the Options icon. Once in **Options**, scroll down to Display and touch or click and then touch or click on Screen Display.

2. At the very bottom you will see **Theme.** If you have more than one theme installed, they will be listed one under the other.

3. If no other themes are listed, just press the **Menu** key and choose Download Themes.

4. Inside the Theme screen, just scroll and Tap on the Theme you want to make **Active**. Your currently selected Theme is shown with the word (Active) next to it.

5. Then press the **Activate** key and then the **Escape** key to get back to the Home screen to check out your new theme.

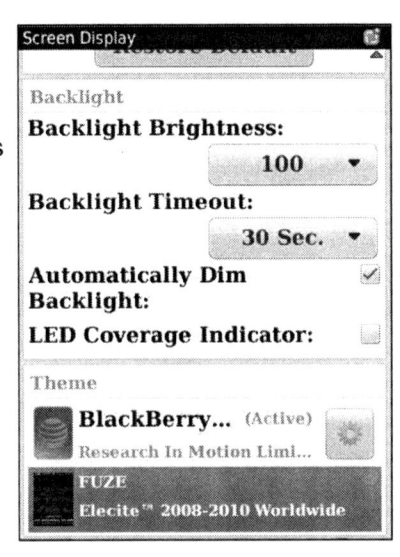

Downloading New Themes

CAUTION: The authors have downloaded many themes on their BlackBerry smartphones. Some themes can cause problems with your BlackBerry, such as the BlackBerry may stop working or freeze, or you may not be able to see everything on the screen. We recommend only downloading themes from a web site you know and trust.

Download from App World

1. Start BlackBerry App World. See Chapter 20 for help getting App World running.

View Video Tutorials and Free Tips at www.MadeSimpleLearning.com 201

2. Click on the Categories soft key in the lower left corner.

3. Scroll down and click on the Themes category.

   Themes (769)

4. Click on the type of theme you would like to download.

5. In this example, we'll choose **Nature**.

6. Now you see all the themes in the **Nature** category.

   Notice that there are FREE and paid themes.

7. Click on any theme to learn more about it, buy it, or download it.

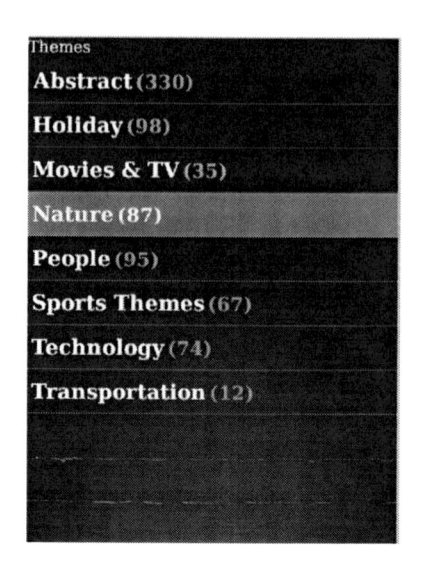

CHAPTER 7: Personalize Your Torch

8. To learn more about the theme, check out the **Screenshots** and **Reviews**.

> **WARNING:** Some of the reviews may have explicit language and many have spelling errors.

9. If you are ready to try the Theme, click the **Download** button.

10. When the theme is downloaded, you will see a message asking if you want to activate it.

11. Click **Yes** to give the new Theme a try right now.

12. Here is an example of the new **Theme** activated.

You can't see it in the book, but the rose is animated. It opens and closes slightly, and the sparkles move around.

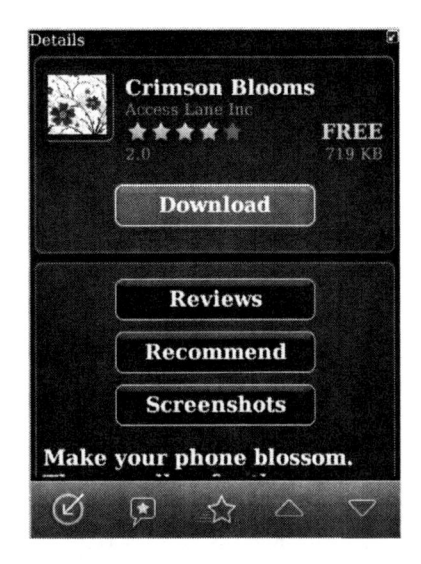

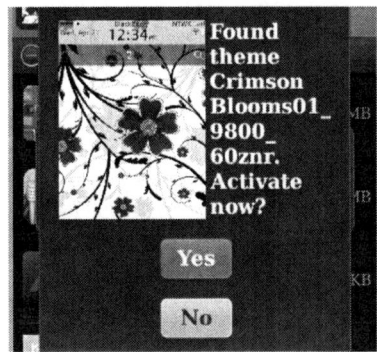

View Video Tutorials and Free Tips at www.MadeSimpleLearning.com   203

Download Themes, Wallpaper, and Ringtones from Other Web Sites

You can also find many Themes from BlackBerry-related web sites. To download these themes, follow the steps below:

1. Tap your **Browser** icon.

2. Tap in the top Address bar to type in one of the BlackBerry community web sites such as:

BlackBerry Mobile Site: http://mobile.blackberry.com

CrackBerry.com : www.crackberry.com

BlackBerry Forums: http://www.blackberryforums.com

3. Click the **Go** key [Go] (where the Enter key usually is located) when using the on screen keyboard or use the Enter key on the keyboard itself.

4. Look for a section on the site that says something like personalize, themes, wallpapers or ringtones.

5. Follow the steps on the site to download the content to your Torch.

> **CAUTION:** Some of the items you find on these web sites may be explicit or offensive. Please view and download with care.

Changing Your Convenience Key

The keys on the bottom of the right side of your Torch is actually a programmable key called a **convenience** key (named so because the key can be set to conveniently open any icon on your BlackBerry, even new third party icons that you added to your BlackBerry).

CHAPTER 7: Personalize Your Torch

1. Touch and click on the **Options** icon (press the **Menu** key and open the tray if you don't see it listed.)

2. Scroll down to **Device** and touch or click on it.

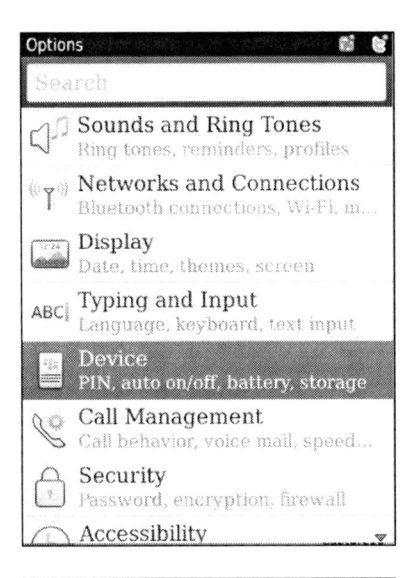

3. Scroll down the screen until you see **Convenience Key** and touch or click on it.

4. To change the application that these keys open, just touch or click on the item to see the entire list. Then touch or Tap on the icon you want.

5. Then press the **Menu** key and select **Save** to save your changes.

> **TIP:** We recommend setting your left convenience key to open the **Camera**. It makes it that much easier to just take a picture when needed.

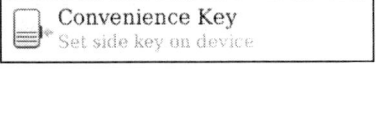

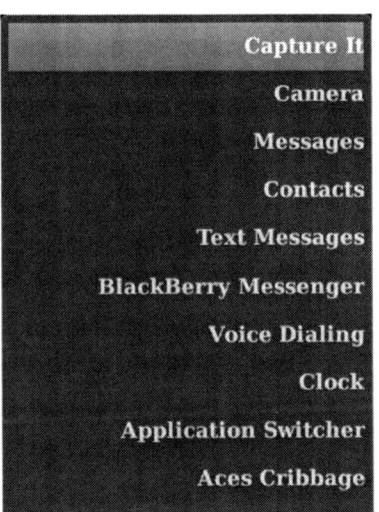

Now give you newly set convenience keys a try.

> **TIP:** The Convenience keys **work from anywhere**, not just the **Home** screen.
>
> **TIP:** You can set your Convenience keys to open any app, even newly installed ones!

View Video Tutorials and Free Tips at www.MadeSimpleLearning.com   205

> After you install new apps, you will notice that they show up in the list of available icons to select in the Screen/Keyboard options screen. So if your newly installed stock quote, newsreader, or game is important, just set it as a convenience key.

The Blinking Light - Repeat Notification

One of the features that BlackBerry users love is the little LED that blinks in the upper right hand corner. It is possible to have this light blink different colors:

- Red when you receive an incoming message (MMS, SMS or Email) or calendar alarm rings
- Blue when connected to a Bluetooth device
- Green when you have wireless coverage
- Amber if you need to charge your BlackBerry or it is charging

Red Flashing Message or Alert LED

By default, whenever you receive a new email message or your calendar or other alarm rings, the red LED on the top of your Torch will also flash. The red light can be pretty bright, especially in the dark, so you might want to turn it off. You can disable the red LED feature by following the steps below:

1. Tap your **Sounds** icon.
2. Scroll down to **Change Sounds and Alerts** and touch or click.
3. Scroll down to **Profile Management** and touch or click.

CHAPTER 7: Personalize Your Torch

4. Tap the profile you wish to adjust. (Most often, **Normal** is the profile that is highlighted.)

   Profile Management (Edit, Add, Dele...
   - Add Custom Profile
   - All Alerts Off
   - Normal ✓
   - Loud
   - Medium
   - Silent
   - Vibrate Only
   - Phone Calls Only

5. To edit email accounts, click on Messages. You will see each of your email accounts listed as Email [(your email address)], and Level 1, PIN, Text, etc.

6. To edit Calendar and other alerts, click on the **Events -Reminders** section.

   Messages - Notifiers
   Email [gary@madesimplelearning.com]
   Email [garymazo1@gmail.com]
   Level 1
   PIN
   Text Messages

7. Click on any account or app you wish to modify.

8. Set the LED to **Off** to disable the red flashing light.

   Visual Alerts
   LED: Off
   Tactile Alerts   On

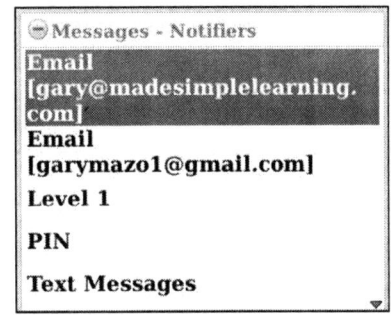

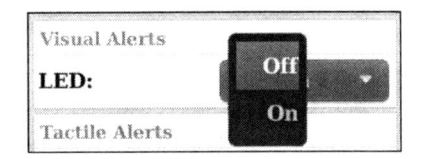

9. Press the Menu key and select Save.

Bluetooth Flashing Blue LED

In order to disable or enable the blue flashing LED when you are connected to a Bluetooth device, follow these steps:

1. From your **Home** screen, touch or click on the **Options** icon (wrench).

2. Scroll to **Networks and Connections** and touch or click and then scroll to **Bluetooth Connections** and touch or click.

3. Press the **Menu** key and select **Options**.

4. To turn on the blue indicator, check the box next to **LED Connection Indicator**. To turn this off, uncheck the same box.

5. Press the **Menu** key and **Save** your settings.

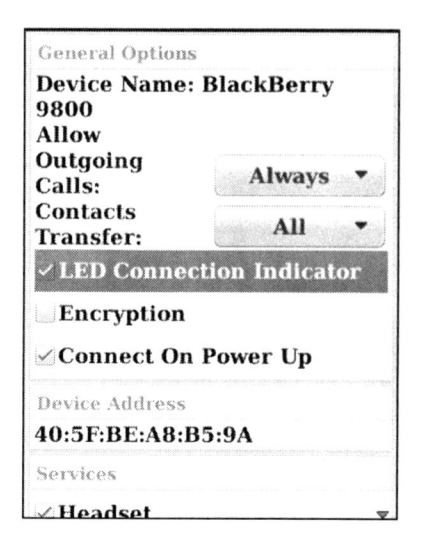

Coverage Flashing Green LED

To enable or disable the green coverage flashing LED, follow these steps:

1. Click on your **Options** icon.

2. Scroll to **Display** and touch or click.

3. Touch Screen Display and then Scroll down to **LED Coverage Indicator** and place a check in the radio box

4. Press the **Menu** key and **Save** your settings.

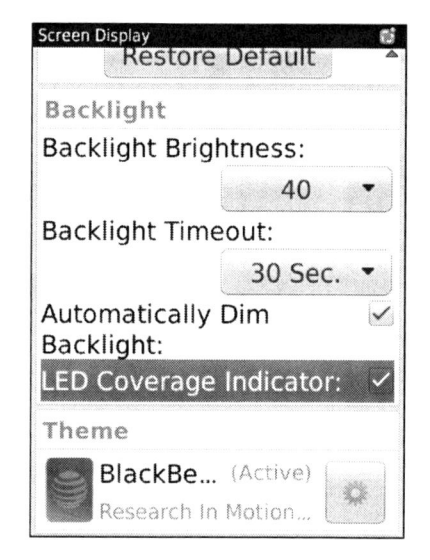

Sounds: Ring and Vibrate

Your Torch is highly customizable—everything from Ringtones to vibrations to LED notifications can be adjusted. Traveling on an airplane but still want to use your calendar or play a game without disturbing others? No problem. In a meeting or at a movie and don't want the phone to ring, but you do want some sort of silent notification when an email comes in? No problem.

Virtually any scenario you can imagine can be dealt with preemptively by adjusting your Sound Profile settings.

Preloaded Sound Profiles

By default, the Torch is set to a **Normal** sound profile, meaning that when a call comes in, the phone rings; when a calendar alarm rings, you will see an alert on the window (without any sound or vibration); and when a message comes in, you hear a sound and the phone may vibrate. The currently selected sound profile has the word "(Active)" next to it. Seven preloaded sound profiles are available on your Torch.

All Alerts Off will turn off all notifications.

Normal is the default Active profile; in it, the phone rings and messages beep and vibrate.

Loud increases the volume for all notifications.

Medium adjusts all notifications to mid-level volume.

Silent will display notifications on the display and via the LED only.

Vibrate Only enables a short vibration for all alerts. (Great for meetings, movies, family dinners, or other places where cell phone rings are discouraged.)

Phone Calls Only will turn off all notifications except incoming calls.

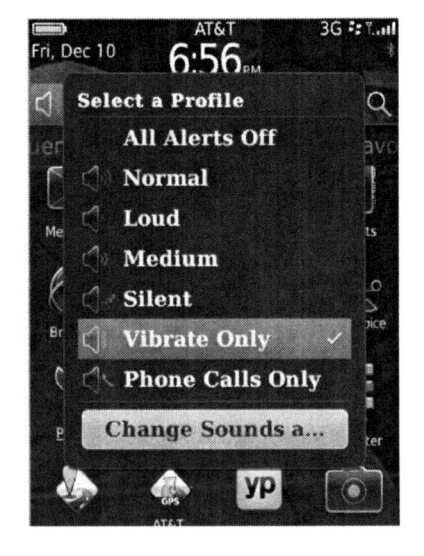

Selecting a Different Preloaded Sound Profile

You can quickly change your sound profile from your **Home** screen.

1. Tap the **Speaker** icon in the upper left portion of the Home screen.

2. Tap to select a new **Sound Profile** (see Figure 7-3).

3. The new profile will be shown by a change in the **Speaker** icon.

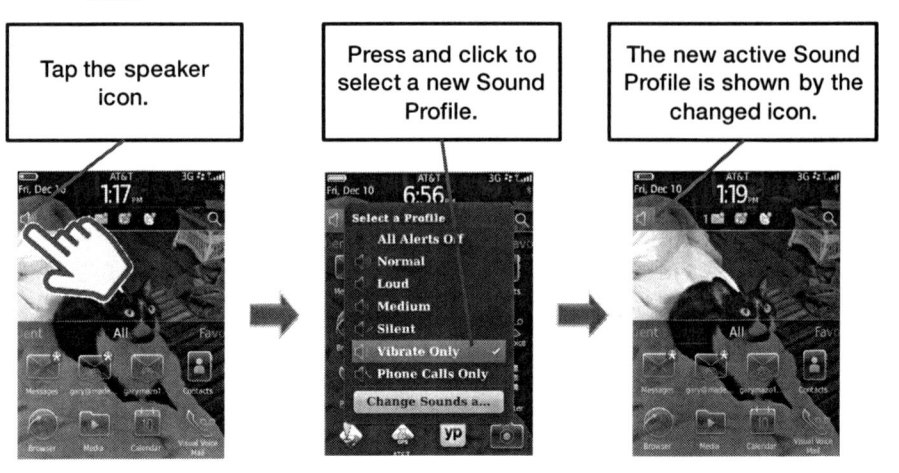

Figure 7-3. Selecting a different preloaded sound profile

If you don't see the speaker icon on your home screen, then follow the steps below to adjust your **Sound Profile**:

1. Press the **Menu** key and open the tray of the Home screen to see the entire list of icons.

2. Scroll down then scroll to the **Sounds** (Speaker) icon and Tap on it.

3. As shown in Figure 7-3, Tap on the new sound profile you wish to activate.

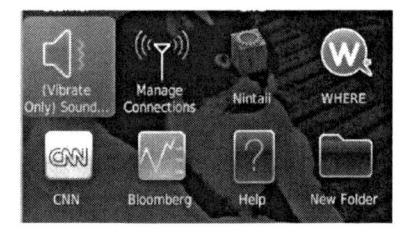

NOTE: You can also scroll to the bottom of the profile menu to where it says Change Sounds and Alerts. Then scroll to Profile Management and touch or click to get to the next screen where you can make individual adjustments for each profile.

Customizing a Sound Profile

There may be some situations where you want a combination of options that one preloaded profile alone cannot satisfy. The Torch is highly customizable, so you can adjust your profile options for virtually any potential situation. The

CHAPTER 7: Personalize Your Torch

easiest way to accomplish this is to choose a profile that is closest to what you need and edit it as shown below.

1. Touch or click on **Sounds** icon (speaker).

2. Select **Change Sounds and Alerts** near the bottom of the list by touching or clicking it.

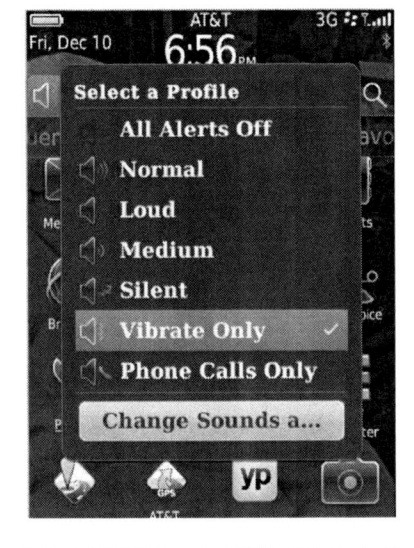

3. Now, scroll down to **Profile Management** and touch or click. Then touch or click on a profile that is close to what you want as a custom profile. In this case, we want to tweak the **Normal** profile.

    If you wanted to adjust a different profile, simply Tap on it instead.

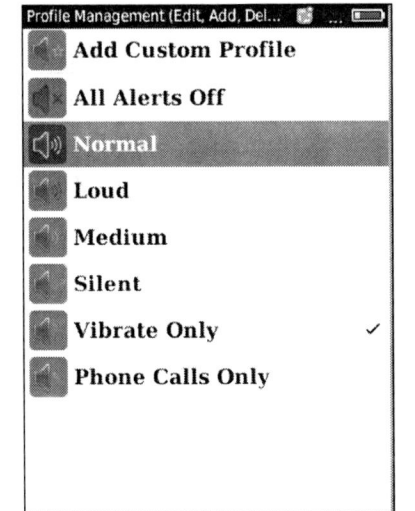

View Video Tutorials and Free Tips at www.MadeSimpleLearning.com   211

4. Tap on the category of alert you wish to edit. In this case, we want to edit the Calendar, so we touch or click on **Events-Reminders** to see the Calendar listed below.

5. Touch or click on **Calendar**.

6. The default for the calendar does not vibrate or ring, and we want to change it. You can adjust virtually every aspect of how the Calendar alerts you when a calendar alarm rings. We changed the **Volume** from **Silent** to **5** and set the **Vibration** to **On**.

7. Press the **Menu** key and select **Save**.

TIP: Make sure to leave the Vibrate with Ring Tone to Off. This will give you a few seconds of vibration before the alarm starts ringing, giving you a chance to keep the phone silent if necessary.

8. If you want to adjust your email notifications, touch or click on **Messages** to expand that section, then click on your **Email** account to see the image shown.

9. For example, choose **Email** and notice that you can make adjustments for vibration both **Out of holster** and **In holster**. (A holster may be supplied with your device or sold separately. This is essentially a carrying case that clips to your belt and uses a magnet to notify your Torch it is **In holster** and that you should turn off the screen immediately, among other things.)

TIP: If you make e-mail sound profile changes **before** you set up individual e-mail accounts, you won't need to adjust each one individually.

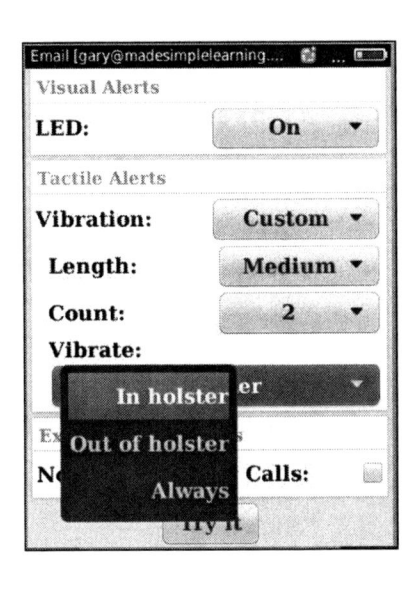

Changing Your Phone Ring Tone

Please see Chapter 8: "Phone and Voice Dialing," to learn how to change your ring tone. See Chapter 16: "Your Music Player," to learn how to use a song on your BlackBerry as a ring tone.

Downloading a New Ring Tone

You may find that your stock ringtones just are not loud enough for you to hear, even when you turn the volume up to loud. Sometimes, you just want a fun Ringtone. We have found that you can download a new ringtone from **mobile.blackberry.com** (among other web sites) to help with this problem.

1. Open your **Web Browser**.
2. If can't see a place to type a web address, then **Menu** key and select **Go To...**
3. Type in **mobile.blackberry.com** and press the **Go** key (where **Enter** was located) on the on screen keyboard or use the **Enter** key on the hardware keyboard.
4. Then scroll down and touch or click on the link called **Personalize**. Then, just Tap on **Ringtones.**
5. When you touch or click on a ringtone, you can **Open** (listen/play it) or **Save** it on your BlackBerry.
6. Go ahead and **Open** a few to test them out. To get back to the list and try more ringtones, press the **Escape** key.
7. If you like the ringtone, then press **Menu** key after you listened to it and select **Save** from the menu.
8. This time, scroll down and check the box at the bottom that says **Set as Ringtone**, then select **Save.**
9. You are done. Next time you receive a phone call, the new ringtone should play.

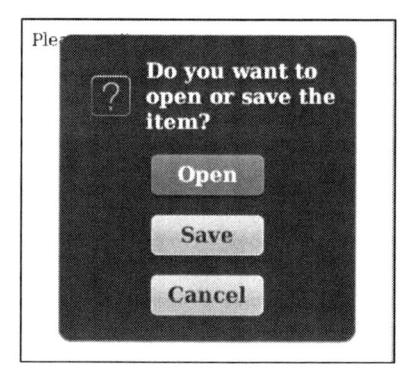

CHAPTER 7: Personalize Your Torch

> **TIP:** New Ringtones are available on many BlackBerry user Websites like:
>
> www.crackberry.com
>
> www.blackberryforums.com
>
> www.blackberrycool.com
>
> View some examples of ringtones for sale (see Figure 7-3).

Figure 7-3. Other sources for ringtones for your Torch.

Setting Different Ring Tones for Contacts

Would you like to know who is calling, emailing, or sending a SMS text message without having to look at the screen?

You can do this by setting what is called a **Contact Alert** on your Torch. You can set these up in two places on your Torch: in the **Sound Profile** app (Speaker icon) and in the **Contacts** app.

Using the Sound Profile App

Since the Sound Profile app is to easy to get to, you might want to use this method:

1. Touch or click on your **Sounds** icon.

2. Scroll down and click on **Change**

View Video Tutorials and Free Tips at www.MadeSimpleLearning.com   215

Sounds and Alerts.

3. Touch and click **Sounds for Contacts** and select **Add Contact Alert** to choose a contact from your Address Book for whom this new profile will apply.

4. Type a name for the Contact alert. In this case, we typed **Gary**.

5. Next, touch or click the box under where is says **Contacts**.

6. Type a few letters of the person's first and last name to find them in your Contact list.

7. Then click on their name to select it.

8. Now, you can adjust the custom alerts for this person for both Messages (e-Mail, SMS, etc.) and the Phone by clicking on **Messages** or **Phone**.

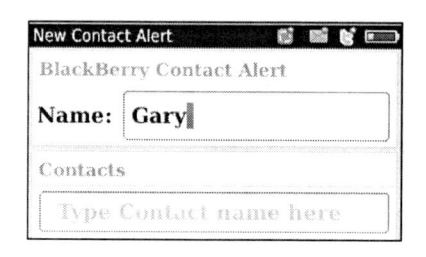

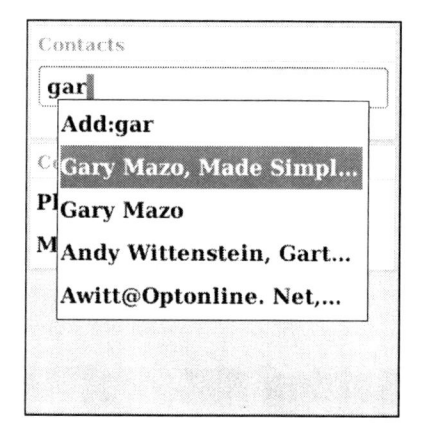

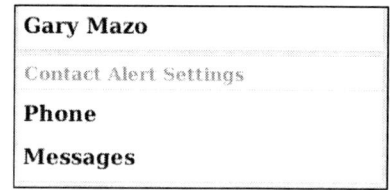

CHAPTER 7: Personalize Your Torch

9. Now, make the adjustments for this person. In this case, we want a different ring tone, so we changed it to **Adventurous** and changed the **Volume** to **10**.

10. We also adjusted the **Vibration** to **On** and the Vibrate with **Ring Tone** to **Off**.

11. Press the **Menu** key and select **Save**.

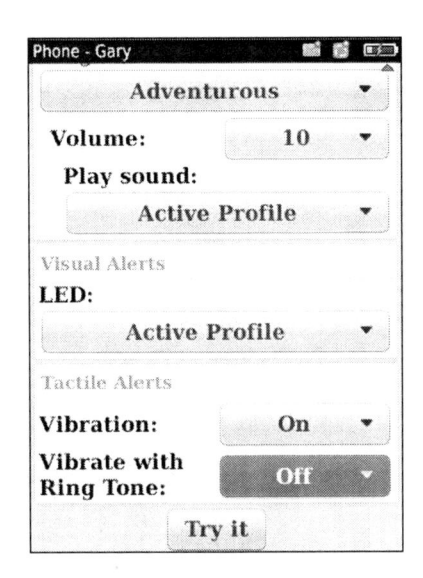

Using the Contact List

You can also set up custom ringtones and alerts for contacts directly in your Contacts icon.

1. Start your **Contacts** app.

2. Type a few letters to find a contact to customize.

3. Click on the contact name you wish to edit to see their details.

4. Touch or click the **Edit** soft key at the bottom of the screen.

View Video Tutorials and Free Tips at www.MadeSimpleLearning.com   217

5. Scroll down until you see the Custom Ring Tones/Alerts section.

6. Click on **Phone** or **Messages** to adjust either one.

7. Make your changes to Ring Tones, volume, vibration, and more on the next screen.

8. Press the **Menu** key and select **Save** to return to the contact edit screen.

9. Press the **Menu** key and select **Save** again to finish editing the contact.

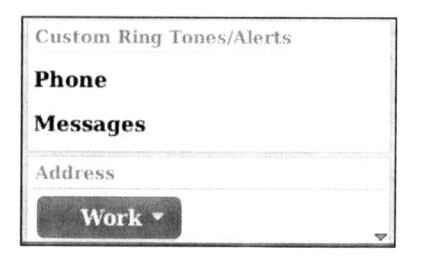

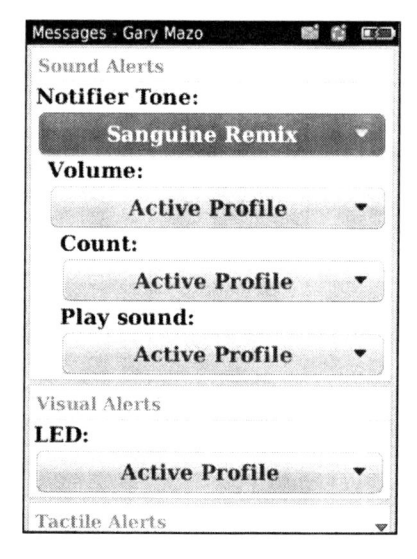

Accessibility Options

Your BlackBerry offers some "tweaks" to the accessibility options – which can change the look and feel of your **Home** screen, color contrast and event sounds.

Touch or click on your **Options** icon and then scroll down to **Accessibility** and click.

CHAPTER 7: Personalize Your Torch

To change the **Home Screen Grid Layout**, just click the dropdown and choose from the default up to 4 columns of icons in the **Home Screen** grid.

The change the contrast (to make the screen more visible) click the dropdown and choose from **Normal, Reverse Contrast** or **Grayscale**. Experiment with this to see that gives you the best viewing experience.

Finally, you can choose to have **Even Sounds** be **On** or **Off**.

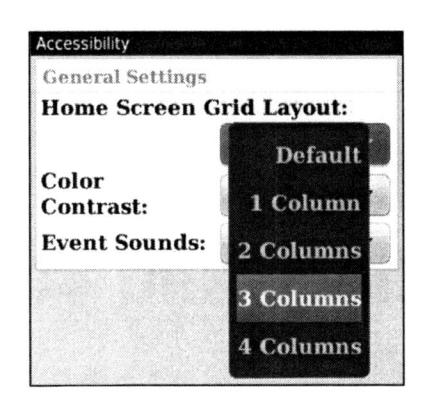

Chapter 8: Phone and Voice Dialing

In this chapter, we dig into the many ways your Torch is a powerful and full-featured phone. You will learn how to quickly make phone calls, use your call logs, and call people from your Contact list. You'll also learn quick ways to dial voice mail, use speed dial, and add new contacts from the phone calls you place or receive. Finally, you will learn how to use voice dialing on your Torch.

Three Main Phone Screens

When you open the phone the first time, you may notice the three soft keys at the top of the phone screen (see Figure 8-1). These soft keys open your **Dial Pad** (left key), **Call Logs** (middle key), and your **Contact** list (right key).

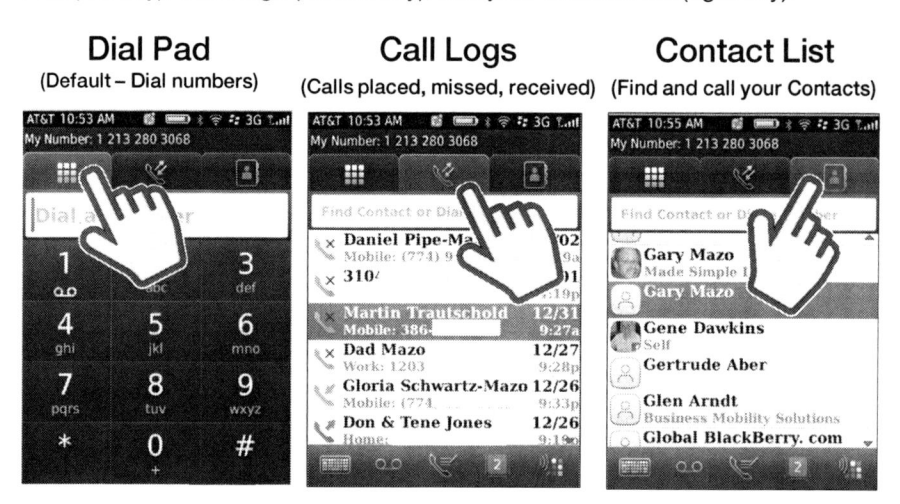

Dial Pad	Call Logs	Contact List
(Default – Dial numbers)	(Calls placed, missed, received)	(Find and call your Contacts)

Figure 8-1. The three phone views on your Torch.

CHAPTER 8: Phone & Voice Dialing

Working with Your Phone

It helps to get a feel for how the phone works by looking at all the keys, buttons, and on-screen icons shown in Figure 8-2. Your main phone keys are the **Green** and **Red Phone** keys on the bottom of the device.

Call Logs
Press to see the phone call logs.

Contact List
Press to bring up the Contact List / Address Book.

Mute key

Your Phone #

Dial Pad
Press for the regular dial pad (shown)

Voice Mail
Press & Hold '1'

Green Phone Key
Start Call

Menu Key
Press & hold to Multi-Task

Volume Up & Down

To Dial Letters:
Slide your keyboard open, press and hold the ALT key and type each letter.

Red Phone Key
End Call

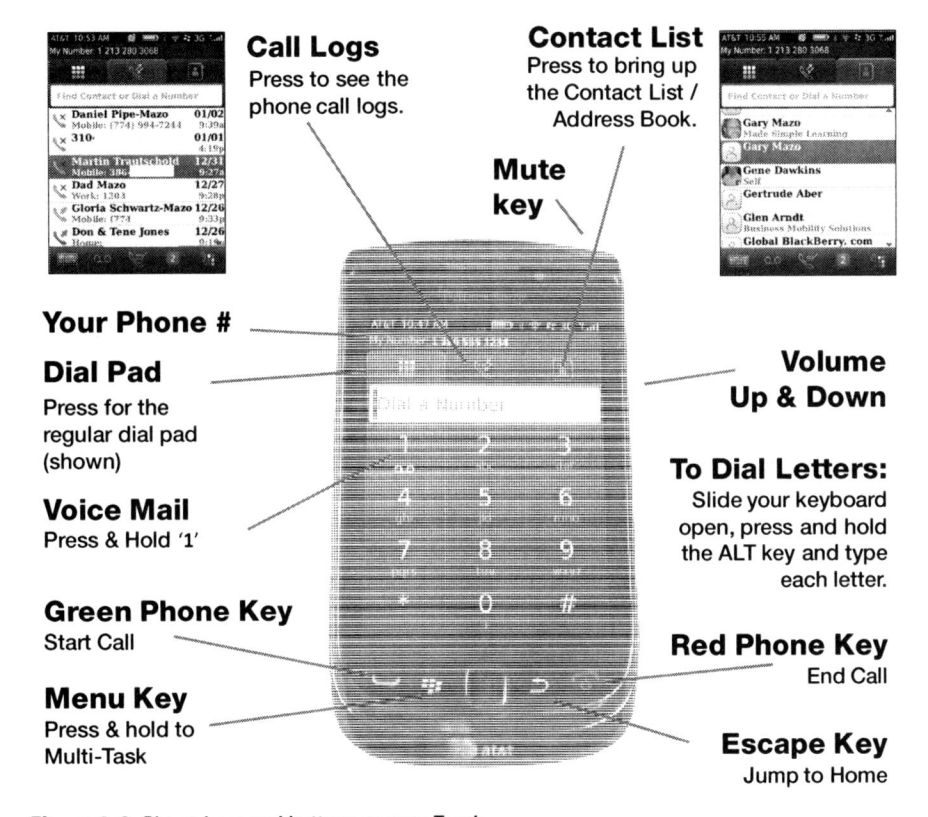

Escape Key
Jump to Home

Figure 8-2. Phone keys and buttons on your Torch.

Placing a Call

Making phone calls is easy. There are many ways to place a call: you can dial a number, use your call logs, or dial by name from your contact list.

View Video Tutorials and Free Tips at www.MadeSimpleLearning.com   221

Dialing a Phone Number

1. Press the **Green Phone** key at any time to get into the **Phone** application.
2. Your default view should be the **Dial Pad**, as shown, but it may be the **Call Log**. If so, just press the **Dial Pad** soft key.
3. Press the number keys to dial the number.
4. Press the **Green Phone** key to start the call.

TIP: To re-dial the last number you called, simply press the **Green Phone** key when on the Dial Pad. If you are not in the Phone, just press the **Green Phone** key three times to dial the last number.

TIP: Press and hold the * key to enter a two-second pause when the phone is dialed. Press and hold the # key to enter a wait (when the phone waits for you to click a button before it continues). These are both useful tricks if you want to dial an access number, then a password.

Dial from Call Log

You can quickly dial from your Call Logs (calls you have placed, received, or missed).

1. Start the **Phone** by pressing the **Green Phone** key.

2. Click the **Call Log** icon at the top.

3. Once in the call log screen, just scroll to the call log you wish to use and touch or click on it to call that number.

TIP: If you want to call a person, but not the number shown on the Call Log (e.g. "Work" is shown, but you want to call "Mobile"), highlight the call log and press the **Menu** key. Then, scroll to Contact (name of contact) and click. Additional phone numbers will then be displayed.

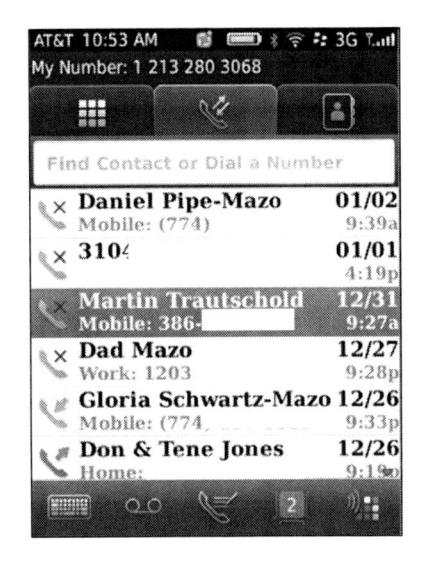

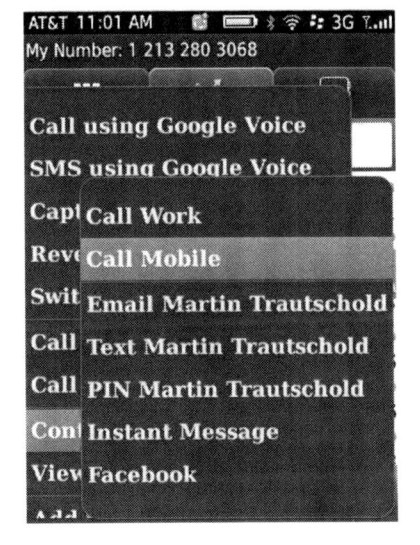

Dialing a Contact by Name

You can quickly dial anyone in your Contact list by following these steps.

1. Start the **Phone** by pressing the **Green Phone** key.
2. Then, click the **Contacts** icon at the top.

> **TIP:** You can always go to your **Contacts** app and dial a contact right from there – see Chapter 10.

3. Once in the **Contacts** screen, start typing a few letters of a contact's first, last, or company name to find them.
4. Tap on the contact's name.
5. If the contact has more than one phone number, you will be prompted to select from the available numbers.

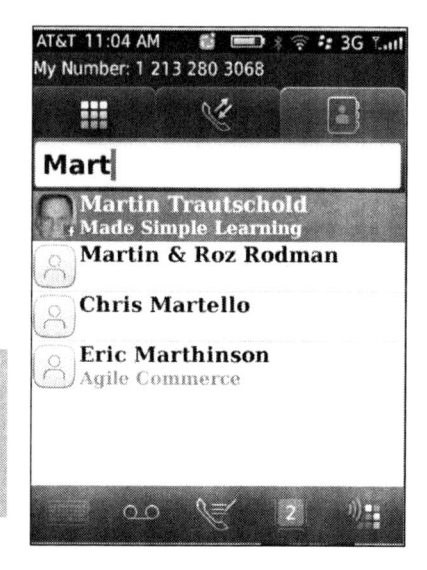

Answering a Call

Answering a call couldn't be easier. When you call comes in, the number will be displayed on the screen. If you have that particular number already in your **Contact** list, the name and/or picture will also be on the screen (if you have entered that information into that particular contact.)

CHAPTER 8: Phone & Voice Dialing

When a call comes in:

Press the **Green Phone** key to answer the call.

If you are using a Bluetooth Headset, you can usually press a button on the headset to answer the call (see Chapter 1)

TIP: Just press the **Volume Down** key to stop the ringing. Press the **Red Phone** key to instantly send the caller to voice mail.

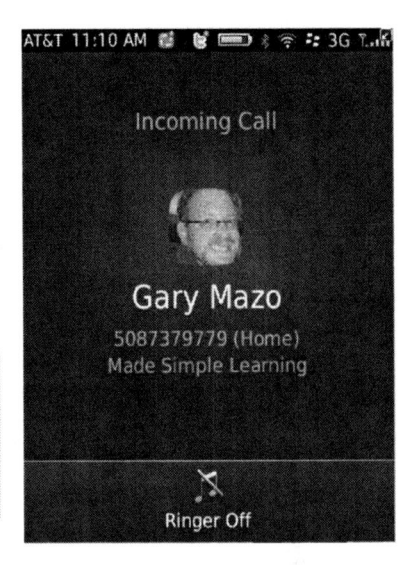

Ignoring Phone Calls

Sometimes you can't take a call. In such cases, you need to make a decision to ignore the call or mute the ringing. Both of these options can be achieved quite easily with your Torch.

Ignoring a Call and Immediately Stop the Ringing:

When the phone call comes in, instead of answering by pushing the **Green Phone** key, simply press the **Red Phone** key to ignore.

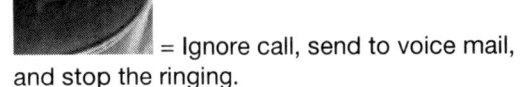

= Ignore call, send to voice mail, and stop the ringing.

TIP: Need to silence the ringer but still want to answer the call?

View Video Tutorials and Free Tips at www.MadeSimpleLearning.com   225

Just Tap the **Ringer Off** button on the bottom of your screen. If the ringing or vibrating had started while your BlackBerry was still in the holster (carrying case), simply pulling the BlackBerry out of the holster should stop the vibrating and ringing, but still give you time to answer.

Ignoring a call will immediately send the caller to your voice mail.

The **Missed Call** icon will be displayed on your Home screen; it's an icon that shows a phone with a red X next to it. The number shows the total number of missed calls.

You can immediately call back the missed caller by doing the following:

1. Press the **Green Phone** key.

2. If you are not already viewing the call logs, press the **Call Log** soft key (middle top).

3. Locate and press on the missed call—usually it will be the top entry, as shown in the image.

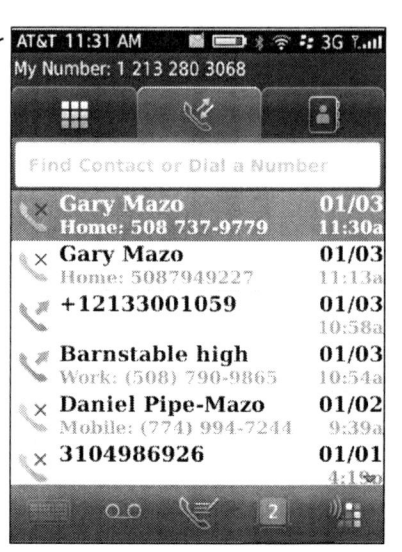

Using the Mute Button to Turn Off the Ringing Phone

Besides pressing the **Ringer Off** button, you can also mute the call via the mute key on the top of your BlackBerry.

If you would prefer not send the call immediately to voice mail and simply let it ring a few times on the caller's end, but you don't want to hear the ring (perhaps you are in a movie theatre or a meeting), press the **Mute** key on the very top right edge of your Torch.

This will silence the ring. You may still pick up the call or let the caller go to voice mail.

CHAPTER 8: Phone & Voice Dialing

Dialing Numbers, Taking Notes, and Jumping to Other Apps

When you are on a call, you will notice some soft keys at the bottom of the screen that allow you to do a number of useful things. (You may want to press the **Speaker** soft key [Speaker] before pressing any of these keys so you can hold the Torch away from your face.)

Press the **Dial Pad** soft key to bring up the dial pad in case you need to dial numbers; for example, when you have received a company directory that asks you to dial someone's extension, first, or last name. If you have to dial the name, use the number keys, just like on a regular phone.

NOTE: You will only see the **Dial Pad** if the Torch is closed. If the keyboard is open, the **Dial Pad** will not be visible and you are to use the number keys on the keyboard.

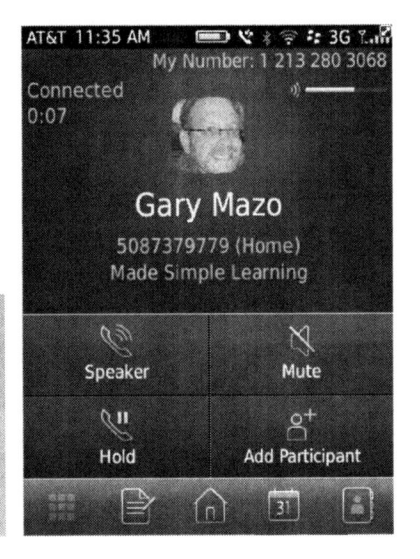

Press the **Notes** soft key to take notes while on the call. Learn more in the "Taking Notes While On a Call" section just below.

Press the **Home** soft key to jump to your Home screen and start any icon you want.

Press the **Calendar** soft key to jump to the Calendar app to check your schedule or book a new event while you are still on your phone.

Press the **Contacts** soft key to jump to your Contacts icon to lookup someone's number, address, or other information.

View Video Tutorials and Free Tips at www.MadeSimpleLearning.com   227

Taking Notes While On a Call

Have you ever hung up the phone and asked yourself, "What was it that they promised to do?" If so, the **Notes** feature on your Torch is a perfect solution to keep track of exactly what was said or promised during a phone call.

You can take notes, save them, and even email them to yourself or others.

The notes you take while on a call are attached to the call log, also known as **Call History**.

1. In order to take the phone away from your ear so you can type notes, either press the **Speaker** key or use your headset.

2. Tap the **Notes** soft key at the bottom.

3. Now, you can type your notes using the keyboard.

4. If you need to get back to the phone screen, then press the **Menu** key and select **Hide Notes**.

5. When done, you can simply hang up with the **Red Phone** key.

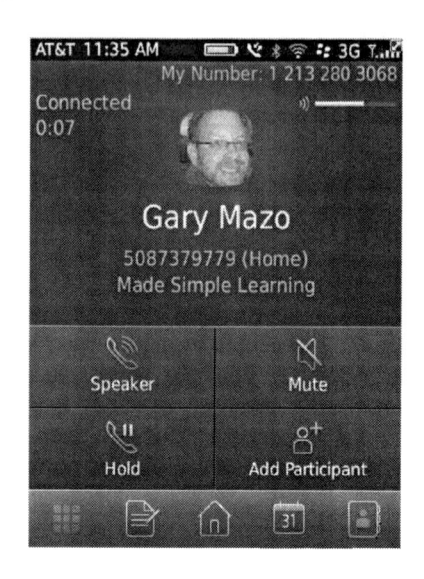

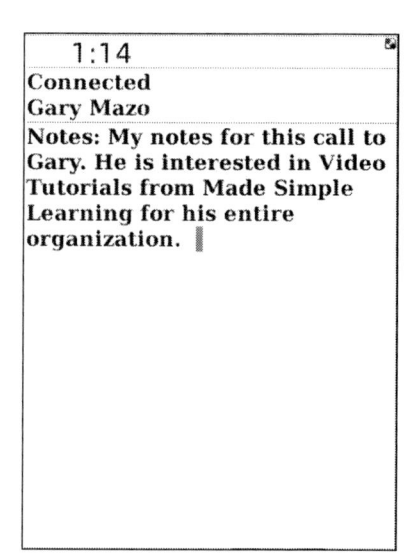

View, Edit, Delete, or Send Call History and Notes

1. From the Call Log screen, highlight the log entry, then press the **View Call Notes** soft key.. You can also press the **Menu** key and scroll to **View** and select **Call Notes**.

   NOTE: You have to use the **View** soft key or the **Menu** because if you touch or click the Call Log entry with your finger, the Torch will start a call to that number.

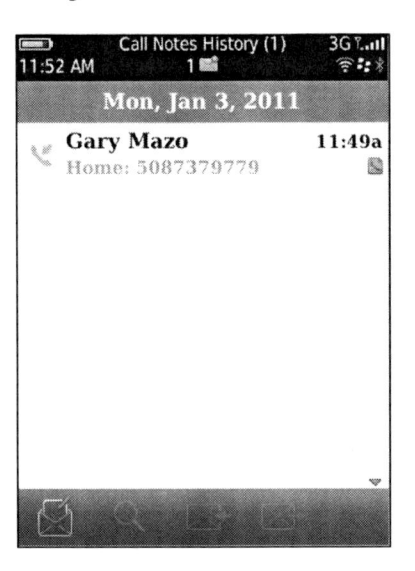

2. Click on the Call History and touch the View Notes soft key and view the entry with the small note icon to view it.

3. From this screen, you can also do the following:

Press the **Compose** soft key to write a note for any call history entry.

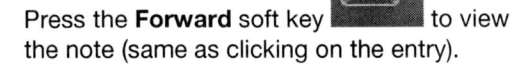

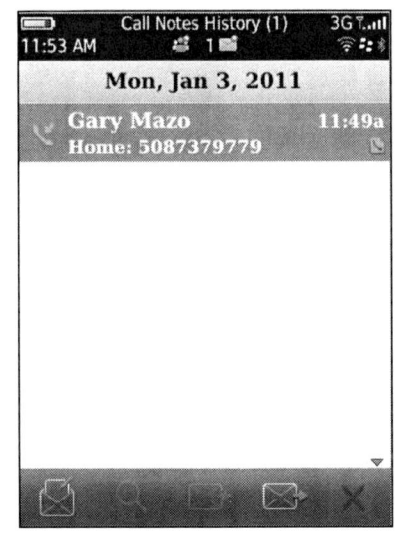

Press the **Forward** soft key to view the note (same as clicking on the entry).

Press the **Delete** soft key to remove the Call History item and related note.

Press the **Menu** key and select **Forward, Open, Call, Compose Email or View Contact information** and then perform one of the desired actions.

Adjusting the Volume on Calls

There may be times when you have trouble hearing a caller. The connection may be poor or you may be using a headset. Adjusting the volume is easy. While on the phone call, simply use the two volume keys on the right hand side of the Torch to adjust the volume up or down. You can also use your finger to adjust the volume slider in the upper right hand corner if that is easier.

What's My Phone Number?

You have your new phone, and you want to give your number to all your friends—you just need to know where you can get your hands on that important information.

1. Press the **Green Phone** key to start the phone.
2. Read your phone number next to **My Number** at the top of the screen.
3. In the image to the right, the number is **1-213-280-3068**. (You may have to wait a second for it to appear.)

230

CHAPTER 8: Phone & Voice Dialing

TIP: Quickly Dialing Voice Mail

When in the Phone Dial Pad screen, press and hold the 1 key.

When in the **Call Log** or **Contacts** screen, press the **Voice Mail** soft key at the bottom to dial your voice mail.

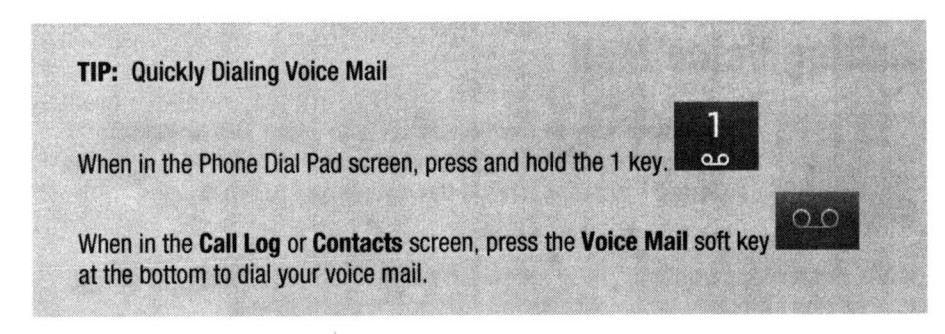

Changing Your Ring Tone

You can change your main phone ring tone inside the Phone app by following these steps:

1. From any Phone screen, press the **Menu** key (see Figure 8-3).
2. Scroll down and select **Phone Ring Tones**.
3. Then click on the button under **Ring Tone** to select a different tune.
4. You can adjust other properties, such as volume ,vibration, length and count.
5. Press the **Menu** key and select **Save**.

Figure 8-3. Changing your ring tone from the Phone app

View Video Tutorials and Free Tips at www.MadeSimpleLearning.com   231

Calling Voice Mail

The easiest way to call voice mail is to press and hold the number 1 key from the dial pad or pressing the 1/W key on the slide out keyboard. This is the default key for voice mail.

If it is not working correctly, read below for some troubleshooting help or call your phone company technical support for help in correcting it.

To setup voice mail, just call it and follow the prompts to enter your name, greeting, password, and other information.

When Voice Mail Doesn't Work

Sometimes, pressing and holding the **1** key or pressing the voice mail soft key will not dial voice mail. This happens if the voice mail access number is incorrect in your BlackBerry.

You will need to call your phone company (wireless carrier) and ask them for the local voice mail access number. This sometimes happens if you move to a different area or change cell phones, then restore all your data onto your BlackBerry.

Once you have the new phone number from the carrier, you need to enter it into your BlackBerry.

1. Start your Phone by pressing the **Green Phone** key.
2. Scroll down to **Options** item and Tap to select it.
3. Tap on **Voice Mail.**
4. Enter the phone number you received into the Voice Mail **Access Number**.

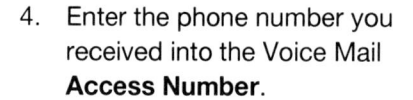

TIP: You can even enter your voice mail password if you like.

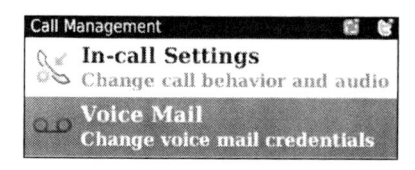

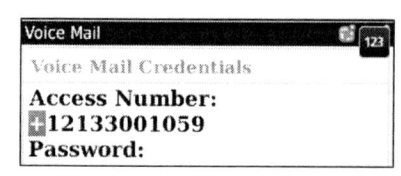

Using Your Call Logs

The Call Log is an especially useful tool if you make and receive many calls. Besides being able to quickly dial them, you can use the call log to add contacts to your Contact list, add notes, forward, and view details.

For instance, you can't remember if you added an individual to your Contact list, but you definitely remember that they called yesterday. Here is a perfect situation: use your Call Log to access the call, add the number into your address book, and place a return call.

Checking Your Call Logs

The easiest way to view your call logs is to do the following:

1. Press the **Green Phone** key.
2. Initially, you should see the Dial Pad.
3. Press the **Call Log** soft key in the middle of the top row.

4. The default setting is to show the most recent calls and then move sequentially backwards, showing calls made and received listed by date and time.

Use the **Trackpad** to highlight a call log entry. Then, either press the **Menu** key to see your options or just touch and hold to bring up a short menu on the screen.

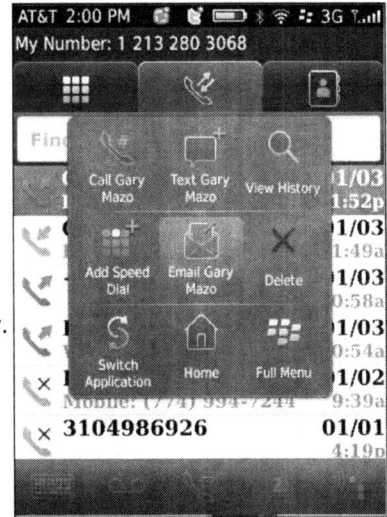

Both Names and Numbers in Call Logs

You will see both phone numbers and names in your phone call logs.

When you see a name instead of a phone number, you know that the person is already entered in your Torch contact list.

It is easy to add entries to your contacts right from the phone call log screen. We show you how below.

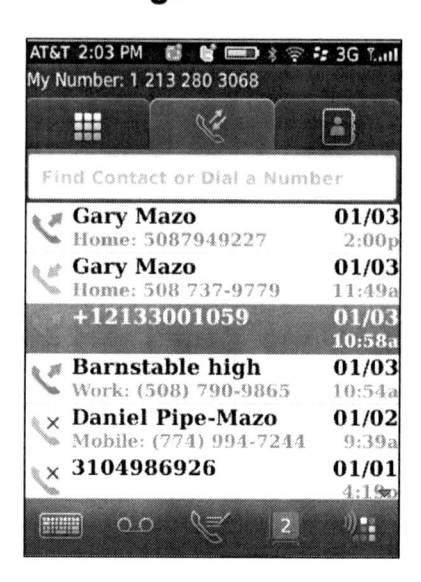

Add New Contact Entries from Call Logs

If you see just a phone number in your call log screen, there is a good chance you will want to add that phone number as a new Contact entry.

> **NOTE:** Call log entries are generated whenever you place, receive, miss, or ignore a call from your Torch.

Get into the call log screen by tapping the **Green Phone** key once.

CHAPTER 8: Phone & Voice Dialing

1. Highlight the call log phone number you want to add to your address book.

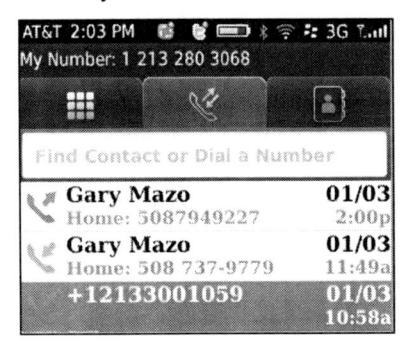

2. Press the **Menu** key and select **Add to Contacts**.

3. Notice that the phone number is automatically placed in the **Work** phone field but you can also move it.

4. To move the phone number to another phone field, follow these steps:

Touch or click the drop down menu next to the phone number.

Touch or click in the new phone number field, e.g. Mobile.

5. Fill in as much information as possible, press the **Menu** key, and select **Save**.

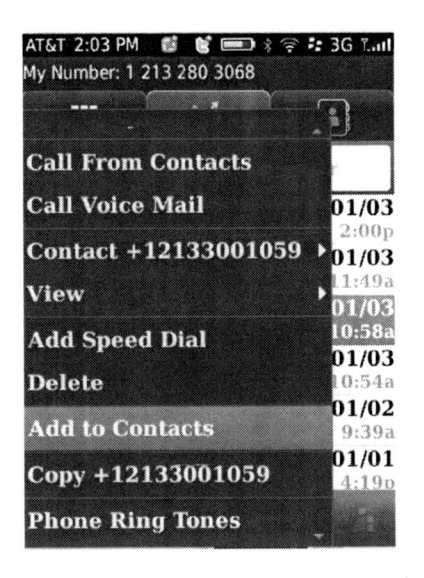

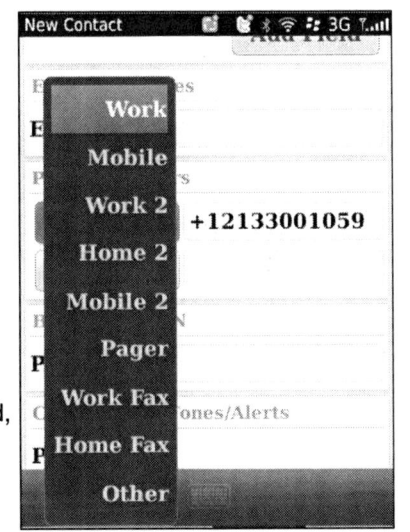

View Video Tutorials and Free Tips at www.MadeSimpleLearning.com   235

Copy and Paste Phone Numbers

For Underlined Phone Numbers:

With underlined phone number like 313-555-1212, or phone numbers in call logs, simply highlight the call log and press the **Menu** key and select **Copy** to copy it.

> **TIP:** This trick works on any underlined email address or PIN as well!

Move to where you want to paste, press the **Menu** key, and select **Paste**.

To Show Your Call Logs in the Messages App

It might be useful to show calls made, received, and missed in your message list (inbox) for easy accessibility. This allows you to manage both voice and message communication in a single unified inbox.

1. Press the **Green Phone** key to see your call logs.
2. Press the **Menu** key.
3. Scroll and click on **Options**.

4. Scroll to **Call Logs and Lists** and touch or click. Under **Show These Call Log Types in Message List** select either:

    - **Missed Calls** (see only missed calls)
    - **All Calls** (see all placed, missed, received)
    - **None** (this is the default; don't see any calls)

5. Press **Menu** and select **Save**.

Benefits of Adding People to Your Contact List/Address Book

- Call any number for this person (that is entered into your address book)
- Send them an email (if this person has an email address entered)
- Send them an SMS text message
- Send them an MMS Message (multimedia message with pictures or other media like songs)
- Send them a PIN message
- View the contact information
- Add a Speed Dial for the contact.

Speed Dial on Your Torch

Speed dialing is a great way to call your frequent contacts quickly. After you setup speed dial, press and hold a number key from the dial pad to have their phone number automatically dialed. There are a couple of ways to set up speed dialing on your Torch.

Use On-Screen Numbers and Keyboard Letters

You can assign the numbers on your on-screen dial pad or the letters on your slide out keyboard as speed dials. If your preference is to have the slide-out keyboard closed, then you'll prefer the on-screen speed dial numbers. If you like having your slide out keyboard open, you have the added choice of using almost every letter on your keyboard. (A few letters are reserved for use like A = Lock the screen, Q = Toggle between Normal and Vibrate Sound profiles, W/1 = Voice Mail)

Set Up Speed Dial from Call Logs

1. Press the **Green Phone** key and then select call logs using the soft key at the top.

2. Use the **Trackpad** to highlight the call log entry (either phone number or name) that you want to add to speed dial and press the **Menu** key.

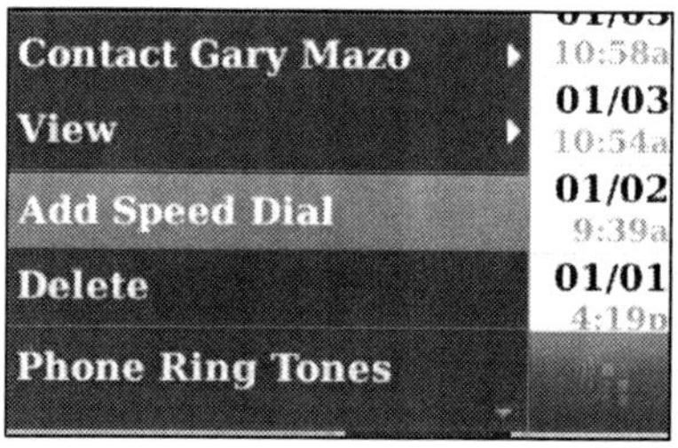

3. Tap **Add Speed Dial**.

4. You will be asked to confirm that you want to add this speed dial number with a pop-up window.

5. In the **Speed Dial** list, the number defaults to the first unused spot or sometimes to the first letter of the name of the contact if that is free.

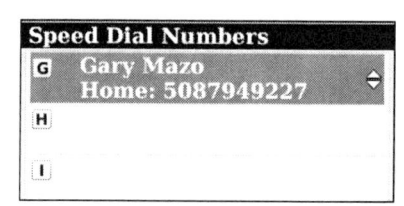

6. Tap on any other number and vacant spot to select that spot instead.

7. Press the **Escape** key to exit the **Speed Dial** list.

Set Up Speed Dial from Dial Pad

If you press and hold **any number key** from the dial pad that has not already been assigned to a speed dial number, you will be asked if you want to assign this key to a speed dial number.

Select **Yes** to assign it.

Then you will be shown your **Contacts** list to select an entry or select **[Use Once]** to type in a new phone number that is not in your Contacts list.

Type the first letters of a name on your keyboard and matching contacts will be displayed; just choose the correct contact for this **Speed Dial** entry.

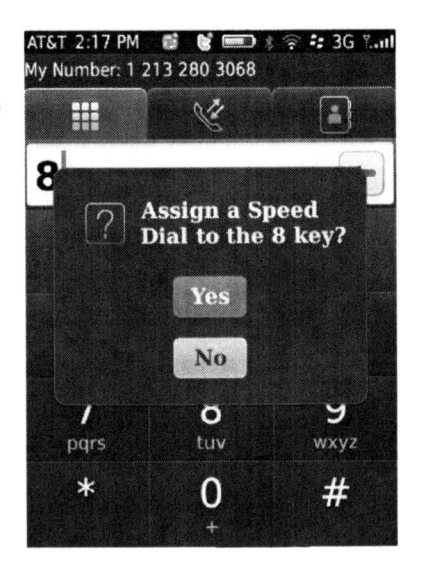

Set Up Speed Dial from Slide-out Keyboard

Very similar to setting up speed dial numbers on the Dial Pad, you can also set up speed dial letters from your slide-out keyboard. Simply press and hold any unassigned letter to assign that letter as a new speed dial entry. Follow the same steps as shown above to complete the process (see Figure 8-4).

1/W Key
Press and hold to speed dial voice mail.

Assign Speed Dial
Press and hold unassigned keys on keyboard to set up or use as Speed Dial letters.

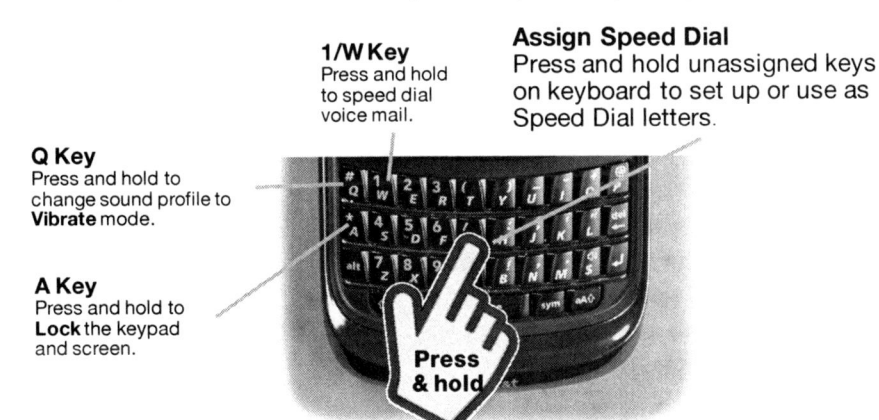

Q Key
Press and hold to change sound profile to **Vibrate** mode.

A Key
Press and hold to **Lock** the keypad and screen.

Figure 8-4. Setting New Speed Dial Entries from the Slide-Out Keyboard

Give your newly created speed dial numbers and letters a try by pressing and holding the same key you just assigned on your dial pad or physical keyboard.

Set Up Speed Dial from Speed Dial Icon

1. Tap the **Green Phone** key and go to any of the three screen views.

2. At the bottom of the screen, touch the **Speed Dial** soft key (second from the right.)

3. Choose any available slot and your BlackBerry will then display your contact list for you to select.

4. Select a contact for the vacant slot.

5. If there is more than one number to choose, pick which number you wish to be in the vacant speed dial slot.

TIP: You can also press the **Menu** key and select **Add Speed Dial** and follow steps to select the speed dial letter as shown above.

Edit, Move or Delete a Speed Dial Number

You can adjust or delete speed dial numbers.

1. Press the **Green Phone** key to start the phone.

2. Press the **Speed Dial** soft key.

3. Highlight the number you want to adjust.

4. Press the **Menu** key and select **Edit** to change the number, select **Move** and tap the space to move to or select **Delete** and confirm your choice to remove this entry.

5. Press the **Escape** key to exit the Speed Dial list.

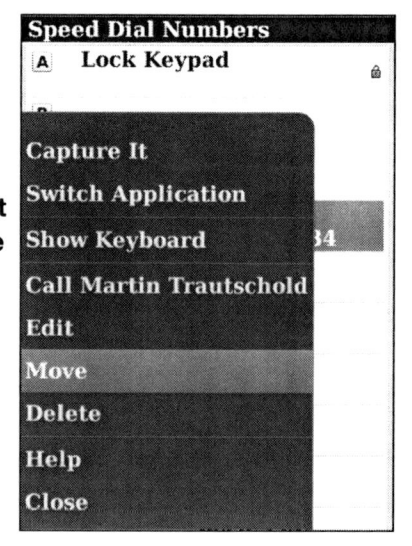

CHAPTER 8: Phone & Voice Dialing

Using Your Speed Dial Entries

To use any speed dial entry, you can press and hold the number on the Dial Pad or press and hold the key on the slide-out keyboard.

Voice Dialing Basics

Now that you know how to use Speed Dialing, we help you understand another way to save time – using the **Voice Dialing** program that allows you to voice dial and perform other simple commands. Voice dialing provides a safe way to place calls without having to look at the BlackBerry and navigate through menus. Voice dialing does not need to be trained like on other smartphones—just speak naturally.

Voice Dialing a Contact

The **Convenience** key is usually set for the **Camera** app but you can also set it for Voice dialing. Try pressing that key. If you have set the key for another program, press the **Green Phone** key and touch the **Voice Dialing** icon in the lower right hand corner.

> **TIP:** We show you how to set or change your convenience keys in the "Convenience Key" section of Chapter 7.

The first time you use this feature, the BlackBerry will take a few seconds to scan your Contact list.

When you hear **Say a Command**, just speak the name of the contact you wish to call using the syntax "Call Martin Trautschold."

If this contact has more than one phone number, you will then be prompted with **Which number?**

Again, speak clearly and say **Home**, **Work**, or **Mobile**

Say **Yes** to confirm the selection and the BlackBerry will begin to dial the number.

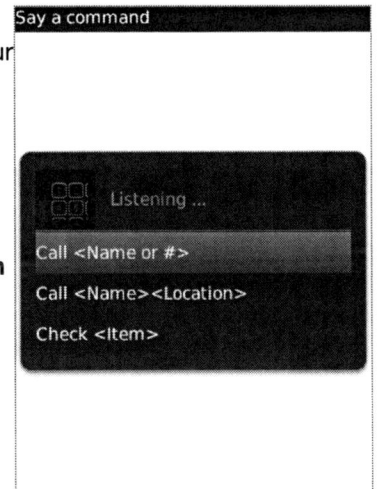

View Video Tutorials and Free Tips at www.MadeSimpleLearning.com   241

Voice Dialing a Number

Start **Voice Dialing** by pressing the icon from the on-screen Dial pad, the icon from the home screen or, if you've set it, the **Convenience** key.

When you hear "**Say a command**," say "**Call**" and the phone number. Example: "**Call 386-506-8224.**"

Depending on your settings, you may be asked to confirm the number you just spoke or it will just start dialing.

Other Voice Dialing Commands

You can use the **Voice Dialing** app to perform other functions on the Torch. These are especially useful if you are in a position where you can't look at the screen (while driving) or in an area where coverage seems to fade in and out.

The most common are:

- **Call Extension** will call a specific extension.
- **Call Martin Home** will call the contact at their home number.
- **Check Battery** will check the battery status.
- **Check Signal** will let you know the strength of your wireless signal and whether you have No Signal, Low Signal, High Signal, or Very High Signal.
- **Turn Off Voice Prompts** will turn off the "Say a command" voice and replace it with a simple beep.
- **Turn On Voice Prompts** turns the friendly voice back on.

Changing Your Voice Dialing Options

You can control various features of Voice Dialing by going into your **Phone Options** by pressing the **Green Phone** key and scrolling to **Options** and selecting **Voice Dialing**.

Change the **Choice Lists** if you do not want to be confronted with lots of choices after you say a command. Your options are **Automatic** (default), **Always On**, or **Always Off**.

Audio Prompts can be set to **No Prompts**, **Basic Prompts** (default), and **Detailed Prompts**. With Detailed Prompts, you will hear more detailed questions and confirmations back from the BlackBerry after you speak. Use Detailed Prompts if you find that you are misdialing a great deal.

Adapt Voice in Voice Dialing

In order to teach the Torch how you speak, you will need to use the **Voice Adaptation** function (which used to be called Adapt Digits on earlier BlackBerry software versions).

1. From the Voice Dialing options screen click the Start button under **Voice Adaptation.**

2. It helps to be in a quiet room and then you will have to repeat 15 sets of prompts (see Figure 8-5).

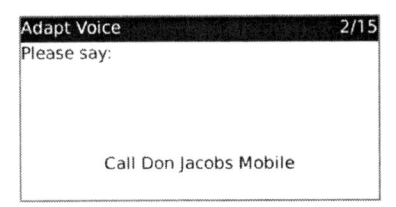

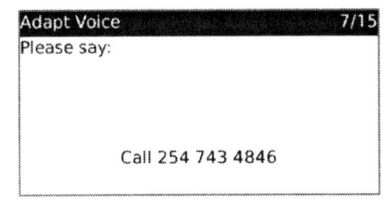

Adapt Voice 2/15	Adapt Voice 7/15
Please say:	Please say:
Call Don Jacobs Mobile	Call 254 743 4846

Figure 8-5. *Training your Voice Dialing*

3. After about 15 screens, you will see a message that the device is working to adapt your voice.

4. When it's done, you will be placed back into the Voice Dialing options. Press the **Red Phone** key to jump back to the Home screen.

Now, give your newly trained **Voice Dialing** a try.

Voice Dialing Tips and Tricks

There are a few ways to speed up the voice dialing process. You can also customize the way Voice Dialing works on the BlackBerry.

Make Voice Dialing Calls More Quickly

When using Voice Dialing, give more information when you place the call. For example, if you say "**Call Martin Trautschold, Home**," the Voice Dialing program will only ask you to confirm that you are calling him at home.

The call will then be placed.

Give your Contacts Nicknames

Make a **Nickname** entry for a contact, especially one with a long name.

In addition to my **Gary Mazo** contact, I might also edit his contact entry to add a Nickname field and then add **GM** as the Nickname.

I would then say: "Call GM."

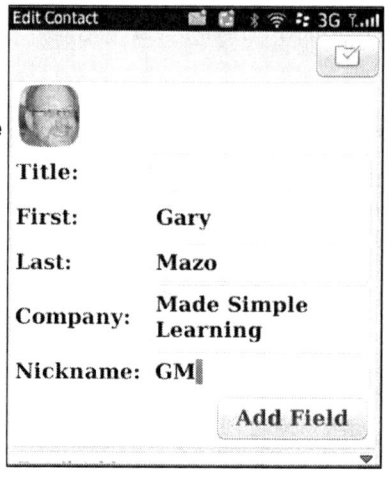

Changing Your Voice Dialing Language

If you need to speak in a different language on Voice Dialing, you can change the Voice Dialing language in the Options icon. Start the **Options** icon (usually a wrench), then click on **Typing and Input** and then click on **Language** to see the screen shown here.

NOTE: If you do not see all of these languages displayed, it is likely they were removed during the Setup Wizard process and will need to be re-installed. If you are a Windows PC user, you can use Desktop Software to add back languages in the Application Loader section. If you are a Mac user, use BlackBerry Desktop Software for the Mac.

Setting your Phone Ring Tone

The BlackBerry supports using many types of audio files as Ringtones. You can set one general ringtone for everyone or set up individual tones for your important callers.

You can change ringtones from various places:

- The **Phone** itself
- The **Sound** profile (Speaker icon) app,
- The **Music** app
- The **Ringtones** app

Important: Place Ringtones in Ringtone Folders

In some BlackBerry smartphones, when you are attempting to set a ringtone for a specific person in the Contact list or in Profiles, you can only browse to the Ringtone folder, not the Music folder. If this is the case, then you must copy your music ringtones to the Ringtone folder using the methods to transfer media found in this book.

Changing Your Ring Tone from the Phone

We will show you a few ways to set your ringtone starting with the **Phone** app itself.

1. Press the **Green Phone** key to start the phone.
2. Press the **Menu** key and select **Phone Ring Tones**.

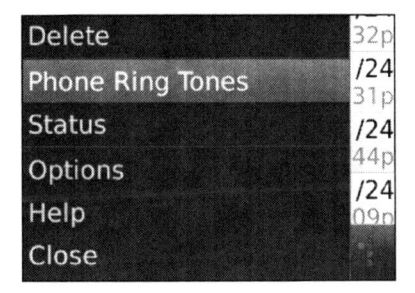

3. Click the drop-down under Ring Tone to select a new ring tone.

4. From this screen you can also adjust the ring tone **Volume** and when it is played (**Play sound**). There are other adjustments below with the **LED** and **Vibration**.

5. When you're done changing things, you can try out your new settings by clicking the **Try It** button at the bottom.

6. Press the **Menu** key and **Save** your settings.

7. Next you will see a question about which Sound Profile to change. You can select **All Profiles** to set this for every profile or **Normal Profile Only** to select this new ring tone only for the Normal sound profile.

Changing Your Ring Tone from Other Places

Music app: To change your ring tone from your **Music** app, highlight to the song you want to use from the list on the screen. Press the **Menu** key and select **Set as Ring Tone**.

Ringtone app: Start the Ring Tone app, select a ring tone you want to use and follow the same procedure as described for music.

Sounds app: Tap the Sounds app (speaker icon). Touch the button at the bottom called Changes Sounds and Profiles. Click the Phone Ring Tone at the top. Tap the Ring tone drop down list and select a new ring tone.

TIP: Unique Ringtones for Callers

Set up unique ringtones for each of your important callers. This way you will know when each of these people is calling without looking at your BlackBerry screen.

Set a Custom Ring Tone for a Caller

We covered how to do this in the "Setting Different Ring Tones for Contacts" in Chapter 7: "Personalize Your Torch."

Call Waiting – Handling a Second Caller

Like most phones these days, the Torch supports call waiting, call forwarding, and conference calling – all useful options in the business world and in your busy life.

If you are on the phone with someone, and a second person calls, you can do any of the following:

- Answer the second caller by pressing the **Green Phone** key. This will put your first caller on hold.
- Ignore the second caller and send them to voice mail by pressing the **Red Phone** key.

Join a Conference Call or Swap

After you have two callers on the phone, you can join them in a conference call or toggle between them (called **Swap**) see Figure 8-6.

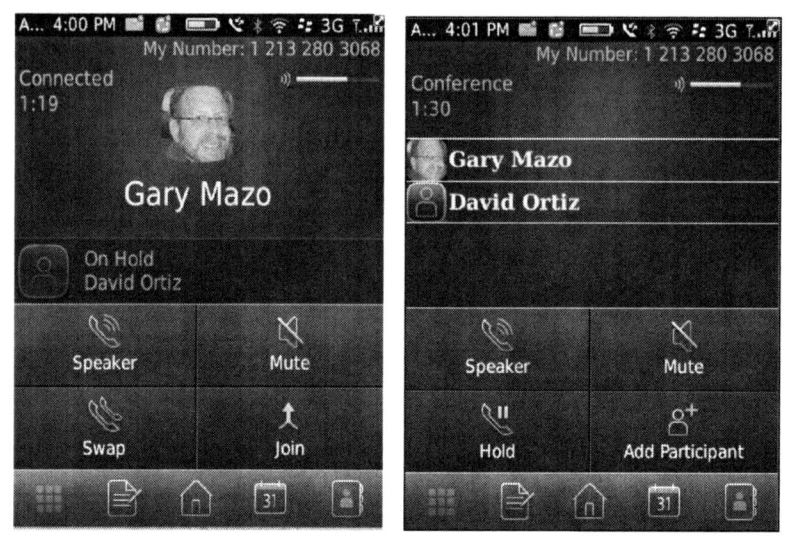

Figure 8-6. Working with two callers on your Torch

1. To swap between the callers (put one on hold and talk to the other), just press the **Swap** soft key (or the **Greeen Phone** key).
2. To join the two callers together in a conference, press the **Join** soft key.
3. To split the call, press the **Menu** key and select **Split Call** from the menu.
4. Then choose which call to separate from the conference call first.

Call Forwarding

Call forwarding is a useful feature when you are traveling or plan on leaving your BlackBerry at home. With call forwarding, you can send your BlackBerry calls to any other phone number you choose.

WARNING: Make sure you know how much your wireless carrier will charge you per call forwarding connection; some can be surprisingly expensive. Also, make sure that your SIM card has been set up by your service provider for this feature.

CHAPTER 8: Phone & Voice Dialing

NOTE: Not all BlackBerry phone companies (service providers) offer this feature.

To Forward Calls Received by Your BlackBerry:

1. Press the **Green Phone** key and then press the **Menu** key.

2. Scroll to **Options** and Tap and then scroll to **Call Forwarding** and Tap. You may see a message stating "Querying call forwarding parameters from the network."

3. Your screen may look a little different from the one below, but the functionality will be very similar. you should see four fields that you can adjust **Conditional, If Busy, If No Reply, If Not Reachable**, then you need to Tap on these items and select whether or not you want to forward each instance.

4. If your screen looks similar to the one below, then you forward your calls conditionally – and set parameters or choose **Always** or **Never** forward calls.

5. Click the drop-down next to Forward Calls to select **Never, Always,** or **Conditional.** The **Conditional** setting (Figure 8-7) allows you to specify different phone numbers or actions based on the situation (Busy, No Reply or Not Reachable).

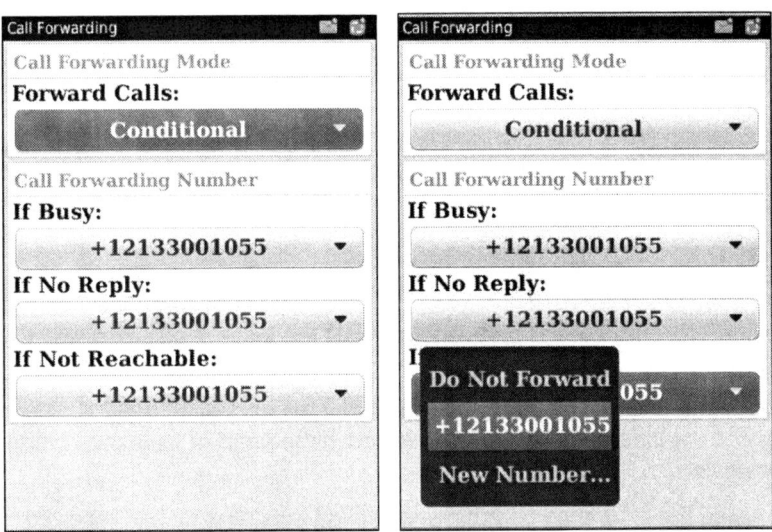

Figure 8-7. The Conditonal option of Call Forwarding

6. Selecting **Always** will allow you to forward all calls, regardless of the condition to a specific number or contact in your **Contact** list.

> **TIP:** To select a new number (one not in your Contact List), select the [Use Once] item at the top of your Contact List and type the new phone number.
>
> **NOTE:** The default set up is to send these calls to your voicemail, the phone number that is most likely already in these fields.

7. To delete a Call Forwarding number, repeat steps 1-4 and Tap **Delete** after you have Taped **Edit Numbers**.
8. Press the **Menu** key and select **Save**.

Conference Calling

Conference Calling is a very useful option for talking with more than one person at a time.

> **TIP:** In your Calendar, you can pre-load conference call dial-in information so you don't have to dial it every time. This works well if you use the same conference call service regularly. See Chapter 11 for details.

Sometimes conferencing several parties together is the faster (and safer) way to transfer necessary information.

For instance, one author was leasing a new car and the car dealer left a voicemail to call the insurance company to approve the proof of insurance being faxed to the dealer.

The author called the insurance company, expecting they already had received the dealer's fax number.

Unfortunately, the insurance company did not have the fax number.

CHAPTER 8: Phone & Voice Dialing

Instead of hanging up and calling the dealer, asking for the fax number, and calling the insurance company back, the author used the BlackBerry conference call feature.

Over the course of the conference call, the dealer's fax number was relayed, along with any special instructions.

To Set Up a Conference Call

Place a call as you normally would.

While on the call, press the **Green Phone** key (or touch the **Add Participant** button on the screen) and either choose a contact from your Contact List or type in a phone number.

Choose which number (if the contact had more than one) that you wish to conference.

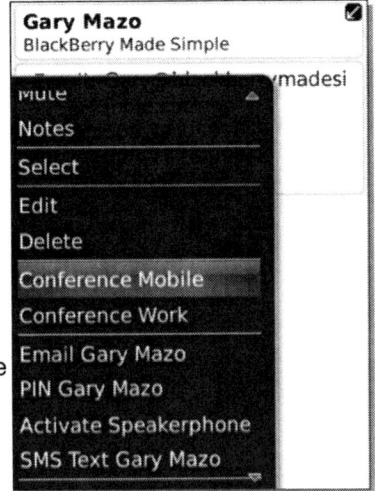

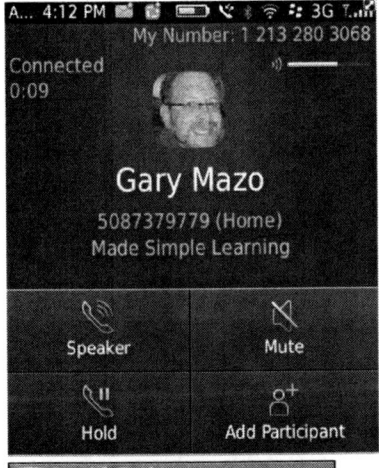

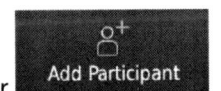

Figure 8-8. Conference call setup

If you add more than two callers to the conference call (Figure 8-8), just repeat the process starting with another **New Call** (press the **Green Phone** key).

Join the calls as you did above. Repeat as needed.

To speak with only one of the callers on a Conference Call:

1. Press the **Menu** key and select **Split Call**.

2. Select the caller to whom you want to speak privately.

3. The other caller will be placed on hold.

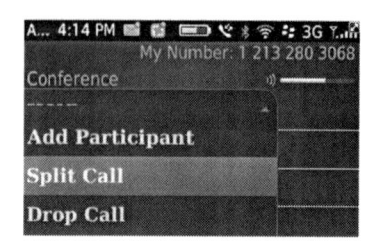

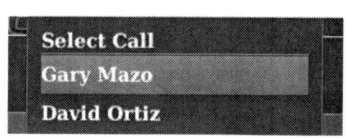

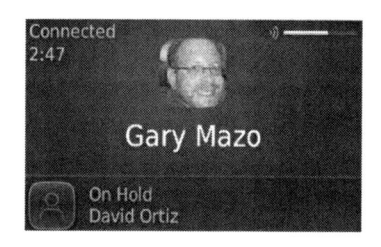

To End or Leave a Conference Call

To hang up on everyone and end the conference call for all, press the **Red Phone** key or press the **Menu** key and select **Drop Call** to just disconnect from one caller.

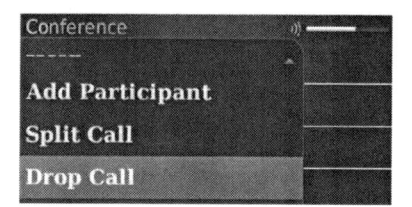

Advanced Dialing (Letters, Pauses, and Waits)

You may want to do some advanced dialing whether you are on the phone or entering a phone number in your Contact list. How to dial letters and how to add pauses and waits when a phone number is dialed are all covered in this section.

Dialing Letters on a Phone Call

Sometimes you call a company with a **Dial by Name** directory that will ask you to "Dial a few letters of the person's last name to look them up."

CHAPTER 8: Phone & Voice Dialing

Simply slide open the physical keyboard, press the ALT key and start dialing letters from the keyboard. It's that easy!

Dialing Phone Numbers with Letters in your Contact List

Typing phone numbers with letters in your Contact list:

If you had to enter **800-CALLABC** into your address book you would follow these steps:

1. Type in **800**
2. Using the on-screen keyboard, press the **ABC** key and hold it until it locks. (You see a small lock icon), then type **CALLABC** using the keyboard normally.
3. Using the slide-out keyboard, press and hold ALT while typing the letters.

Adding Pauses and Waits in Phone Numbers

Sometimes you want to type a phone number that has preset pauses and waits.

A **Pause** is a 2-second pause, and then dialing continues. If you need more than 2 seconds, just put more than one pause in the phone number.

Press and hold the * key.

A **Wait** waits for you to press the screen before continuing to dial.

Press and hold the # key.

View Video Tutorials and Free Tips at www.MadeSimpleLearning.com   253

If you need a variable amount of time to wait, then you should use a **Wait** instead of a **Pause**.

TIP: This is a great way to enter a voice mail access number to quickly check voice mail on a work or home number.

TIP: When you are editing a contact entry in your Contact List, and you are using the on screen keyboard, you actually see buttons to add Pauses and Waits.

More Phone Tips and Tricks

Like most features on the BlackBerry, there is always more you can do with your Phone. These tips and tricks will make things go even quicker for you.

- To dial the previous call, press and hold the **Green Phone** key when in the **Phone** app.
- To insert a **plus** sign when typing a phone number, hold the number zero 0.
- To add an extension to a phone number, press the X key and then type the extension number. It should look like this: **8005551212 x 1234.**
- To check your voice mail, press and hold the number 1.
- To view the last phone number that you dialed, scroll to the top of the Phone screen, and press the **ENTER** key. Press the **Send** key to dial the number.

Chapter 9: Email Like a Pro

The Torch, even though small and stylish, is a BlackBerry to the core—a powerful email tool. This chapter will get your up and running with your email. In minutes, you will be an emailing pro.

> **NOTE:** If you have not yet set up your email accounts, please see the "Set Up Email Accounts" section of Chapter 1: "Getting Started."

Composing Email

The Torch, like all BlackBerry smartphones, gives you the freedom to email on the go. With the cellular network, you are no longer tied to a Wi-Fi hotspot or your desktop or notebook; email is available to you at all times almost anywhere in the world.

Send Email from Your Messages App

This first option is perhaps easiest for learning how to initially send an email.

1. Tap on the **Messages** icon. You will have an icon for each email address that you have set up in addition to the overall **Messages** icon.

2. You have a few ways to start composing a new message:

- Press the **Compose** soft key ![compose icon] in the bottom left hand corner of your **Messages** list screen, then select the type of message you want to compose.

NOTE: You will see options based on installed apps on the BlackBerry. So, if you have **Facebook** and **Twitter** installed, they should appear in the **Compose** list.

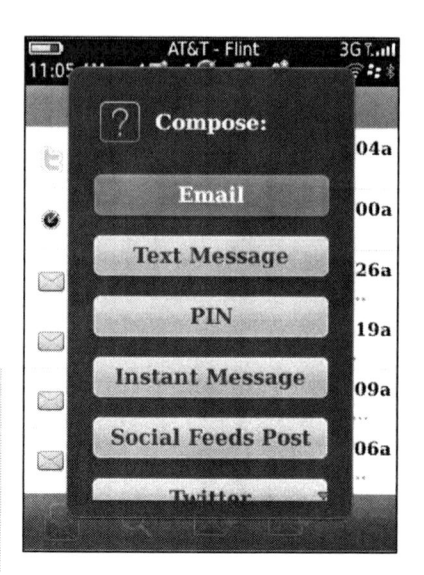

- Press the Email Shortcut key letter **C** (for Compose) on your slide-out keyboard.

TIP: Learn all the shortcut keys in Part 4 at the end of this book.

- Touch or click on a date row separator gray bar and select **Compose** or other message type from the pop-up window (see image to above and right).

- Press the **Menu** key and scroll down to select **Compose Email**.

3. Type in the recipient's name or email address in the **To** field. If the Torch finds a match between what you are typing and any **Contacts** entries, they are shown in a selectable drop-down list.

4. Select the correct name by tapping it. If there is more than one email address stored for that particular contact, just choose the correct address.

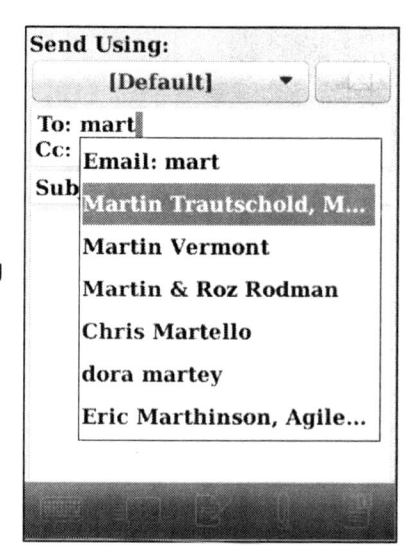

256

CHAPTER 9: Email Like a Pro

> **TIP:** Press the **space** key for the @ and "." in the email address. EXAMPLE: To type susan@company.com, you would type "susan" **space** "company" **space** "com"

5. Repeat to add additional **To:** and **Cc:** addressees.

6. If you need to add a Blind Carbon Copy (Bcc:), then press the **Menu** key and select **Add Bcc:**

7. Then type the **Subject** and **Body** of your email message.

8. When you are done, press the **Send** button in the upper right corner or press the **Menu** key and select **Send**. That's all there is to it.

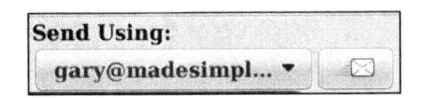

Send Email from Your Contacts App

After you have entered or synced your names and addresses to your BlackBerry, you may send emails directly from your **Contacts** list.

1. Click on your **Contacts** icon.

   Begin to type a few letters from your person's first or last name to find them. Then touch their name to see the contact details.

2. To compose a new message you could do any of these actions:

   - Tap on the **email address** to compose a new message.
   - Tap on the **Email** soft key  at the bottom of the screen.
   - Press the **Menu** key and select **Email (the person's name)**.

View Video Tutorials and Free Tips at www.MadeSimpleLearning.com   257

Selecting a Different Email Address to Send From

If you have several email addresses integrated with your BlackBerry, you can select which one to send your email from.

1. Touch or click on the field that shows **your email address** or **[Default]** at the top of email compose screen under **Send Using**.

2. Scroll up or down to tap on another **Email Account** to use.

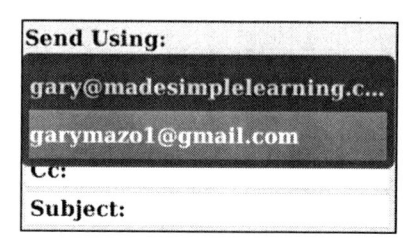

See the Email Address

When you receive email on your BlackBerry, you will often see the person's name (e.g. Margaret Johnson) and not their email address in the **From** field.

You can see the email address by doing either of the following actions:

- Hover your finger over the person's name (without pressing and clicking) or just highlight using the **Trackpad** to see their email address in a little pop-up window.

- From your slide-out keyboard highlight the name/email address and press the letter **Q**.

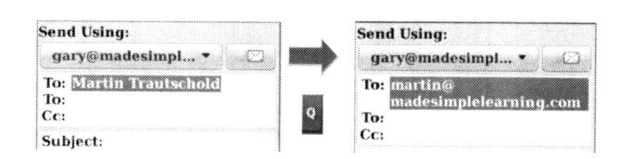

CHAPTER 9: Email Like a Pro

Replying To Messages

Once you get the hang of emailing on your Torch, you will quickly find yourself checking your email and wanting to respond quickly to your emails. Replying to messages is very easy on the Torch.

1. Tap on the **Messages** icon.
2. Tap on the message you wish to reply to.
3. Click on the **Reply** soft key in the bottom row of soft keys.

4. The recipient is now shown in the **To** field.
5. Type in your message.
6. Click the **Send** button in the top right corner to send your email. You can also use the **Send** button at the bottom or press the **Menu** key and select **Send**.

Navigating Around and Other Tidbits

There are a few useful things to know when you are working with your **Messages** inbox. You can get around by using the soft keys at the bottom or the swipe gesture. You may also notice some blank spots in email messages

View Video Tutorials and Free Tips at www.MadeSimpleLearning.com 259

you receive where pictures should be located. In this section, we show you how to retrieve those images. There are times you want to change the importance level from **Normal** to **Low** or **High** depending on the situation. Also, you will want to know how to turn on **Spell Checking** for every email you compose and send from your Torch.

Email Shortcut Keys

You can use special keys on your slide-out physical keyboard to quickly move around your inbox. For example, pressing the letter **T** will jump to the top of your inbox, **B** will jump to the bottom, **N** will go to the Next day and **P** will go to the previous day. There are many more shortcut keys listed at the end of this book.

> **TIP:** See the all **Email Shortcut keys** in Part 4 of this book at the very end.

Email Soft Keys

There are a series of soft keys that you can use while in the Email program.

To view what a soft key does, lightly hover your finger over that soft key and see a little pop-up window show you the name of the key. Simply tap the soft key to activate that function.

The soft keys are:

Reply – Touch or click to reply to an e-mail.

Forward – Touch or click to forward e-mail to another contact.

Reply All – Touch or click to reply to all recipients who were sent the email.

Delete – Touch or click to delete the message.

CHAPTER 9: Email Like a Pro

Swipe to Navigate Your Inbox

You can use the swipe motion with your finger on the screen to move around in your email inbox.

- To move to the next message, swipe from right to left.
- To move to the previous message, swipe from left to right.
- Once you get back to the first message, swiping from left to right will bring you back to the message list.

Getting Rid of Blank Spots in Emails You Receive

On some email messages, you may see blank spaces where images should be. If you see this, follow these steps to see all your images:

1. Press the **Menu** key.

2. Select **Get Image** to retrieve just one image or **Get Images** to retrieve them all.

3. You may see a warning message about exposing your email address; you need to Tap **Ok** or **Yes** in order to get the image.

TIP: To set your BlackBerry so that it automatically loads the images every time, you can set an option. Start **Messages**, press the **Menu** key and select **Options**. Select **Email Preferences** and make sure to check the box next to **Download Images Automatically** for each email account.

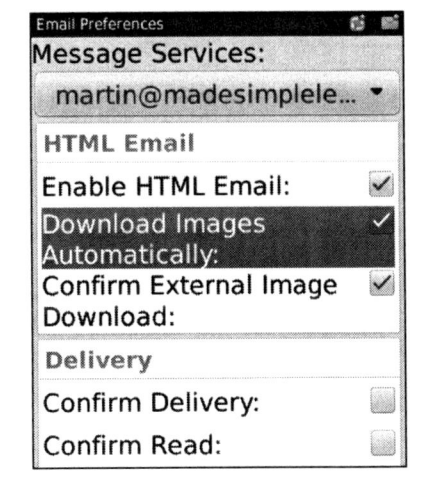

View Video Tutorials and Free Tips at www.MadeSimpleLearning.com   261

Setting the Importance Level of the Email

Sometimes you need your email to be noticed and responded to immediately. The BlackBerry lets you set the importance level so that the recipient can better respond.

Begin composing a new email message, then press the **Menu** key.

Select the **Options** menu item.

> **TIP:** Pressing the letter key that matches the first letter of the menu item a couple of times will jump you down to that item.

In the **Options** screen, you will see a line that says **Importance** and the default **Normal** at the end of the line. Tap on the word **Normal** and you see the options **High** or **Low**.

Just Tap on the appropriate option. Press the **Menu** key and **Save** your choice to return to the email message. Then press the **Menu** key and **Send** the message.

Finally, you will see high importance and low importance messages marked with special icons in your messages list (Inbox):

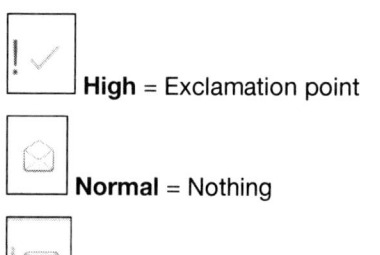

High = Exclamation point

Normal = Nothing

Low = Arrow pointing down

Spell Checking Your Email Messages

Please see later in this chapter about how to enable spell checking on email messages you type and send. The spell checker may not be turned on when you take your BlackBerry out of the box the first time.

CHAPTER 9: Email Like a Pro

Flag for Follow Up

After receiving and email message, have you ever found yourself thinking:

"This is important, but I can deal with it now."

"I need to spend more time on this email and get back to them later."

"I need to give them a call about this message on Wednesday."

If so, then the **Flag for Follow Up** can be a great feature for you.

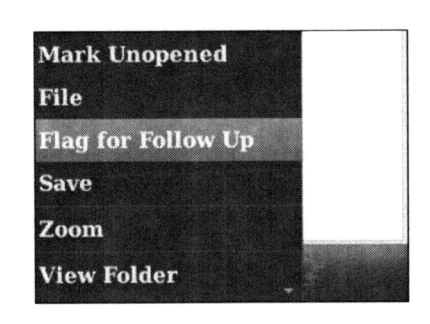

You can set flags of many types and colors with or without due dates from the main email inbox or while you are reading a particular message.

Setting a New Flag

To set a flag, follow these steps:

1. Press the **Menu** key and select **Flag for Follow Up**.
2. You can set various **Flag Properties** on this screen. As you set these properties, keep in mind that you can use the messages **Search** command to find flags or specific colors of flags.

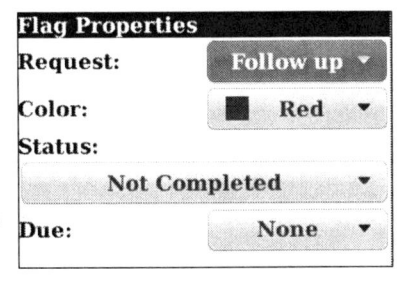

263

3. **Request type**: There are quite a few types of follow-ups that you can select. Click on the Request field to see them all. In some cases, you might even want to select **No response necessary**.

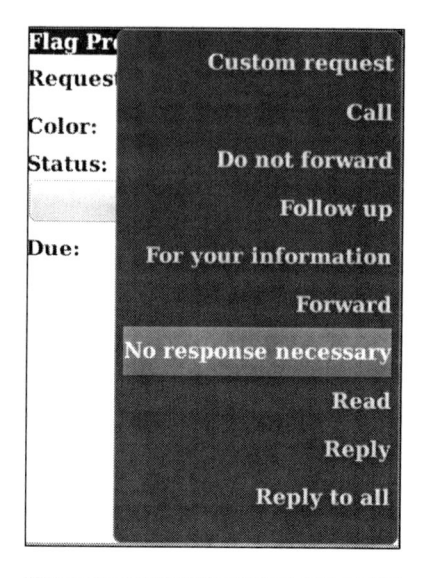

4. **Color:** You can change the color of the flag, which can help you locate the flagged items later with the email search feature.

5. **Status**: Is either **Not Completed** or **Completed**

6. **Due**: Is either **None** or **By Date** and it lets you specify a due date.

7. Press the **Menu** key and select **Save**.

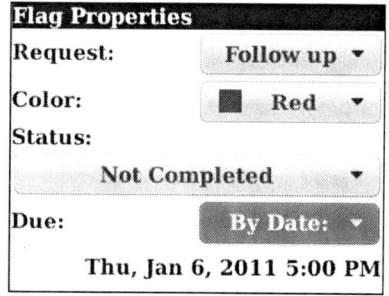

Changing or Editing a Flag

Once you set the flag, you will see it in the top of the message under the addresses and subject information.

1. Tap on the flag itself in an email or press the **Trackpad** or the **Menu** key and select **Flag Properties**.

2. This will allow you to change all the properties.

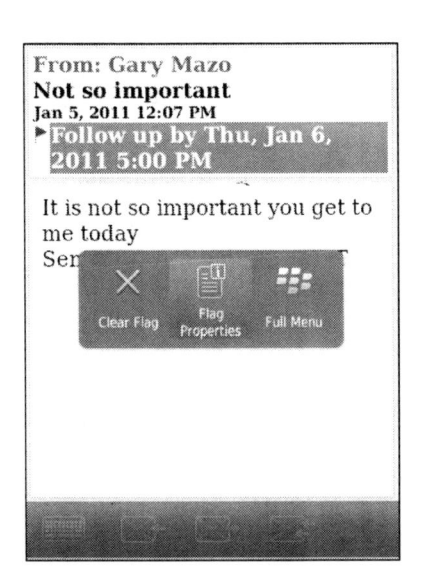

Finding Flagged Items

1. In your email inbox (Messages), press the **Menu** key and select **Search**

2. Scroll down to the bottom where it says **Type** and select **Email With Flags**.

3. Notice you can also select a specific **Color** or **All colors**.

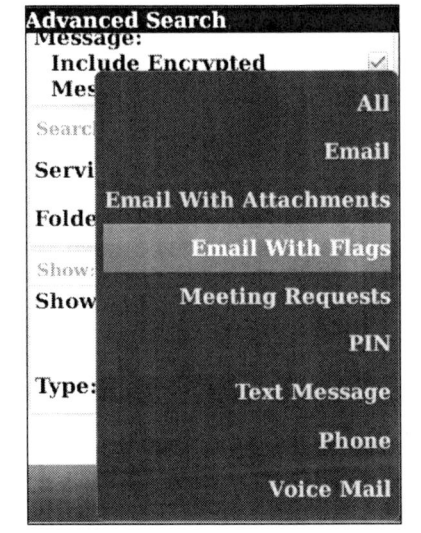

TIP: Set a Hotkey for your Search

Before you execute your Search, press the **Menu** key and select **Save**. This allows you to name your search and set a hotkey combination ALT + some letter.

Try **ALT + Q**

Now, in your email inbox, you can quickly find all flagged items by pressing **ALT + Q**.

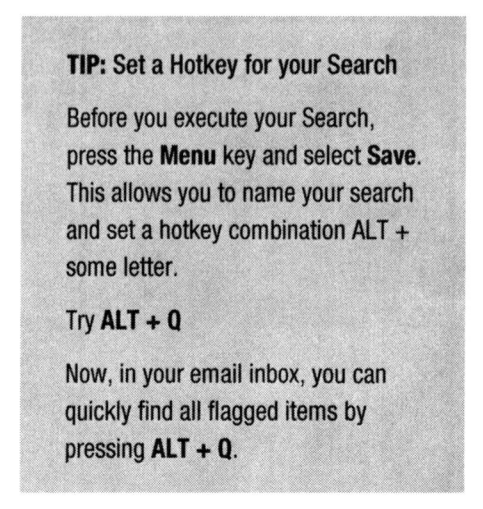

Flag Alarms

When a flagged item alarm rings, you will see a pop-up screen similar to the one shown.

You will also notice a flag on your top status bar with a number next to it showing how many flag due dates have passed.

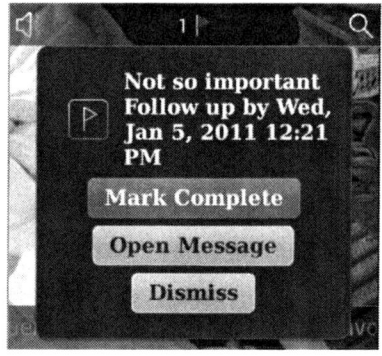

Attaching Contacts, Files, and Pictures

There may be times you want to send a Contact entry to someone who asked for your colleague's mailing address or phone number. There are other times you want to share a picture, a word processing file, or a spreadsheet file. For this, you would attach a file to a message. In this section we show you how to do these things.

Attaching a Contact Entry

If you need to send someone an address that is contained in your **Contacts** list:

CHAPTER 9: Email Like a Pro

1. Start composing an email.
2. Press the **Menu** key and click on **Attach** and then choose **Contact** from the menu.

3. Either scroll to the contact you want to add or use the keyboard to type in the first few letters of the desired contact and then Tap the highlighted name.

4. You will now see the attached contact shown as a little address book icon at the bottom the main body field of the email.

Attaching a File, Picture, Video, or Voice Note

There may be times you want to send a spreadsheet, document, picture, video, music or voice note from your Torch via email. For this, you will need to attach a file (much like you would do on your computer) to the email you send from your Torch.

1. Start composing an email message and press the **Menu** key.
2. Select **Attach** and choose **File** from the menu.

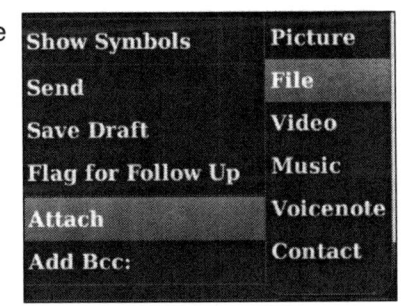

View Video Tutorials and Free Tips at www.MadeSimpleLearning.com   **267**

3. Next, locate the directory in which the file is stored. Your two initial options are **Device Memory** or **Media Card**.

4. Navigate to the folder where the file is stored. Once you find the file, simply Tap on it and it will appear in the body of the email at the bottom of the screen.

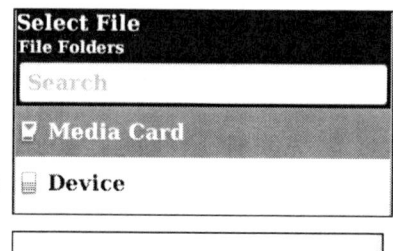

Working with Email Attachments

One of the things that makes your Torch more than just another pretty Smartphone is its serious business capabilities. Often, emails arrive with attachments of important documents: Microsoft Word files, Excel spreadsheets or PowerPoint presentations. Fortunately, the Torch lets you open and view these attachments and other common formats wherever you might be.

Your Torch also comes with the **Documents to Go** program from DataViz. This is an incredibly comprehensive program that allows you to not only view but also edit Word, PowerPoint and Excel documents and it preserves the native formatting. This means that the documents can open on your BlackBerry and look just like they do on your computer.

Supported Email Attachment Formats

Your Torch can open and view the following document and image formats:

- Microsoft Word (DOC)
- Microsoft Excel (XLS)
- Microsoft PowerPoint (PPT)
- Corel WordPerfect (WPD)
- Adobe Acrobat PDF (PDF)
- ASCII text (TXT)
- Rich Text Format files (RTF)
- HTML
- Zip archive (ZIP)
- (Password protected ZIP files are not supported)
- MP3 – Voice Mail Playback (up to 500Kb file size)

- Image Files of the following types: JPG, BMP, GIF, PNG, TIFF

NOTE: Multi-paged TIFF files are not supported.

NOTE: Additional file types may be supported in newer versions of the system software running on your BlackBerry.

Features available in attachment viewing:
- Images: Pan, Zoom, or Rotate.
- Save images to view later on your BlackBerry.
- Show or hide tracked changes (e.g. in Microsoft Word).
- Jump to another part of the file instead of paging through it.
- Show images as thumbnails at the bottom of the email message.

Knowing When You Have an Attachment

How do you know if you have an email attachment?

You will see an envelope with a paperclip, as shown.

= Has Attachment

= No Attachment (or it has an attachment that cannot be opened by the BlackBerry)

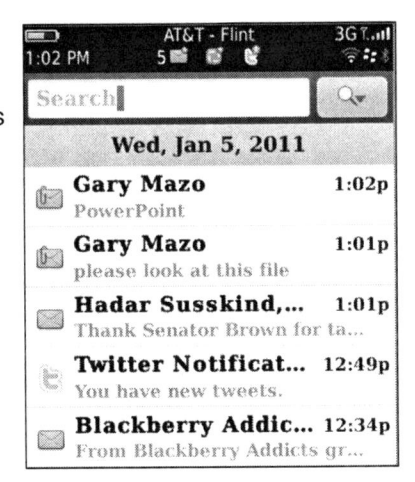

Opening Attachments

Navigate to your message with the attachment icon showing [icon] (paperclip on envelope) and Tap on it.

BlackBerry Torch Made Simple

At the very top of the email, in parenthesis, you will see **[1 Attachment.]** or **[2 Attachments]**, depending on number of attachments.

1. Touch or click on the Attachment (or click the Trackpad) and select **Open Attachment(s)**.

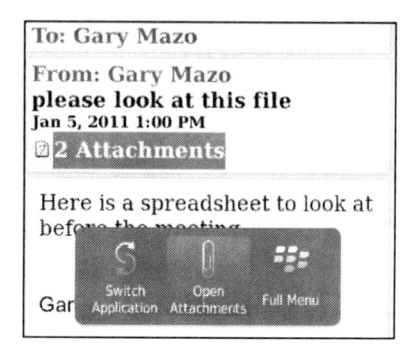

2. If the document is a Microsoft Office document format, it will then be presented with the option of **View** or **Edit with Documents to Go**.

    - For a quick view without true document formatting, select **View**.

    - To really see the document the way it was meant to be seen, we suggest you select **Edit with Documents to Go**.

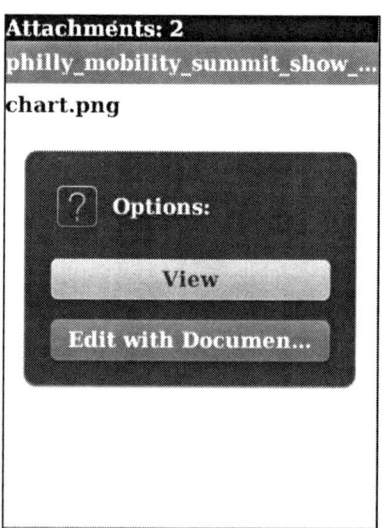

3. If you get an error message such as **Document Conversion Failed**, it is very likely that the attachment is not a format that is viewable by the BlackBerry Attachment Viewer. Check out the list of supported attachment types up above.

Editing Attachments with Documents to Go

Once you select **Edit with Documents to Go** the document will open on your screen.

1. You can scroll through just like you were reading a Word Document on your computer.

CHAPTER 9: Email Like a Pro

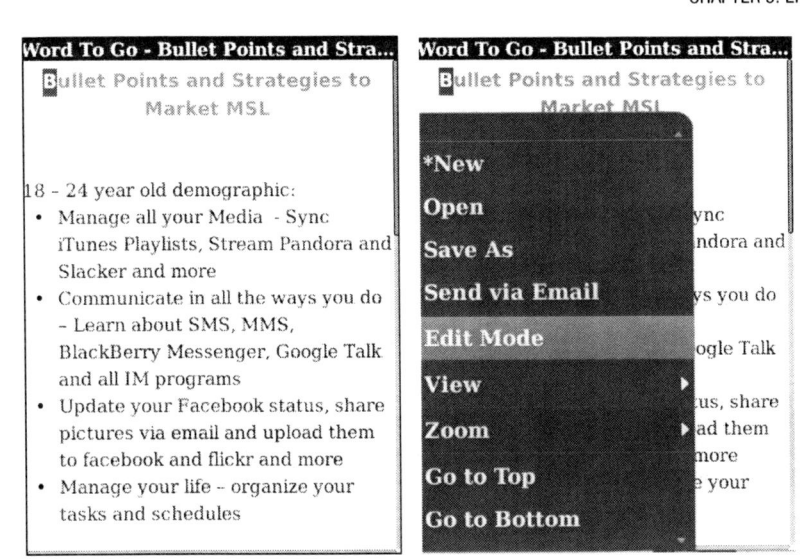

Figure 9-1. Editing a document with Documents to Go

2. If you want to make changes to the document, just press the **Menu** key and select **Edit Mode** from the menu (Figure 9-1).

3. If you want to adjust the Formatting of the document, just press the **Menu** key and select **Format** (right image of Figure 9-2).

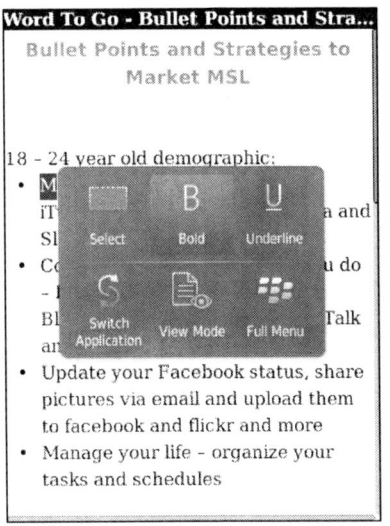

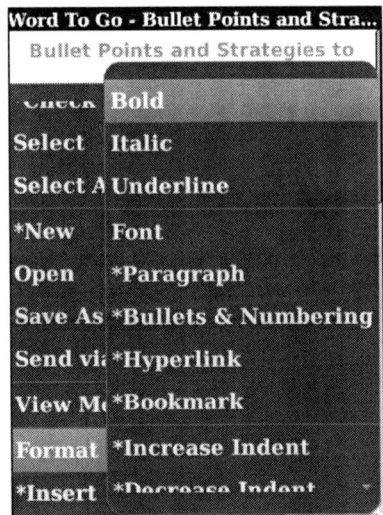

Figure 9-2. Formatting options in Edit Mode

4. Alternatively, you may press and hold text or selected text to see the formatting pop-up menu (see left portion of Figure 9-2).

View Video Tutorials and Free Tips at www.MadeSimpleLearning.com   **271**

Using the Standard Document Viewer

You may decide you don't want to use the **Documents to Go** app.

1. Click on the attachment in the email message.
2. Select the **View** option (Figure 9-3).
3. You will see the document open with the standard document viewer. The document won't have the same look, but you will be able to navigate through it quickly.

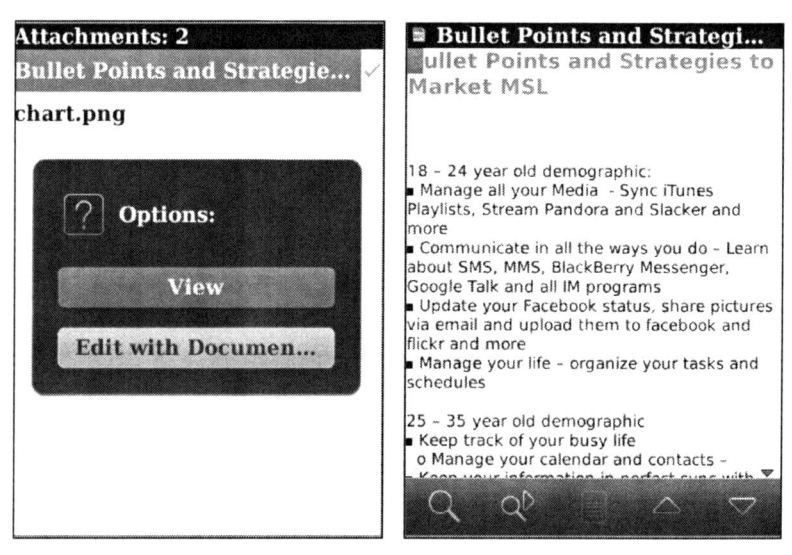

Figure 9-3. Using the standard document viewer to view a Word document

Using Sheet to Go and Slideshow to Go

Following the same steps you did earlier when you opened the word processing document you can edit a spreadsheet with **Edit with Documents to Go** and a presentation file with **Edit with Documents to Go** (see Figure 9-4).

CHAPTER 9: Email Like a Pro

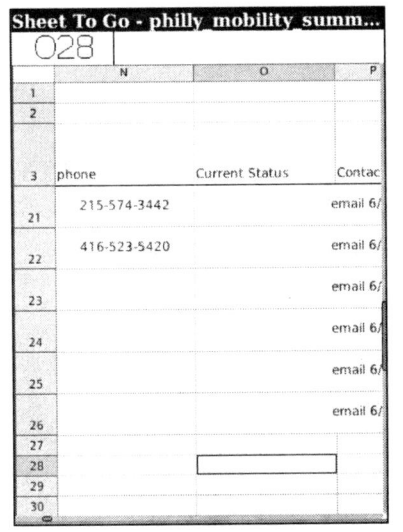

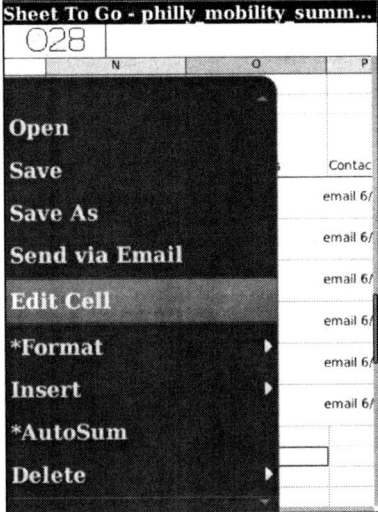

Figure 9-4. Viewing and editing a spreadsheet with Documents to Go

To Open a Picture

1. Open a message with pictures attached.

2. Touch or click on the **[1 Attachment]** or **[2 Attachments]**, etc. at the top of the email message.

3. Select **Open Attachment** or **Download Attachment** (to save it on your BlackBerry).

4. Then Tap on the image file names to open them.

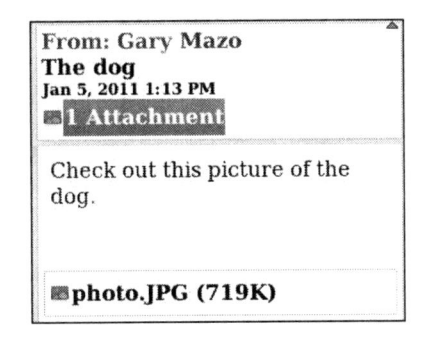

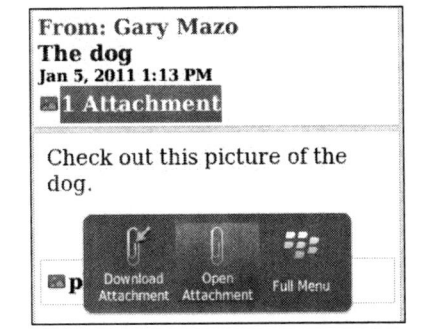

View Video Tutorials and Free Tips at www.MadeSimpleLearning.com   273

> **TIP:** Once you have opened the pictures, the next time you view that email, you will see the thumbnails of all the pictures at the bottom of the message. You can then just scroll down to them and Tap on them to open them.

5. To save the picture, press the **Menu** key and Tap on **Save Image**. The picture will be saved where you specify, either on your Media Card or the main Device Memory (Figure 9-5).

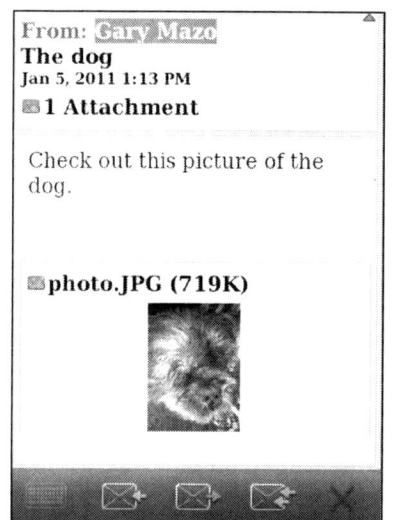

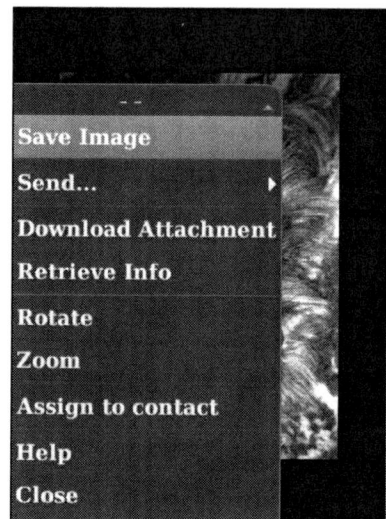

Figure 9-5. Viewing and saving an image on your Torch

6. Other menu options include the following:
   - **Zoom** - to expand the image
   - **Rotate** - to rotate the image
   - **Send** - Then choose how you would like to send it; Email, Text Message, Messenger Contact, Twitter, Group or Facebook.
   - **Retrieve Info** - Retrieves image type, size and other pertinent information.

CHAPTER 9: Email Like a Pro

7. To save the image as a Caller ID picture, follow these steps:

- Press the **Menu** key and select **Assign to Contact**.

- Begin to type in the contact name to find the contact.

- Navigate to the correct contact and save as prompted.

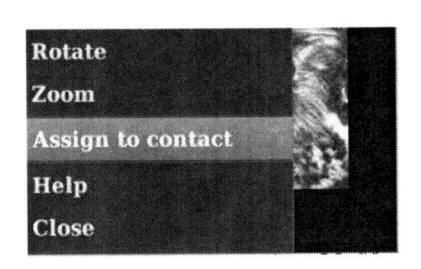

Searching for Messages (Email, SMS, MMS)

You might find that you use your messaging so often, since it is so easy and fun, that they really start to collect on your BlackBerry.

Sometimes you need to find a message quickly, rather than scroll through all the messages in your in box. There are three primary ways to search through your messages; searching the entire message through any field, searching the sender, and searching the subject (see Figure 9-6).

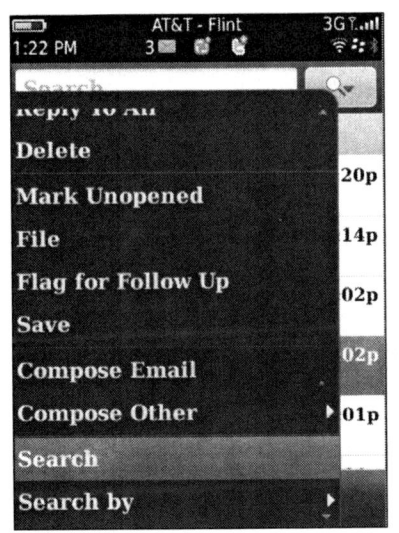

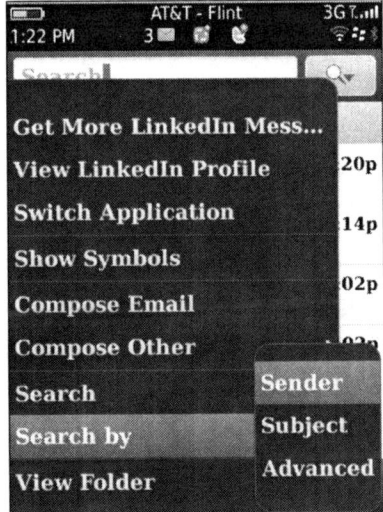

Figure 9-6. Various ways to search your Messages Inbox

View Video Tutorials and Free Tips at www.MadeSimpleLearning.com   275

The General Messages Search Command

This is the easiest way to search for a message if you are not sure of the subject or date:

1. Tap on your **Messages icon** and press the **Menu** key.
2. Select **Search**.
3. Enter in information in the **Search** box at the top of the screen. When you are done, click the **Trackpad** or press the **Enter** key on your slide-out keyboard.

The messages that match the search criteria are then displayed on the screen.

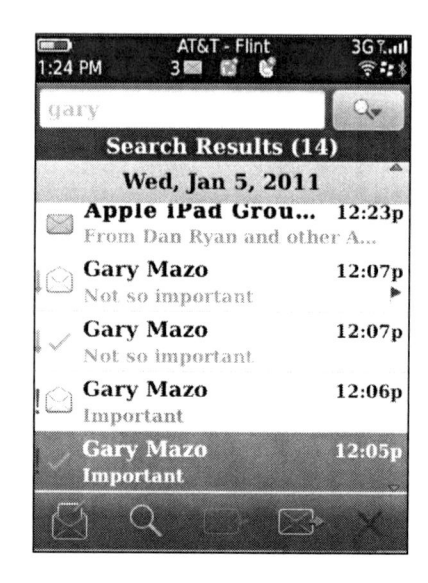

Search Sender or Search Recipient

The easiest way to search through your emails is to search by either **Sender** or **Subject**. The **Search By** command from the Menu makes this very easy to accomplish.

Search Subject

Find an email with a subject you wish to search for in other messages. Press the Menu key and scroll down to Search by and choose Subject.

The search results will then be displayed as shown in (Figure 9-7).

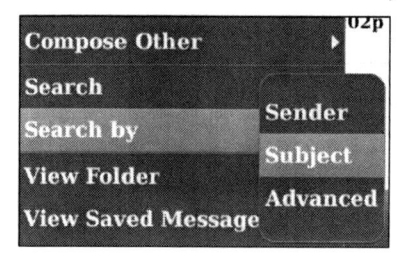

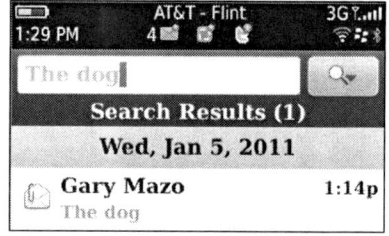

Figure 9-7. Search by subject.

CHAPTER 9: Email Like a Pro

Search Sender or Recipient Menu Command

TIP: The Search, Search Sender, Search Recipient, and Search Subject work on SMS messages, e-mail, MMS —anything in your Messages Inbox!

Sometimes, you only want to see the list of your communication with a particular individual.

From the messages list, scroll to any message from the person you wish to search and press the **Menu** key.

Say that you want to find a specific message from Gary and you have lots of messages from Gary on your device. Just highlight one of the messages and then press the **Menu** key.

Scroll to Search by and choose Sender from the options.

All the messages from Gary will now be displayed.

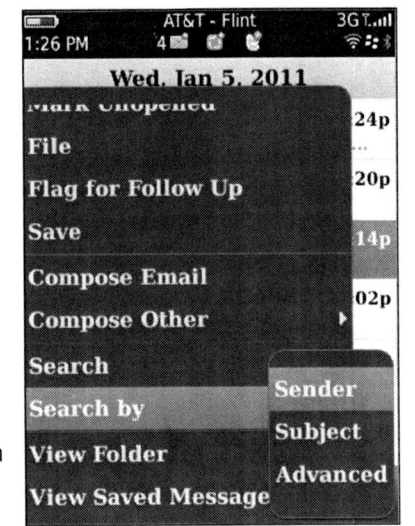

Only the list of messages sent by that particular person (in this case, Gary) is now displayed. Just scroll and find the particular message you are looking for.

View Video Tutorials and Free Tips at www.MadeSimpleLearning.com

277

Chapter 10:
Your Contact List

Your BlackBerry excels as a contact manager. You will most likely turn to your **Contacts** app more than any other app on the device. From a contact, you can email, send a text message, call, fax, or even "poke" on Facebook.

One rule of thumb you will hear from us often is to add anything and everything to your contacts. Whenever someone calls, add him or her to your contacts. Add contacts from emails and messages—that way they are always available for you later.

The Heart of Your BlackBerry

Your address book is really the heart of your BlackBerry. Once you have your names and addresses in it, you can instantly call, email, send text (SMS) messages, PIN-to-PIN BlackBerry messages, or even pictures or Multimedia Messages (MMS). Since your BlackBerry came with a camera, you may even add pictures to anyone in your address book, so that when they call, their pictures show up as Picture Caller ID (as we see here with Gary's picture).

Ways to Get Your Addresses on Your BlackBerry

You can manually add contact addresses one at a time as we show you in this chapter. You can also sync your computer's contacts with your BlackBerry.

If your BlackBerry is tied to a BlackBerry Enterprise Server, the synchronization is wireless and automatic, see Chapter 1 for steps to connect to an "Enterprise Account."

Otherwise, you will use either a USB cable or Bluetooth wireless to connect your BlackBerry to your contacts and calendar on computer to keep it up to date. For Windows PC users, see Chapter 2: "Windows PC Set Up;" Apple Mac computer users, see Chapter 4: "Apple Mac Set Up."

CHAPTER 10: Your Contact List

If you use **Gmail** (Google Mail), you can use the Google Sync program to wirelessly update your contacts on your BlackBerry with your address book from Gmail for free!

Learn how in the "Using Google Sync for Contacts and Calendar" section at the end of Chapter 11: "Manage Your Calendar."

When Is Your Contact List Most Useful?

Your Contacts program is most useful when two things are true:

- You have **many names** and addresses in it.
- You can **easily find** what you need.

Our Recommendations

We recommend keeping two rules in mind to help make your contacts most useful.

> **Rule 1:** Add anything and everything to your contacts.

You never know when you might need that obscure restaurant name/number, or that plumber's number, etc.

> **Rule 2:** As you add entries, make sure you think about ways to easily find them in the future.

We have many tips and tricks in this chapter to help you enter names so that they can be instantly located when you need them.

Transfer SIM Card Contacts into Your Contact List

If you are using your SIM (Subscriber Identity Module) card from another phone in your BlackBerry and have stored names and phone numbers on that SIM card, it's easy to transfer your contacts into your contact list.

Start your **Setup app**; it should be in the **All** Home Screen.

279

Scroll down to Personalization

Touch or click the **SIM Contacts Sync** icon.

The BlackBerry will then sync the SIM card contacts to your device **Contacts** app.

> **TIP:** Your SIM card contains only the bare minimum—**name** and **phone number**. You should review your imported contacts and add in email addresses, mobile/work phone numbers, and home/work addresses to make your BlackBerry more useful.

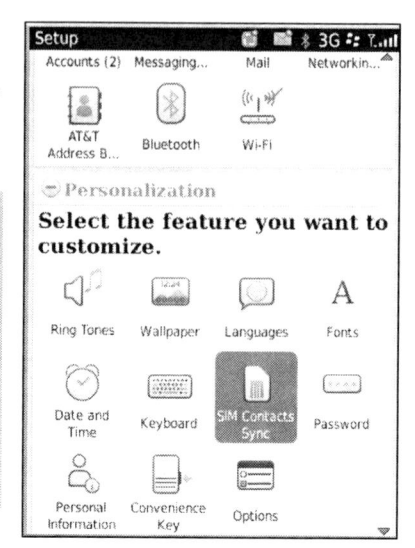

How to Easily Add New Addresses

On your BlackBerry, since your address book is closely tied to all the other icons (Messages/Email, Phone, and Web Browser) you have many methods to easily add new addresses:

- **Option #1:** Add a new address inside the **Contacts** app.
- **Option #2:** Add an address from an email in **Messages**.
- **Option #3:** Add an address from a phone call log in the phone.
- **Option #4:** Add a new address from an underlined email address or phone number anywhere (web browser, email, tasks, MemoPad, etc.).
- **Option #5:** (If your BlackBerry is tied to an Enterprise Server) Perform a Remote Lookup of your organizations main address list and add the resulting contacts to your **Contacts** list.

Option #1: Type an Address into Contacts

The most obvious way of adding a contact is to use New Contact and type in the information. To do this:

1. Click the **Contacts** icon.

CHAPTER 10: Your Contact List

OR

2. You can either tap the **New Contact** tab at the top of the list or press the **Menu** key and select **New Contact**.

3. Add as much information as you can because the more you add, the more useful your BlackBerry will be!

TIP: Press the **Space** key instead of typing @ and "." in the email address.

TIP: If you add someone's work or home address, you can easily map his or her address and get directions right on your BlackBerry.

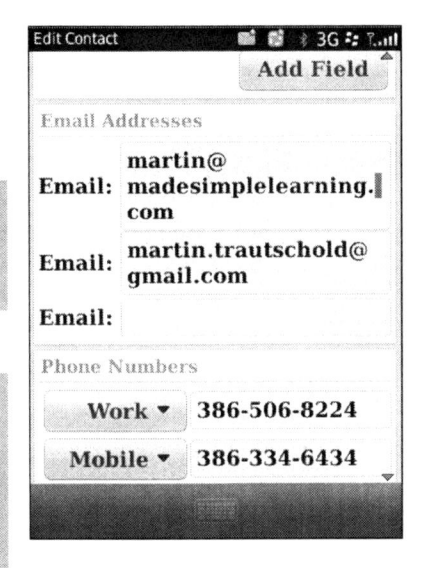

TIP: Finding Restaurants

Whenever you enter a restaurant into your contact list, make sure to type the word **Restaurant** into the **Company Name** field, even if it's not part of the name. Then when you type the letters "**rest**," you should instantly find all your restaurants!

View Video Tutorials and Free Tips at www.MadeSimpleLearning.com 281

Up to Three Email Addresses

While you are adding or editing their contact entry, just touch or click the **Email** field (there are three of them.)

Add up to three email addresses per contact.

Be sure to save your changes by pressing the **Menu** key and selecting **Save**.

Typing Phone Numbers with Letters

Some business phone numbers have letters, like "1 800-REDSOX1." These are easier than you might think to add to your BlackBerry address book (or type while on the phone).

Here are two steps to do this easily:

1. Slide open your BlackBerry to reveal the keyboard. Press and hold the **Alt** or **Shift** Keys while you type letters.

2. Or, use the on screen keyboard and press and hold your **ABC** key until you see the little lock sign. Then use the full keyboard to type the letters as you see them.

TIP: Do you need to enter a phone number with pre-set **pauses** or **waits** (like the pause before entering a voice mail password)? See the "Adding Pauses and Waits" section of Chapter 8: "Phone and Voice Dialing" for more information.

CHAPTER 10: Your Contact List

Option #2: Add Contacts from Email Addresses

Another easy way to update your address book is to simply add the contact information from email messages that were already sent to you.

1. Open up an email message in your inbox.
2. Highlight any email address for the contact you want to add.
3. Press the **Menu** key and select **Add to Contacts**.
4. Add the information in the appropriate fields, press the **Menu** key, and **Save**.

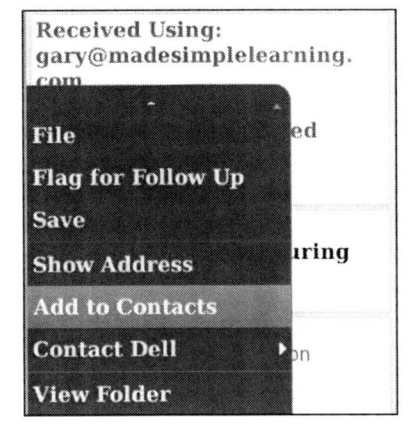

Option #3: Add Contacts from Phone Call Logs

Sometimes you will remember that someone called you a while back, and you want to add his or her information into your address book.

1. Press the **Green Phone** button to bring up your call logs.
2. Use the Trackpad to highlight the number you want to add to your address book.
3. Press the **Menu** key and select **Add to Contacts**.
4. Add the address information, press the **Menu** key, and select **Save**.

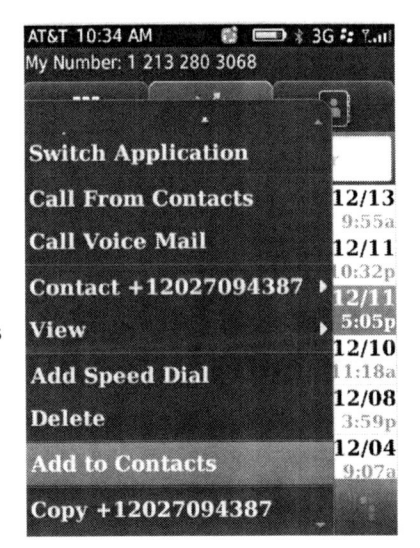

View Video Tutorials and Free Tips at www.MadeSimpleLearning.com   283

Option #4: Add Contacts from Underlined Numbers, Email Addresses and Street Addresses

One of the very powerful features of the BlackBerry is that you can really add your contacts from just about anywhere.

Whenever you see an underlined phone number, email address or physical street address, you can tap it to bring up a short menu and select **Add to Contacts**.

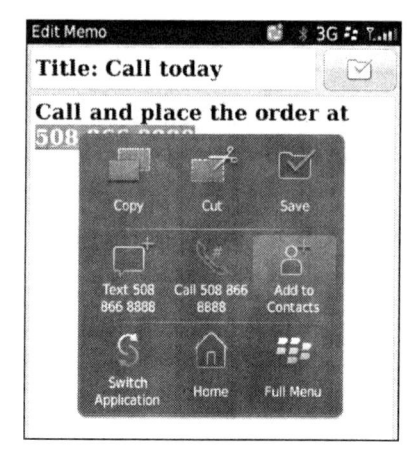

Here is an example with an underlined street address.

Notice that with this address, the menu also allows you to:

- View On Map
- Navigate to Here

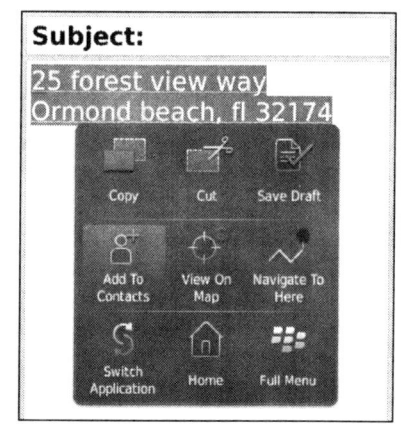

Option #5: Perform a Remote Lookup (If your BlackBerry is tied to an Enterprise Server)

If your BlackBerry is connected to a BlackBerry Enterprise Server (See "Enterprise Account Set Up" of Chapter 1), then you can perform a **Remote Lookup**. This allows you to search your corporate or enterprise-wide database stored on the central server for names of people in your organization. Once you locate these people, you can easily add them to your local **Contacts** list with a menu command.

CHAPTER 10: Your Contact List

TIP: This **Remote Lookup** also works from the **Universal Search** on the Home screen. All you need to do is type a few letters of someone's first or last name in the **Search** window and tap on the **Remote Contact Lookup** icon (shown here).

1. Click on your **Contacts** icon.
2. Type a few letters of someone's first name (you can also type a few letters of their last name if you end up finding too many people).
3. Tap **Remote Lookup** as shown.
4. Now, the BlackBerry will search the remote list of all the names in your organization for matches. In this example, there were 23 matches for "Gary." Press the **Menu** key and select **Get More Results**.
5. To add one name to your **Contacts** list, select **Add to Contacts** from the menu. To add everyone of the names listed, select **Add All to Contacts.**
6. Notice that you can also **Delete the Lookup** from the menu (lookups are saved until you delete them).

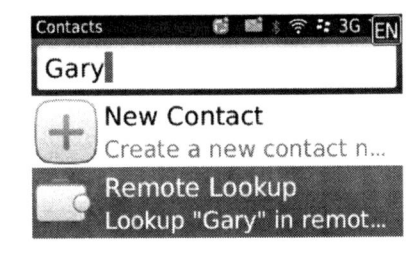

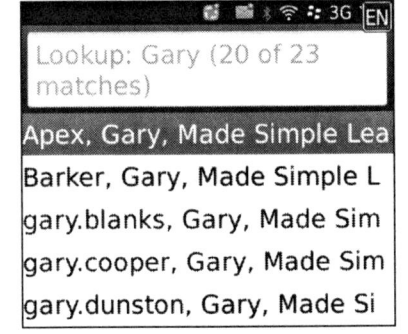

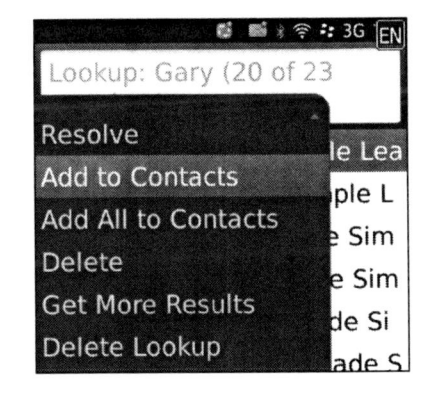

How to Easily Find Names and Addresses

Once you get the hang of adding contact information to the **Contacts** app, you will begin to see how useful it is to have all that information at your fingertips. The tricky part can be actually locating all the information you have input into the BlackBerry.

Using the Find Feature in Contacts

Contacts has a great Find feature at the top of the Contacts startup page that will search for entries that match the letters you type into one of these fields:

- **First Name**
- **Last Name**
- **Nickname**
- **Company Name**

1. Inside **Contacts**, just type a few letters of a person's first name, last name, nickname, and/or company name (separated by spaces) to instantly find that person.

2. Press the letter **M** to see only entries where the first name, last name or company name start with the pressed letter. Press more letters to get a closer match.

3. You can also press one letter and then the **Space** key and type another letter, like **T**, to further narrow the list to people with an **M** and a **T** starting their first, last, or company name.

CHAPTER 10: Your Contact List

4. In this case, I can easily find what I am searching for: "Martin Trautschold."

Finding and Calling Someone in the Phone

Sometimes, you just want to make a phone call to someone in your contacts.

1. Press the **Green Phone** key to start your phone.

2. Tap the **Contacts** or **Call Logs** soft keys in the upper row.

3. Type a few letters of someone's first name, last name, nick name, or company name, and the BlackBerry immediately starts searching for matching entries from your contacts.

4. Tap the person's name and a call is started to them. If they have more than one phone number, you are asked to select a number.

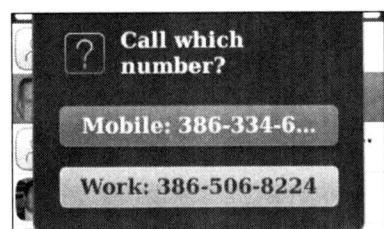

Managing Your Contacts

Sometimes, your contact information can get a little unwieldy, with multiple entries for the same individual, business contacts mixed in with personal ones, etc. There are some very powerful tools within the **Contacts** application that can easily help you get organized.

Editing Contacts

One of the first things to do is to make sure that all the correct information is included in your contacts. To do this, you will follow the steps to select and edit your contact information.

1. Tap the **Contacts** icon.

2. Type a few letters of the first, last, or company name to find the contact, or just scroll through the list.

3. Highlight the contact you want to manage by touching the screen.

4. Press the **Menu** key and choose **Edit** to access the detailed Contact screen and add any information missing in the fields.

For Facebook Users

If you use Facebook, once you load it on your BlackBerry (see Chapter 14) all your Facebook contacts will have their Facebook profile photos loaded, and you can do some cool things right from your contact list to your Facebook friends.

Friend Photos Simply Appear

All your Facebook contacts will automatically have their Facebook profile photos placed in their contact entries. You know it is a Facebook photo because of the little f in the lower right corner of the picture.

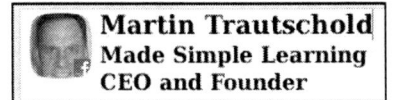

Poke!, Send a Message, or Write on a Wall

You can also Poke! your Facebook friends, send them a Facebook message, or even write on their walls right from your BlackBerry contact list!

To do this, highlight a Facebook friend in the contact list, press the **Menu** key and select **Facebook.**

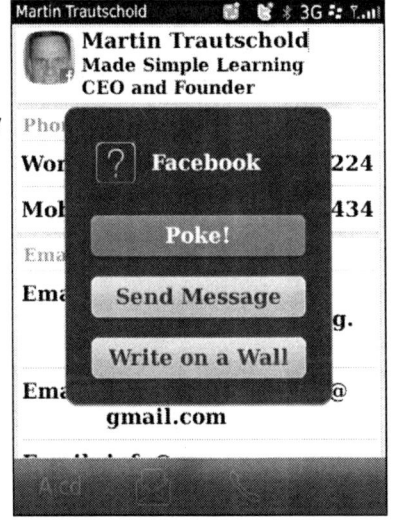

Adding a Picture to the Contact for Caller ID

Sometimes it is nice to attach a face with the name. If you have loaded pictures onto your media card or have them stored in memory, you can add them to the appropriate contact in your contacts. Since you have a BlackBerry with a camera, you can simply take the picture and add it as a **Picture Caller ID** right from your camera.

1. Select the contact to edit as you did previously.
2. Scroll down to the **Picture** icon, and click or touch the screen.
3. Choose the folder containing the picture you would like to use. You will see **Camera Pictures**, **Picture Library**, **Avatars** and possibly **BlackBerry Messenger**.

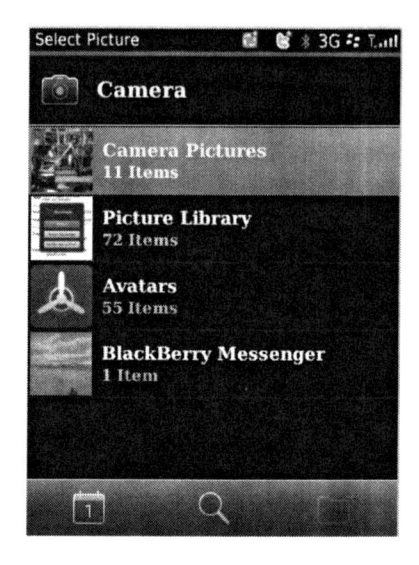

You can also choose Camera at the top to take a new picture with the built in camera.

If you want to use a stored picture, then navigate to the folder in which your pictures are stored by scrolling the screen up/down and clicking the correct folder.

Once you have located the correct picture, click it, and then click Crop and Save.

> **NOTE**: If you choose a small icon like an Avatar, you won't have to crop and save.

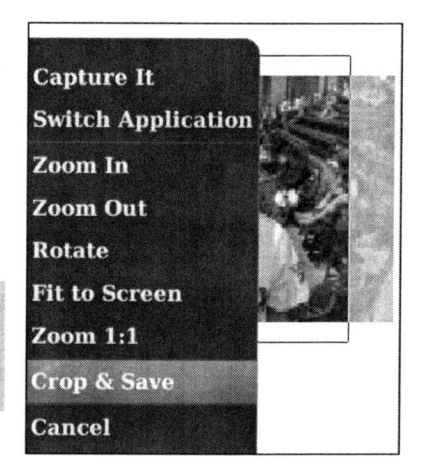

You can use the camera instead to take a picture right now. To do this, click the camera, and take the picture. Move the viewing box to the center of the target of the photo, click, and select **Crop and Save**.

The picture will now appear in that contact whenever you speak to him or her on the phone.

Changing the Way Contacts Are Sorted

You can sort your contacts by first name, last name, or company name.

1. Click your **Contacts**.
2. Press the **Menu** key and select **Options**.
3. Click **Contact Display Actions**.
4. In the **Sort By** field, click the First Name selection (it may say Last Name or Company if you have changed it) and choose the way you wish your contacts to be sorted. You may also select whether to allow duplicate names and whether to confirm the deletion of contacts from this menu.

Using Categories

Sometimes, organizing similar contacts into categories can be a very useful way of finding people. What is even better is that the categories you add, change, or edit on your BlackBerry are kept fully in sync with those on your computer software.

BlackBerry Torch Made Simple

1. Find the contact you want to assign to a category, click to view the contact, click it again, and select **Edit** from the short menu.

2. Click the **Menu** key and select **Categories**.

3. Now you will see the available categories. (The defaults are **Business** and **Personal**.) In you have a contact app like LinkedIn installed (See chapter 14) you may see that as well.

4. If you need an additional category, just press the **Menu** key and choose **New**.

5. Type the name of the new category, and it will now be available for all your contacts.

6. Scroll to the category to which you wish to add this contact, and click.

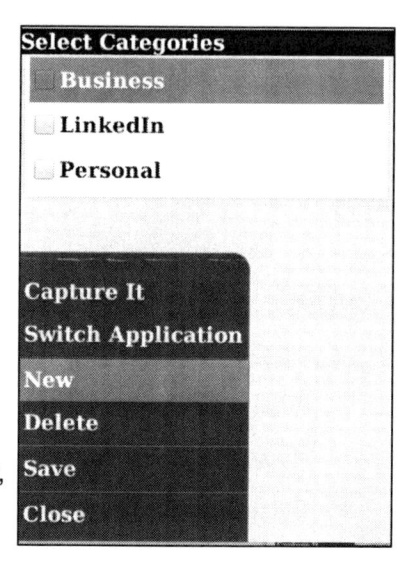

TIP: You can assign a contact to as many categories as you want!

Filtering Your Contacts by Category

Now that you have your contacts assigned to categories, you can filter the names on the screen by their categories. So, let's say that you wanted to quickly find everyone who you have assigned to the business category.

1. Click your **Contacts**.

2. Press the **Menu** key, scroll up, and select **Filter**.

3. Click on any category to select it.

4. Once you do, only the contacts in that category are

292

available to scroll through.

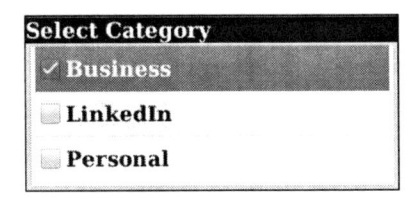

Knowing when Your Contacts Are Filtered

You will see a black bar at the top of your Contact list with the name of the category. In this case the category is **Business**.

> **TIP:** If you filter by a category where you have no contacts assigned, or just a few assigned, it may appear as if you have lost your entire address book. To fix this issue, see how to un-filter the Contact list in the next section.

Un-Filtering Your Contacts

Unlike the Find feature, you cannot just press the **Escape** key to un-filter your categories. You need to reverse the filter procedure.

Inside your contacts, press the **Menu** key. Select **Filter**, scroll down to the checked category, and uncheck it by clicking it or pressing the **space** key when it is highlighted.

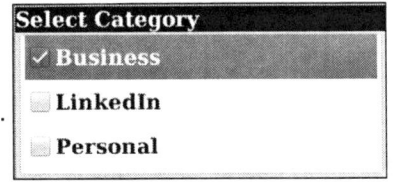

Using Groups as Mailing Lists

Sometimes, you need even more organizing power from your BlackBerry. Depending on your needs, grouping contacts into mailing lists might be useful so that you can send mass mailings from your BlackBerry.

Examples

- Put all of your team members in a group to instantly notify them of project updates.

View Video Tutorials and Free Tips at www.MadeSimpleLearning.com

BlackBerry Torch Made Simple

- Let's say you're about to have a baby. Put everyone in the notify list into a "New Baby" group. Then you can snap a picture with your BlackBerry and instantly send it from the hospital!

Creating and Using a Group Mailing or SMS List

1. Start the contact list (address book) by clicking the **Contacts** icon.
2. Press the **Menu** key, scroll to **New Group**, and click. 
3. Type a name for your new group.
4. Press the **Menu** key again, and click **Add Member**.

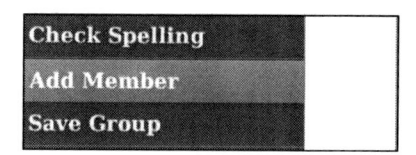

TIP: You can make both an SMS text group and an email group. Make sure each email group member has a **valid email address** and each SMS text group member has a **valid mobile phone number**, otherwise you will not be able to contact them from the group.

5. Scroll to the contact you want to add to that group and click.
6. Choose which Address to use.
7. Continue to add contacts to that group or make lots of groups and fill them using the previous steps.

TIP: You can add either mobile phone numbers (for SMS groups) or email addresses for email groups. We recommend keeping the two types of groups separate. In other words, have an SMS-only group and an email–only group. Otherwise, if you mix and match, you will always receive a warning message that some group members cannot receive the message.

Sending an Email to the Group

Just use the group name as you would any other name in your address book. If your group name was "My Team" then you would compose an email and address it to "My Team." Notice that after the email is sent, there is a separate **To** field for each person you have added to the group.

Chapter 11: Manage Your Calendar

The BlackBerry calendar is both intuitive and powerful. In this chapter we will show you how to add events, schedule individual and recurring appointments, accept meeting invitations, and search and utilize all the features of your **Calendar** app.

Organizing Your Life with Your Calendar

For many of us, our calendar is our lifeline. Where do I need to be? With whom am I meeting? When do the kids need to be picked up? When is Martin's birthday? The calendar can tell you all these things and more.

The calendar on the BlackBerry is really simple to use, but it also contains some very sophisticated options for the power user.

Ways to Get Your Calendar on Your BlackBerry

You can manually add calendar appointments one at a time as we show you in this chapter. You can also sync your computer's calendar with your BlackBerry.

If your BlackBerry is tied to a BlackBerry Enterprise Server, the synchronization is wireless and automatic, see Chapter 1 for steps to connect to an "Enterprise Account."

Otherwise, you will use either a USB cable or Bluetooth wireless to connect your BlackBerry to your contacts and calendar on computer to keep it up to date. For Windows PC users, see Chapter 2: "Windows PC Set Up;" Apple Mac computer users, see Chapter 4: "Apple Mac Set Up."

If you use **Gmail** (Google Mail), you can use the Google Sync program to wirelessly update your calendar on your BlackBerry with your address book from Gmail for free!

Learn how in the "Using Google Sync for Contacts and Calendar" section at the end of this chapter.

Switching Views and Days in the Calendar

The calendar is where you look to see how your life will unfold over the next few hours, days, or weeks (see figures 11-1 and 11-2). It is quite easy to change the view if you need to see more or less time in the Calendar screen.

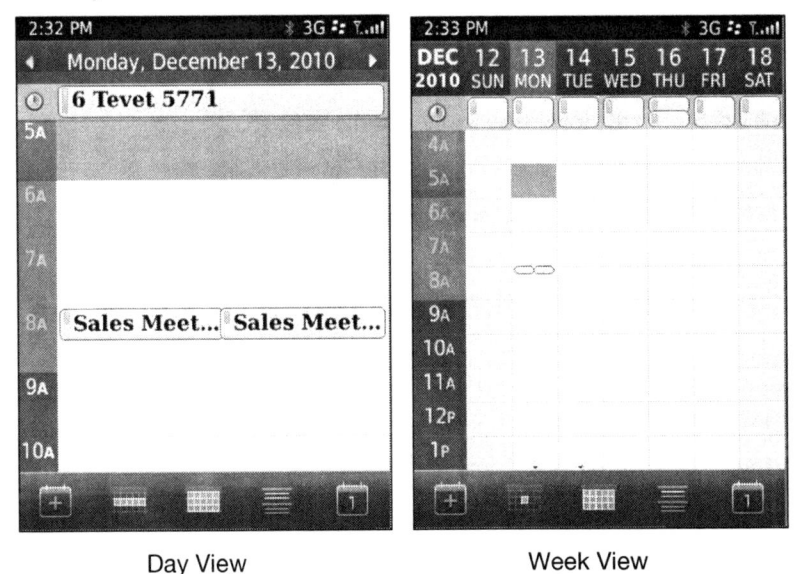

Day View — Week View

*Figure 11-1. Day and week views in the **Calendar** app*

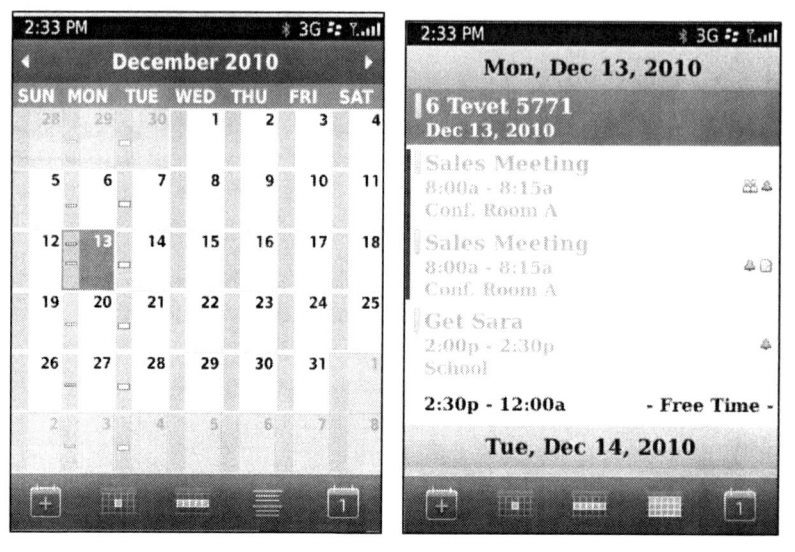

Month View   Agenda View

Figure 11-2. Month and agenda views in the Calendar app

Swiping to Move Between Days

Navigate to your **Calendar** icon, and click. The default view is the Day view, which lists all appointments for the current calendar day.

Swipe your finger left or right to a previous day or an upcoming day.

Notice the date changes at the top of the screen.

Using the Calendar Keyboard Shortcuts

You can enable one key shortcuts that allow you to switch between views and do other functions on your calendar. If you like using your slide-out keyboard, then you will want to turn on these shortcuts.

1. Start your **Calendar**.
2. Press the **Menu** key and select **Options**.
3. Select **Calendar Display and Actions**.
4. Scroll down and uncheck the box next to **Enable Quick Entry**.

Now, you can use all the slide-out keyboard one-letter shortcuts. Here are a few of the more common ones: **D** = Day View, **M** = Month View, **A** = Agenda View, **W** = Week View, **N** = Next Day, **P** = Previous Day.

CHAPTER 11: Manage Your Calendar

> **TIP:** Check out the **Calendar Keyboard Shortcuts** in Part 4 at the end of this book for a complete listing of these **Calendar** shortcuts as well as other useful shortcuts on your Torch.

Using the On-Screen Soft Keys

The calendar has some soft keys at the bottom (or the top) of the screen (Figure 11-3). In the Day View screen, you will see keys for scheduling a new appointment, viewing the month, looking at today, or moving to the previous or next day. Just hover to see the function of the soft key revealed and then click.

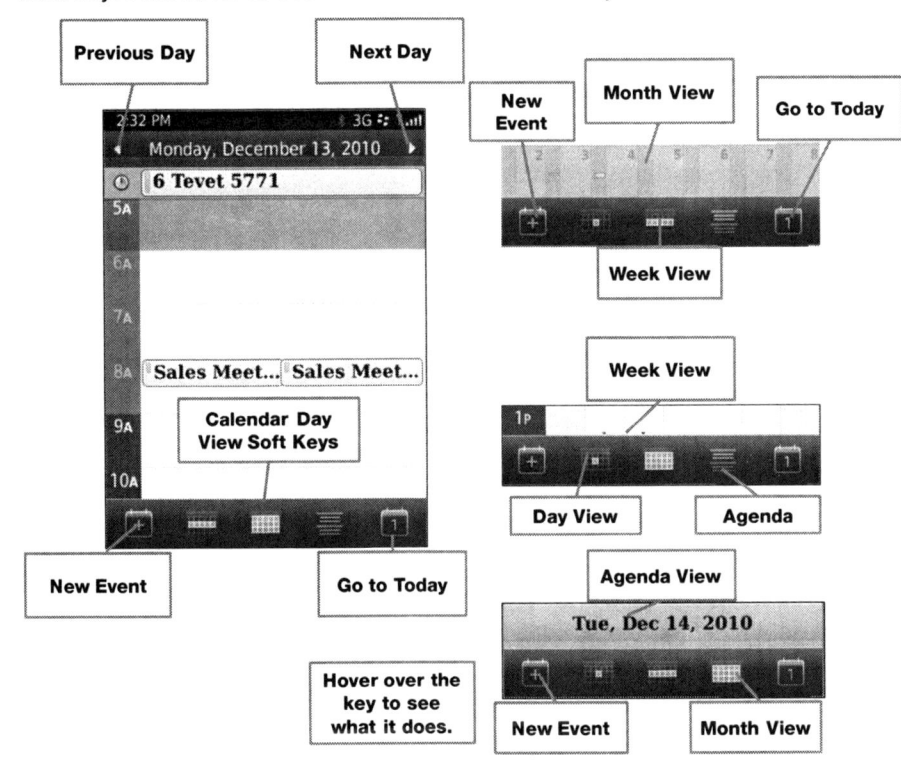

Figure 11-3. Soft keys in the Calendar app

Scheduling Appointments

Putting your busy life into your BlackBerry is quite easy. Once you start to schedule your appointments or meetings, you will begin to expect reminder alarms to tell you where to go and when. You will wonder how you lived without your BlackBerry for so long!

Quick Scheduling

It is amazingly simple to add basic appointments (or reminders) to your calendar.

1. In Day view, move the cursor to the correct day and time by swiping and gently touching (but don't click). The **Trackpad** can help you move the selected hour.

2. Start Quick Scheduling by pressing the **Enter** key on your slide-out keyboard.

3. Simply type your event in the Day View screen.

4. When you're done, press the **Enter** key to save the new event.

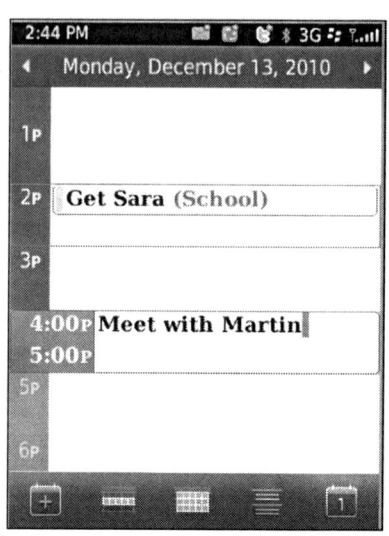

TIP: Quick Scheduling is so fast you can even use your calendar for reminders like:

Pick up the dry cleaning, Pick up Chinese food, Pick up dog food

Detailed Scheduling

1. Click the **Calendar** icon.

2. Touch or click on a time slot to start scheduling. Or, in any view, press the **Menu** key and select **New Appointment** to get into the New Appointment screen.

3. Type the subject and optional location. Click **All Day Event** if it will last all day, like an all-day conference.

CHAPTER 11: Manage Your Calendar

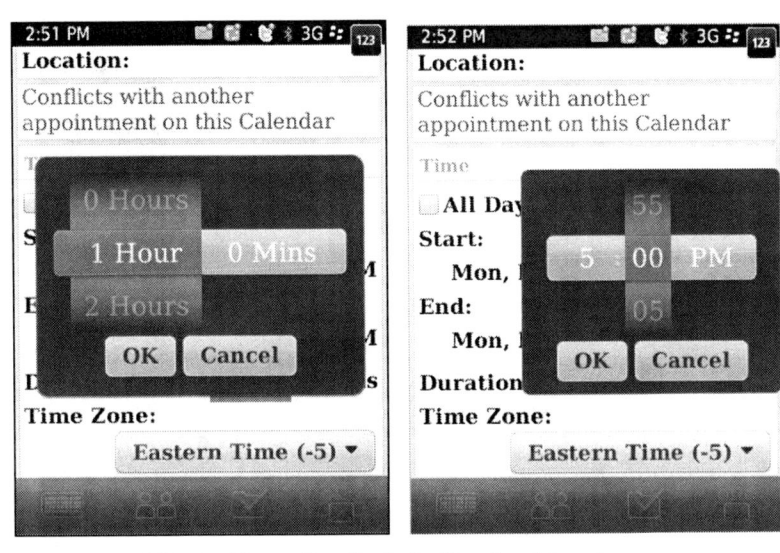

Figure 11-4. Adjusting Time or Duration in the New Appointment screen

4. Click the field you need to change, and then just swipe your finger or move the **Trackpad** up or down to change the field (see Figure 11-4).

TIP: Just slide out your keyboard and type numbers to set specific dates or times. You can also use the on-screen keyboard if it is visible.

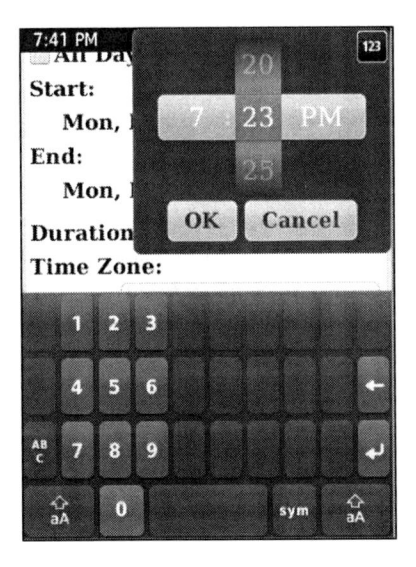

5. Alternatively, you can skip changing the end time of the appointment, and instead just change the length of the appointment by scrolling to the **Duration** field and putting in the correct amount of time.

View Video Tutorials and Free Tips at www.MadeSimpleLearning.com 301

6. Set a reminder alarm by clicking **Reminder** and setting the reminder time for the alarm from five minutes prior to nine hours prior.

TIP: The default reminder time is usually 15 minutes, but you can change this by going into your Calendar Options.

7. If this is a recurring appointment, do the following: Click **Recurrence** and select **None**, **Daily**, **Weekly**, **Monthly**, or **Yearly**. Adjust the additional settings below like how many days per week, interval of the recurrence as well as when the meeting stops recurring.

8. Mark your appointment as Private by selecting the **Mark as Private** check box.

9. If the meeting is a conference call, just put a check in the conference call box. You can then add the phone number and access code as well as participants.

10. If you would like to include notes with the appointment, simply input them at the bottom of the screen.

11. Press the **Menu** key and select **Save**.

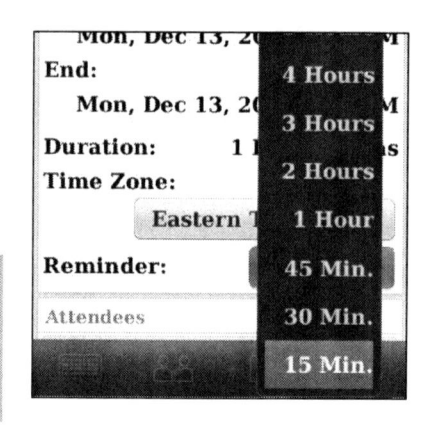

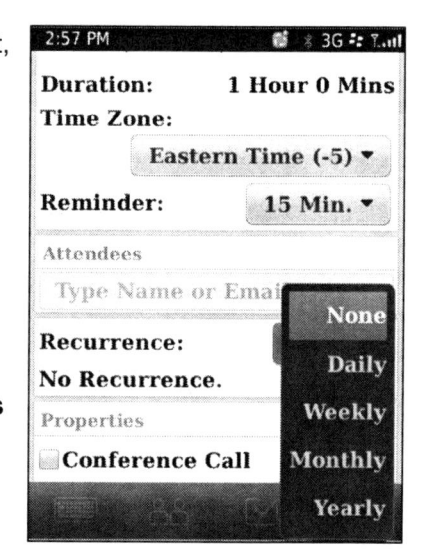

Customizing Your Calendar with Options

You can change a number of things to make your calendar work exactly the way you need using Calendar Options.

1. From the **Calendar**, press the **Menu** key and select Options.

2. Next, touch or click **Calendar Display and Actions** at the top of the screen.

Changing Your Start and End Time on Day View

If you are someone that has early morning or evening appointments, the default 9am–5pm calendar will not work well. You will need to adjust the **Start of Day** and **End of Day** fields in the Options screen. These options are up at the top of the screen, under **Formatting**.

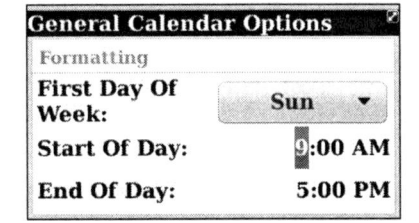

TIP: If you slide out your keyboard, then you may also use the number keys on your keyboard to type the correct hours (e.g., type **7** for 7:00 and **10** for 10:00).

Changing Your Initial View

If you prefer the **Agenda** view, **Week** view, or **Month** view, instead of the default **Day** view when you open your calendar, you can set that in the Options screen. Click the drop-down list next to the **Initial View** field to set these.

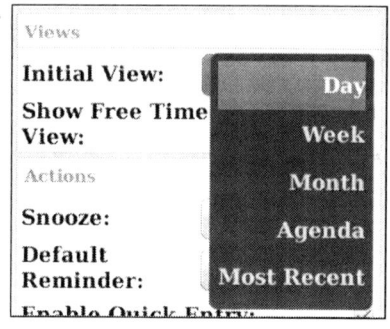

Default Reminder (Alarm) and Snooze Times

If you need a little more advanced warning than the default 15 minutes, or a little more snooze time than the default 5 minutes, you can change those also in the **Options** screen.

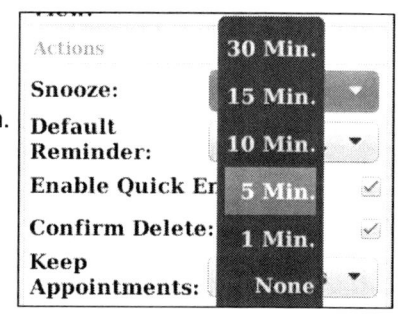

Scheduling Conference Calls

The BlackBerry has some very useful built-in features for scheduling and joining conference calls. You can actually pre-load the conference call participant or moderator dial-in and access code numbers so when the alarm rings as shown, you can simply click the **Join Now** button.

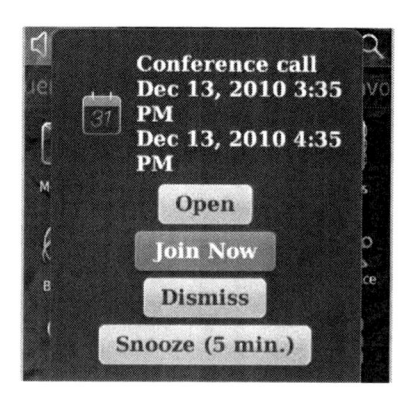

CHAPTER 11: Manage Your Calendar

How do you make this happen?

It's easy—in the **Appointment Details** screen when you schedule a new appointment, select the **Conference Call** check box. Then you can type the access numbers as shown.

> **TIP:** You might want to set the reminder at "**0 Min.**" so you are reminded to dial-in right on time, instead of 15 minutes early (the default reminder).

> **TIP:** If you use the same conference call dial-in service regularly, then you should pre-load the information in your Calendar Options screen.
>
> From the calendar, press the **Menu** key and select **Options**.
>
> Then select **Conference Calling**.
>
> Type in the moderator and participant basic information. This is used to pre-load the conference call numbers on all scheduled appointments. If it changes, then you can alter the individual appointment.

Copying and Pasting Information into Your Calendar

The beauty of the Blackberry is how simple it is to use. Let's say that you wanted to copy part of a text in your email message and paste it into your calendar. A few good examples are:

- Conference call information via email

View Video Tutorials and Free Tips at www.MadeSimpleLearning.com   305

- Driving directions via email
- Travel details (flights, rental cars, hotel) via email

First, scroll to or compose an email message from which you want to copy and open it.

Method #1: Menu > Select

1. You can use this method only if you are typing or editing text.
2. Press the **Menu** key and choose **Select**. Then highlight (by touching the screen) the text you wish to copy.
3. Drag the handles that appear to adjust your selection.
4. Once it is highlighted, press the **Menu** key again and choose **Cut** or **Copy**.

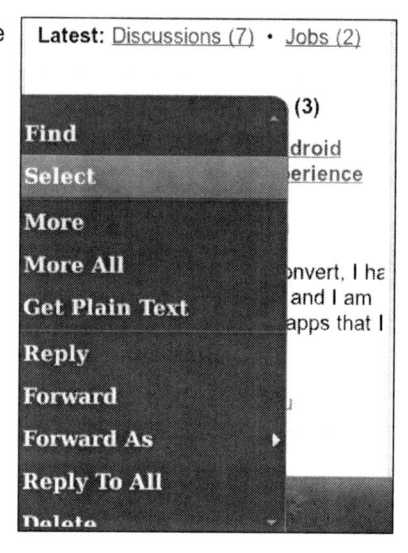

Method #2: Multi-Touch

The multi-touch method allows you to touch the screen at two points and then highlight the text in between. It takes a little practice, but once you master it you will be copying like a pro!

1. Drag the handles that appear to adjust your selection (see Figure 11-5).
2. We recommend placing your thumb at one end of the line you wish to copy and your forefinger at the other.
3. Press the **Copy** soft key at the bottom to copy the text.

CHAPTER 11: Manage Your Calendar

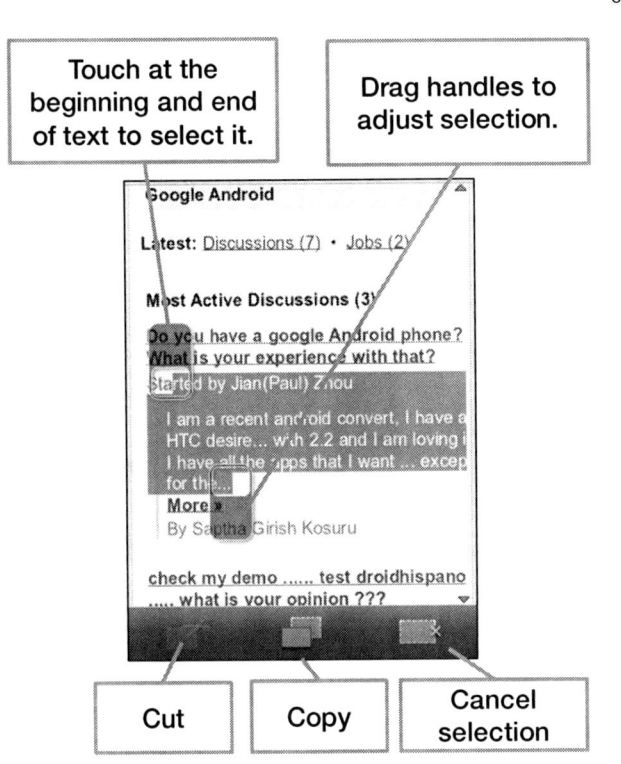

Figure 11-5. Select by Multi-Touch, Copy and Paste

4. Press the **Red Phone** key to jump back to your **Home** screen and tap the **Calendar** icon to start it.

View Video Tutorials and Free Tips at www.MadeSimpleLearning.com **307**

5. Schedule a new appointment by clicking a time slot in the Day view.

6. Move the cursor to the field in which you want to insert the text that is on the clipboard. Then, click the **Paste** icon at the bottom of the screen or press the **Menu** key and select **Paste**.

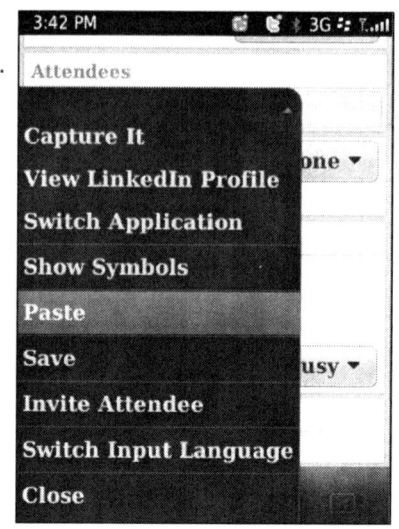

7. Finally you will see your information pasted into the calendar appointment.

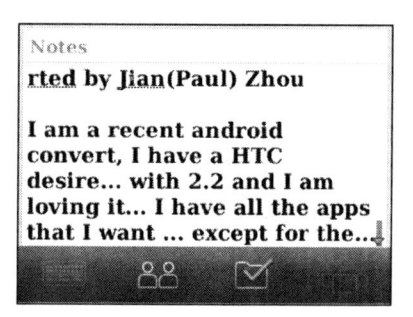

8. Press the **Menu** key and **Save** your appointment.

Now your text is in your calendar. Now it's available exactly when you need it. Gone are the days of hunting for the conference call numbers or driving directions, or asking yourself "What rental car company did I book?"

CHAPTER 11: Manage Your Calendar

Dialing a Scheduled Conference Call

What is really great is that if you put a phone number into the conference call field of a calendar Appointment

If you want to call the number before the alarm rings, open the event and then click the underlined phone number. Or select Call from the menu as shown.

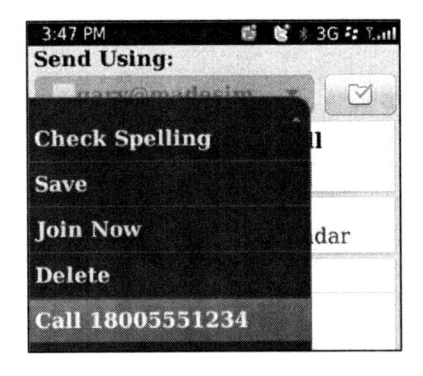

If you place the phone number in the conference call field and wait for the alarm to ring, then you can click the **Join Now** to call the number.

TIP: You can also set a calendar event as a conference call so that you get a **Join Now** menu item.

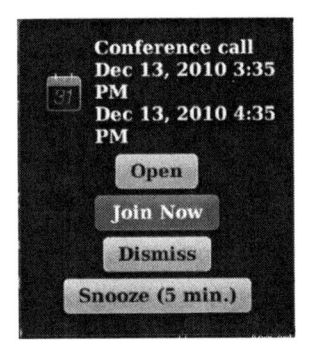

Changing the Calendar Alarm Sound

1. Click the **Sound Profile** icon in the upper left corner of your Home screen.

2. Tap the **Change Sounds and Alerts** button at the bottom and touch or click.

3. Tap **Sounds for Selected Profile.**

4. Tap **Events – Reminders** to see the **Calendar** and tap on **Calendar**.

View Video Tutorials and Free Tips at www.MadeSimpleLearning.com   309

5. Adjust the settings for sounds, LED, Vibration and more using this screen.

TIP: Check out our "Sounds" section of Chapter 7: "Personalize Your Torch" for more information on adjusting **Sound Profiles**.

6. Press the **Menu** key and select **Save**.

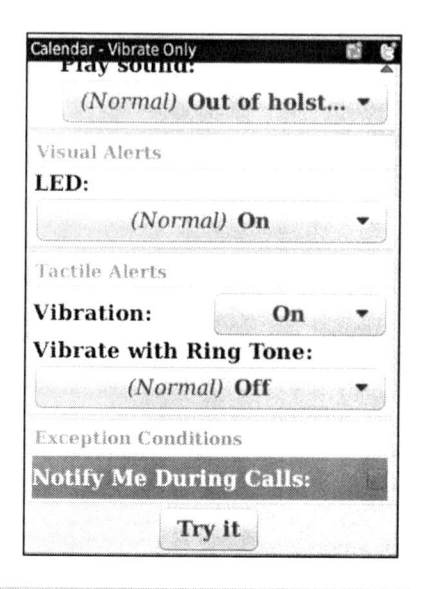

Tip: We highly recommend setting **Vibration** to **On** and **Vibrate with Ring Tone** to **Off** as shown. This allows you to grab your BlackBerry while it's vibrating and (usually) take some action before it starts to make noise.

This can save you some embarrassment if you are in a location where noise would be a problem.

Snoozing a Ringing Calendar or Task Alarm

When a calendar or task alarm rings, you can open it, dismiss it or snooze it.

To make sure you have the Snooze option active:

1. Open the **Calendar** application (see Figure 11-6).
2. Press the **Menu** button and scroll down to **Options**.
3. Click **Calendar Display and Actions** .
4. Look under the **Actions** sub-heading and click in the **Snooze** field.

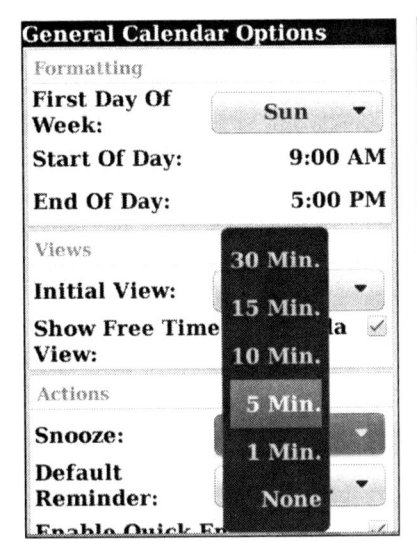

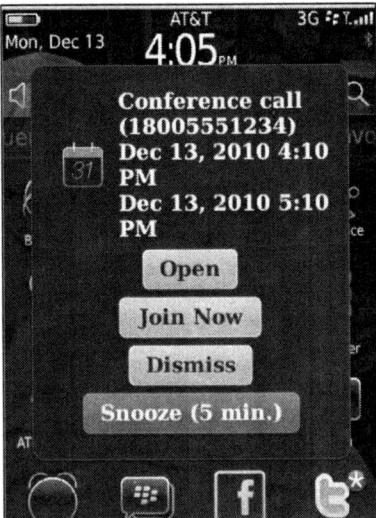

Figure 11-6. Snoozing a Calendar appointment

If you don't see a **Snooze** button as in the right screen of Figure 11-6, then you need to change the setting in your Calendar or Task options screen to some other value than **None**.

Clicking the **Snooze** option does just that—it will snooze five minutes, ten minutes, or whatever you have set in the Calendar or Task options screen.

Working with Meeting Invitations

Now with your BlackBerry you can invite people to meetings and reply to meeting invitations. If your BlackBerry is connected to a BlackBerry Enterprise Server ("BES"), you may also be able to check your invitees' availability for specific times as well.

Inviting Someone to Attend a Meeting

1. Create a new appointment in your calendar or open an existing meeting by clicking it.

2. Touch or click in the field under **Attendees**, follow the prompts to find a contact, and click that contact to invite them. Just start typing their name and click on the appropriate email address.

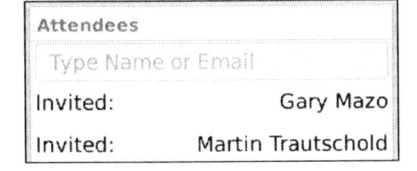

3. Follow the same procedure to invite more people to your meeting.
4. Click the **Menu** key and select **Save**.

Viewing Availability of Invitees

If your BlackBerry is connected to a BlackBerry Enterprise Server, then you will also see a button under the Attendees called **View Availability**.

1. Tap **View Availability** to see when the people you want to invite are available (based on their corporate calendars on the server).

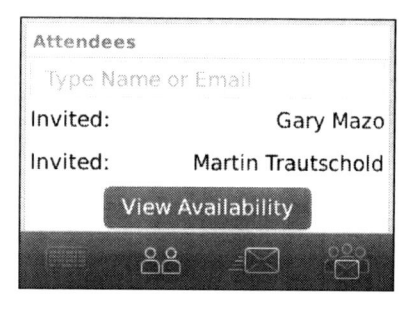

2. From this screen, you will see a list of the availability of all the invited attendees. From this screen you can:

- Scroll left or right with your finger or using the **Trackpad** to see more of the time line. Scheduled events are shown as blocks in the time line for each person.
- Tap **Next Available Time** to search the calendars to see the next time everyone might be free.

3. When done, tap **Save**.

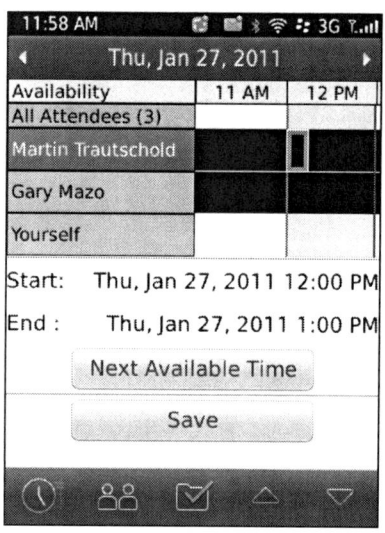

Respond to a Meeting Invitation

You will see the meeting invitation in your messages (email inbox).

Open the invitation by clicking it, and then press the **Menu** key.

Several options are now available to you. Click either **Accept** or **Accept with Comments**, or **Tentative** or **Tentative with Comments**, or **Decline** or **Decline with Comments**.

Changing the List of Participants for a Meeting

1. Click the meeting in your **Calendar** application.
2. Navigate to the **Accepted** or **Declined** field and click the contact you wish to change.
3. The options **Invite Attendee, Change Attendee,** or **Remove Attendee** are available. Just click the correct option.

Sending an Email to Invitees

1. Navigate to the meeting in your calendar and click it to open it.
2. Click **Email all Attendees** and compose your email message.
3. Click the **Menu** key and select **Send**.

Using Google Sync for Contacts and Calendar

Google Calendar BETA   Gmail by Google BETA

> **CAUTION:** You can sync your Google contacts and calendar using the BlackBerry Internet Account Email Setup (see Chapter 1). We would strongly suggest that you **use the Google Sync app instead** of the BlackBerry Internet Account Setup to sync your Google Contacts and Calendar. We have seen problems with the Contacts sync portion of the BlackBerry Internet Account sync (in our case the first line of the street address disappeared from all our Google Contacts!).

The great thing about using **Google Sync** for BlackBerry is that it provides you a full two-way wireless synchronization of your BlackBerry Calendar and Contacts with your Google Calendar and Address Book. What this means is anything you type in your Google Calendar/Addresses "magically" (wirelessly and automatically) appears on your BlackBerry Calendar/Contacts in minutes! The same thing goes for contacts or calendar events you add or change on your BlackBerry—they are transmitted wirelessly and automatically to show up on your Google Calendar and Contacts.

> **NOTE:** Calendar events you have added on your BlackBerry prior to installing the Google Sync application don't get synced. However, contacts on your BlackBerry prior to the install are synced.

This wireless calendar update function has previously been available only with a BlackBerry Enterprise Server (whether in-house or hosted).

Getting Started with Gmail and Google Calendar

First, if you don't already have one, you must sign up for a free Google mail (Gmail) account at www.gmail.com.

Then, follow the great help and instructions on Google to start adding address book entries and creating calendar events on Gmail and Google Calendar on your computer.

Installing the Google Sync Program

1. Click your BlackBerry **Browser** icon to start it.

2. Type this address in the address bar on the top: http://m.google.com/sync.

3. Then click the **Install Now** link on this page.

> **NOTE:** This page may look slightly different when you see it.

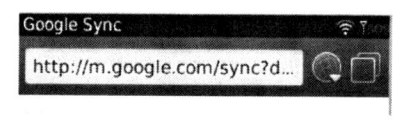

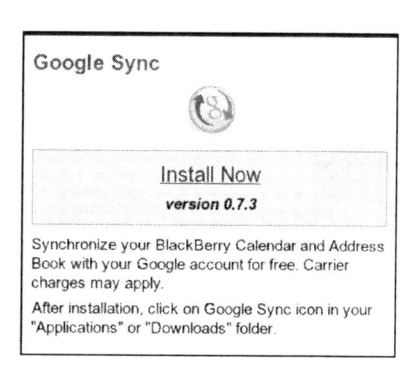

CHAPTER 11: Manage Your Calendar

4. And click the **Download** button here.

> **NOTE:** The version numbers and size will probably be different when you see this page.

After you download and install it, you can run it or just click **OK**. If you clicked **OK** and exited the browser, you may need to check in your **Downloads** folder for the new **Google Sync** icon.

After you successfully login, you will see a page similar to this one describing what Google Sync will sync to and from your BlackBerry.

You will see some sync status screens.

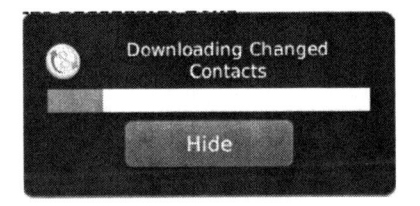

Notice at the bottom of this screen, you will see details on which fields (or pieces of information) from your BlackBerry will be shared or synchronized with your Google address book.

Finally, click **Sync Now** at the bottom of this page.

View Video Tutorials and Free Tips at www.MadeSimpleLearning.com   **315**

Then, finally, you will see a screen like this. Click **Summary** to see the details of what was synced.

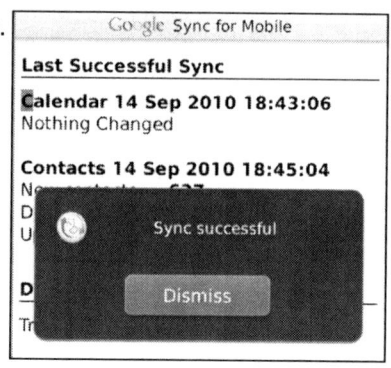

From this point forward, the **Google Sync** program should run automatically in the background. It will sync every time you make changes on your BlackBerry or at a minimum every two hours.

From both the BlackBerry **Calendar** and **Contacts** icons, you will see a new **Google Sync** menu item. Select that to see the status of the most recent sync.

You can also press the **Menu** key from the Sync status screen to **Sync Now** or go into **Options** for the sync. If you are having trouble with the sync, check out Chapter 23: "Troubleshooting" or view Google's extensive online help.

How The Sync Looks

Here are views of the same calendar items in Google and BlackBerry.

Google Calendar on the computer:

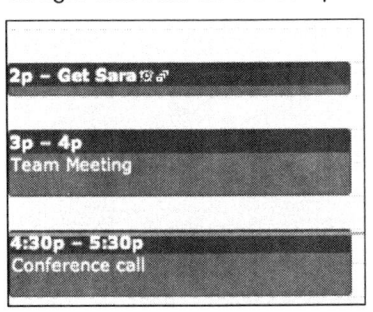

BlackBerry Calendar:

Notice the calendar events from Google Calendar are now on your BlackBerry. Anything you add or change on your BlackBerry or Google calendar will be shared both ways going forward. Automatically!

Google Address Book on your computer:

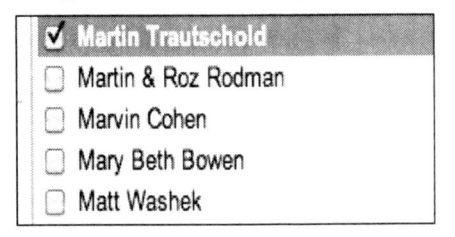

BlackBerry Contact List:

Notice all contacts from your BlackBerry contact list are now in your Gmail address book. Also, all contacts from your Gmail address book are now in your BlackBerry.

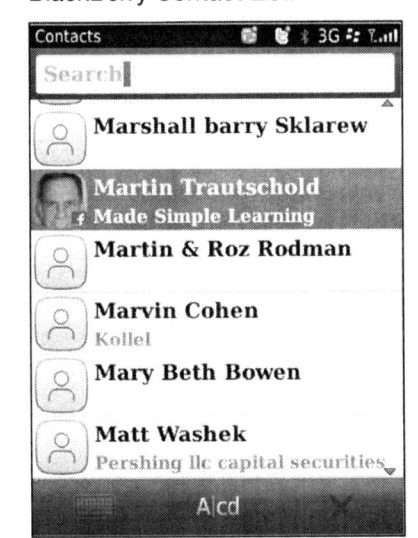

Chapter 12:
Text and MMS Messaging

As you may be aware, a key strength of all BlackBerry devices is their messaging abilities. We have covered email extensively and now turn to Text and MMS messaging. In this chapter, we will show you how to send and view both text messages (SMS) and multimedia messages (MMS.) We will show the various apps from which you can send a message and offer some advice on how to keep your messages organized.

Text and Multimedia Messaging

SMS stands for Short Messaging Service (text messaging) and MMS stands for Multimedia Messaging Service (sometimes called "Picture Messaging" or "Video Messaging"). MMS is a short way to say that you have included pictures, sounds, video, or some other form of media right inside your email message (not to be confused with regular email when media is an attachment to an email message). BlackBerry is beautifully equipped to use both of these services—learning them will make you more productive and make your BlackBerry that much more fun to use.

> **TIP:** SMS and MMS sometimes cost extra! Many phone companies charge extra for text messaging and MMS multi-media messaging, even if you have an "unlimited" BlackBerry data plan. Typical charges can be $0.10 to $0.25 per message. This adds up quickly!
>
> The solution is to check with your carrier about bundled SMS/MMS plans. For just $5–10 per month you might receive several hundred or thousand or even unlimited monthly SMS/MMS text messages.

SMS Text Messaging on Your BlackBerry

Text messaging has become one of the most popular services on cell phones today. You are hard pressed to find a teen ager (or their parents) who do not communicate via text messages these days!

The concept is very simple. Instead of placing a phone call, send a short message to someone's handset. It is much less disruptive than a phone call, and you may have friends, colleagues, or co-workers who do not own a BlackBerry—so email is not an option.

One of the authors uses text messaging with his children all the time—this is how their generation communicates. "R u coming home 4 dinner?" "Yup." There you have it—meaningful dialogue with an eighteen-year-old—short, instant, and easy.

Composing Text Messages

Composing an SMS message is much like sending an email. The beauty of an SMS message is that it arrives on virtually any handset and is so easy to respond to.

Sending from the Text Messages or Messages App

1. Click your **Text Messages** icon. (You can also send from the **Messages** app.)

2. Press the **Compose** soft key at the bottom left of the screen . (Select **Text Message** if prompted.)

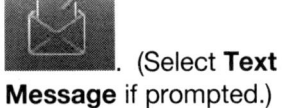

3. You can also press the **Menu** key and choose **Compose Text Message**.

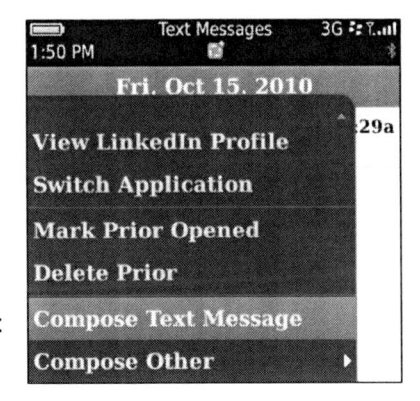

4. Begin typing in a contact name or simply type someone's mobile phone number. If you are typing a name, when you see the contact appear, touch or click it.

5. If the contact has multiple phone numbers, the BlackBerry will ask you to choose which number.

6. In the main body (where the cursor is) just type your message like you were sending an email message.

> **CAUTION:** SMS messages are **limited to 160 characters** by most carriers. If you go over that in the BlackBerry, two separate text messages will be sent.

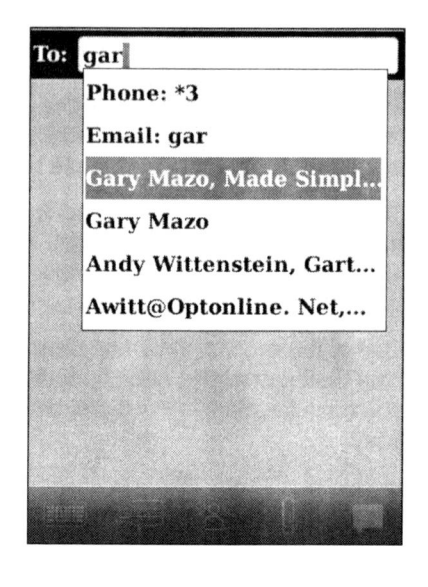

CHAPTER 12: Text and MMS Messaging

7. If you want to add some fun to your text message, click the smiley face to bring up a list of potential emoticons.

8. Tap the emoticon you want to use. Note at the top of the screen there are also the text shortcuts for each emoticon. For example, the shortcut for the "Cool" face is **B-)**.

9. To send the message, just press the **Enter** key or touch the **Send** soft key at the bottom to send the message. That's all there is to it.

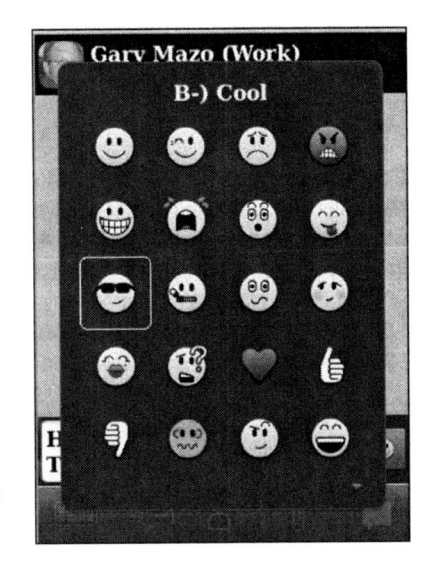

Sending from the Contact List

1. Click your **Contacts** icon.

2. Type a few letters to find the person to whom you want to send your SMS message.

3. With the contact highlighted from the list, press the **Menu** key and you will see one of your menu options is **Text**, followed by the contact name.

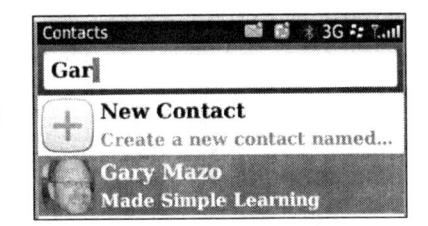

Text Menu Commands

As with the email feature, there are many options via the menu commands in **Text** messaging.

The BlackBerry adds an **Text Messages** icon to the Home screen. One way to initiate SMS messaging is to click the icon.

View Video Tutorials and Free Tips at www.MadeSimpleLearning.com   **321**

BlackBerry Torch Made Simple

> **NOTE:** On some other BlackBerry models, the icon says **SMS and MMS**.

On the Compose Text screen, when you are typing your text message, press the **Menu** key. The following options are perhaps to most popular ones you would use shown in Figure 12-1 are available to you.

Switch Application	Brings up the Switch Applications pop-up window so you can select another icon for multitasking (see Chapter 1)
Add Smiley	Add an emoticon to your Text message.
Attach	Touch or click to add a Picture, Video, Location, Audio, Voice Note, Contact or appointment.
View Contact	Touch or click to see the full contact information for the recipient.
Edit Recipients	Allows you to change or add recipients to the message.
Options	Brings up the SMS option menus
Switch Input Language	Allows you to type your SMS message in another language
Close	This is the same as pressing the **Escape** key. It closes your **Text** window and asks if you want to save changes.

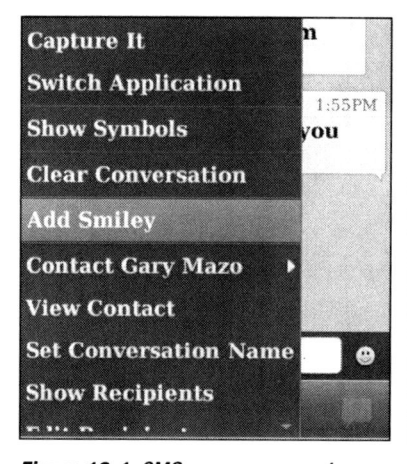

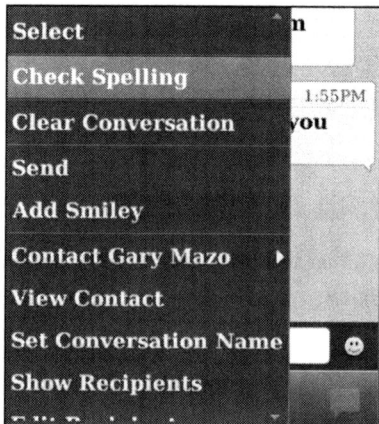

Figure 12-1. SMS menu commands

322

Once you actually start typing your Text message and press the **Menu** key, a new option appears—**Check Spelling**—giving you the power of the BlackBerry spelling checker in your SMS messages.

Opening and Replying to SMS Messages

Opening your Text messages couldn't be easier—the BlackBerry makes it simple to quickly keep in touch and respond to your messages.

1. Navigate to your waiting messages from either the **Text Messages** icon on the **Home** screen or your **Messages** app, and click the new Text message.

2. If you are in the midst of a dialogue with someone, your messages will appear in a threaded message format, which looks like a running discussion.

3. Press the **Menu** key if you want to edit or change or add recipients to the message.

4. The cursor appears in a blank field—type your reply, click the **Menu** key, or the trackpad, and select **Send**.

> **TIP:** Need to find a Text or MMS message? Go to the "Searching for Messages" section of Chapter 9: "Email Like a Pro."

MMS Messaging (Picture Messaging)

MMS stands for multimedia messaging, which includes pictures, video, and audio.

> **NOTE:** Not all BlackBerry devices or carriers support MMS messaging, so it is a good idea to make sure that your recipient can receive these messages before you send them.

Sending MMS from the Message List

Perhaps the easiest way to send an MMS message is to start the process just like you started the Text Message process earlier:

1. Compose a new text message from the **Messages**, **Text Messages** or **Contacts** apps.

2. Press the **Menu** key and select **Attach**. Now you can select from one of the following: **Picture**, **Video**, your **Location**, **Audio**, **Voice Note**, **Contact** or **Appointment**.

3. If you selected a picture, video, audio or voice note, then you will need to navigate to the item you want to attach and select it.

4. Here we see the selected picture of the new car has been sent in this multi-media message.

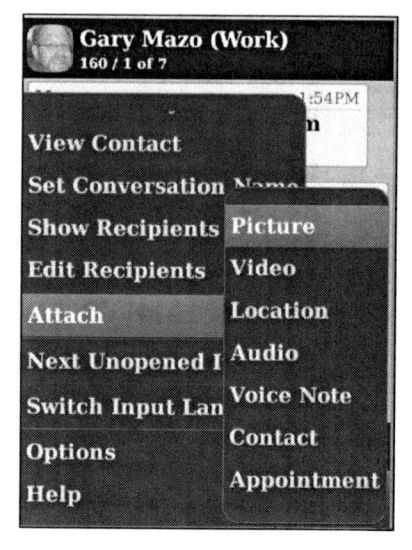

CHAPTER 12: Text and MMS Messaging

5. You can add a subject and text in the body of the MMS. When finished, just press the **Menu** key and click **Send** or press the trackpad and select **Send**.

Sending a Media File from the Media Icon

This might be the easier and more common way for you to send media files as MMS messages.

1. Scroll to the **Media** section of your **Home** screen Where you will see your Music, Videos, Pictures and other media files.

2. Touch or click **Pictures**, and find the picture either on your device or media card that you wish to send.

3. Choose the folder that contains the picture you are looking for. Just highlight the picture (no need to click it) and press the **Menu** key.

4. Scroll down to **Send**, and click. You will then be directed to choose the method for sending the picture; either **Email**, **Text Message**, **Messenger Contact**, **Twitter** or other services that may be installed.

5. Once you select the method of delivery, choose your recipient as you did above.

TIP AND CAUTION: If you do not have a specific MMS or Picture Messaging plan, then you may be charged significantly more than a basic Text Message to send your Picture message. (Could be $0.50 or $1.00 for each picture message sent).

View Video Tutorials and Free Tips at www.MadeSimpleLearning.com

6. Type a subject and any text in the message, press the **Menu** key or the trackpad, and send it.

Advanced MMS Commands

While you are composing your MMS message, press the **Menu** key and scroll down to **Options**.

Scroll down to MMS.

From this screen you can set your phone to always receive multimedia files by setting the **Multimedia Reception** line to say **Always**.

You can also set your **Automatic Retrieval** to occur **Always** or **Never**.

You can adjust the importance at the bottom of the option screen to **Low**, **Normal** or **High**.

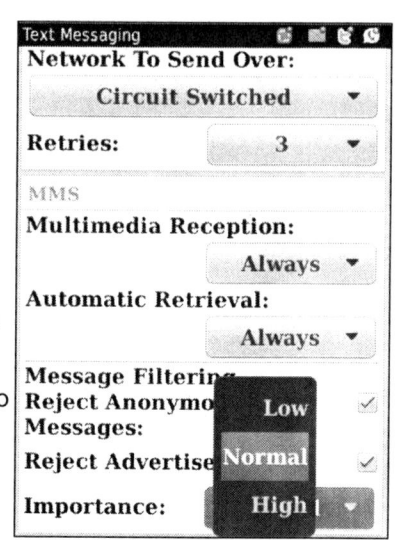

You can also select to **Reject Anonymous Messages** (messages without a return phone number, these are usually ads) and **Reject Advertisements** by checking both boxes.

Messaging Troubleshooting

Here are a few simple things you can do to help get your messaging going again. If they don't work, then try out some more techniques in Chapter 23: "Troubleshooting."

Host Routing Table "Register Now"

These troubleshooting steps will work for MMS, SMS, email, web browsing—anything that requires a wireless radio connection.

CHAPTER 12: Text and MMS Messaging

Host Routing Table—Register Now:

1. From your Home screen, click the **Options** icon, then scroll down to **Device** and touch or click.

2. Click **Advanced System Settings**, and then scroll to **Host Routing Table**.

3. Highlight the one listing in the routing table that is in **Bold** print.

4. Press the **Menu** key, and then click **Register Now**.

NOTE: This image shows "GPRS US." You will see your own wireless carrier name on the screen.

Perform a Soft Reset (Reboot)

Press and hold three keys simultaneously (ALT + CAP + DEL) until you see your BlackBerry screen go off for at least 10 seconds. This will reboot or reset your BlackBerry and many times will help re-establish network connectivity to fix issues with texting and picture texting.

Perform a Hard Reset (Battery Pull)

While the BlackBerry is still on, do a battery pull. Take off the back of the casing, remove the battery, wait 30 seconds, and then re-install it. Once the BlackBerry reboots, you should be all set for Text and MMS messaging.

TIP: Both Chapter 23: "Troubleshooting" and the Quick Start Guide at the beginning of the book has more detailed steps with pictures showing how to remove the battery.

View Video Tutorials and Free Tips at www.MadeSimpleLearning.com   327

Chapter 13: BlackBerry Messenger, PIN Messaging and More

You already know that your BlackBerry is an amazing messaging device. What you may not realize is that there are so many more ways to use "messaging" on the BlackBerry. In this chapter, we will look at the famous **BlackBerry Messenger** (BBM), **PIN** messaging (using the unique identifying PIN for each BlackBerry) and other instant messaging.

> **TIP:** PIN messages are always free, whereas other SMS text and MMS (multimedia messages) may be charged under an extra service plan by your phone company.

BlackBerry Messenger

BlackBerry Messenger is really designed from the ground up to be both a messaging application and a group collaboration software. We show you the group functionality later in this chapter.

> **TIP:** If you cannot find the **BlackBerry Messenger** icon, what to do.
>
> First, tap the Search icon and type Messenger to see if it appears. Then touch and hold the icon to make it a Favorite so you don't lose it again!
>
> If the icon does not appear, then go to BlackBerry App World and download it. Check out Chapter 20 for help with this.

Setting Up BlackBerry Messenger (BBM)

BlackBerry Messenger offers you a little more secure way of keeping in touch quickly with fellow BlackBerry users. Setup is very easy.

If you don't see the **Blackberry Messenger** icon, then press the **Menu** key, look for the **Instant Messenger** folder, and click it.

1. Click the **BlackBerry Messenger** icon after accepting the legal agreement.

2. Type your **Display Name** (the name that others will see) and click OK.

3. You will then be asked which is your contact your **Contact** list. This allows better functionality in BBM.

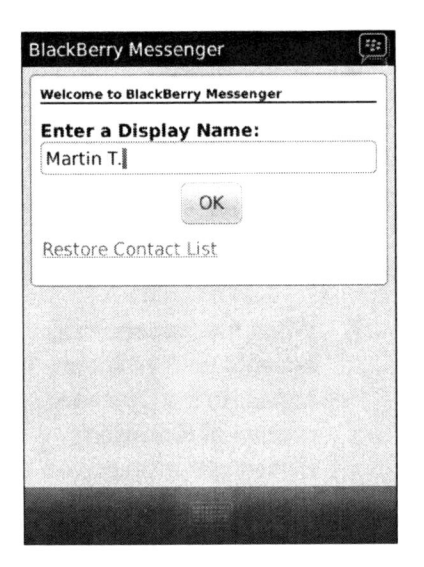

Adding Contacts to BlackBerry Messenger

Once your user name is setup, you need to add contacts to your BlackBerry Messenger group. In BlackBerry Messenger, your contacts are fellow BlackBerry users who have the BlackBerry Messenger program installed on their handhelds.

1. Navigate to the main **BlackBerry Messenger** screen and press the **Menu** key.

2. Select **Invite Contact**.

3. Then you will be asked if you want to invite the person via PIN or email message, scan a barcode or add a contact by text messaging. (We show you the barcode scanning later in this chapter.)

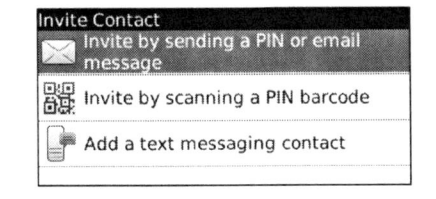

4. Start typing someone's name or email address to invite them. Adjust your **Send using** email account, if you desire.

5. Tap **Send**.

6. The message request shows up in your **Pending** group, under **Contacts**.

7. When the person finally accepts your invitation, they appear in the Contacts section of BlackBerry Messenger main screen as shown with **Gary M** here.

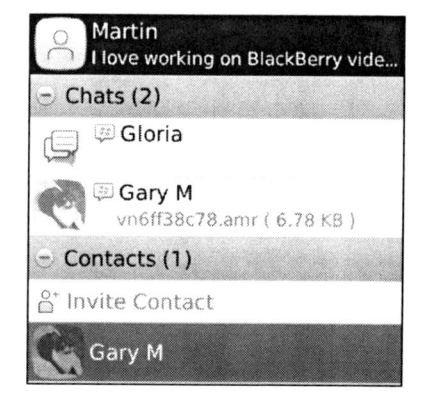

Invite Using Barcodes

Perhaps the most innovative feature of the new BlackBerry Messenger is the use of a unique barcode as a means of connecting to someone as a Messenger contact.

Each BlackBerry can generate its own unique barcode. To find your barcode, just start up the **Messenger** application, press the **Menu** key, and select **View My Profile**.

NOTE: You can also just tap on your display name at the top of your main screen.

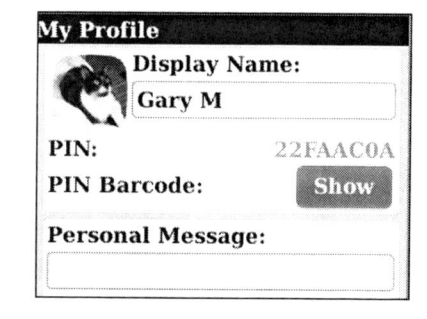

330

Scroll to **PIN Barcode** and click on **Show**. The screen will now show your unique barcode. Any other BlackBerry user (with BlackBerry Messenger 5.0 or 6.0 installed *and* a camera on his or her BlackBerry) can snap a picture of the barcode.

What needs to be done is for the other BlackBerry Messenger user to choose **Invite Contact** from their menu and then choose **Invite by scanning a PIN barcode**. This will activate their camera and they can take a picture of your code.

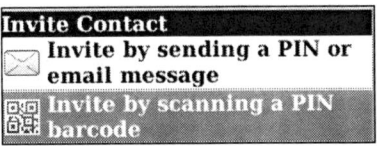

Responding to BlackBerry Messenger Invitations

You may be invited to join another BlackBerry user's Messaging group. You can either accept or decline this invitation.

You will receive your invitations via email or you can see them directly in BlackBerry Messenger.

1. Click your **BlackBerry Messenger** icon.

2. Scroll to your **Requests** group, highlight the invitation, and click.

3. A menu pops up with three options: **Accept**, **Decline**, or **Remove**.

4. Click **Accept** and you will now be part of the Messaging group. Click **Decline** to deny the invitation or **Remove** to no longer show the invitation on your BlackBerry.

BlackBerry Messenger Menu Commands

Open BlackBerry Messenger, go to your main screen, and press the **Menu** key.

Switch Application – Press this to multitask or jump to another application while leaving the **Messenger** application running.

New Multiperson Chat – Choose several BBM contacts to chat with at one time.

New Broadcast Message – Broadcast a quick message to as many BBM Contacts as you wish.

New Contact Category – Create a New Category for one or more BBM contacts.

New Group – Click to add a new messaging group such as Work, Family, or Friends.

Scan a Group Barcode – Member can display a group barcode from the **Group Details Screen.** You can scan the code to join the group.

Collapse All – This hides all group members.

Invite Contact – Use this to add people to your BBM directory.

Expand – This simply expands the dialogue screen, if the screen is already open. The menu command reads **Collapse All**.

> **TIP:** Clicking the main conversation screen does the same thing.

View My profile – View and/or change your name, change barcode settings, change status and display what you are listening to..

Options – Click to bring up your Options screen.

Close – Exit the **Messenger** application.

BlackBerry Messenger Options and Backup, Restore

If you wanted even more control of Messenger, you would press the **Menu** key and select **Options** (See Figure 13-1). On the Messenger Options screen, you can set the following:

1. Whether your BlackBerry will vibrate when someone **PING**s you (the default is **Yes**).

2. Replace BlackBerry Contact pictures with BBM Pictures.

CHAPTER 13: BlackBerry Messenger, PIN Messaging and More

3. Whether your requests can be forwarded by other people (the default is **yes.**)

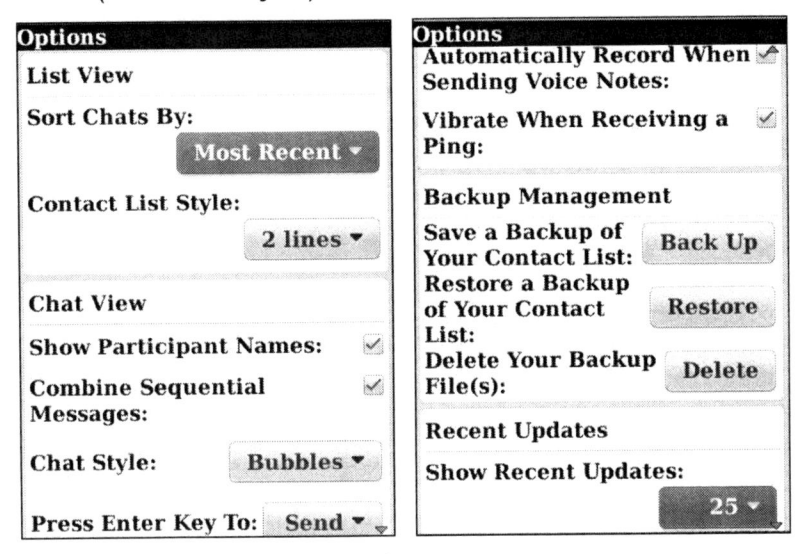

Options	
List View	
Sort Chats By:	Most Recent ▼
Contact List Style:	2 lines ▼
Chat View	
Show Participant Names:	✓
Combine Sequential Messages:	✓
Chat Style:	Bubbles ▼
Press Enter Key To:	Send ▼

Options	
Automatically Record When Sending Voice Notes:	✓
Vibrate When Receiving a Ping:	✓
Backup Management	
Save a Backup of Your Contact List:	Back Up
Restore a Backup of Your Contact List:	Restore
Delete Your Backup File(s):	Delete
Recent Updates	
Show Recent Updates:	25 ▼

Figure 13-1. BlackBerry Messenger options

Finally, you can also set **Backup Management** options. You can **Backup**, **Restore** contacts from a previous backup or **Delete** backup files from the **Options** menu.

Starting or Continuing Conversations and Emoticons

While messaging is a lot like text messaging, you actually have more options for personal expression and the ability to see a complete conversation with the Messaging program.

BlackBerry Torch Made Simple

Your conversation list is in your Main screen. Just highlight the individual with whom you are conversing and Tap them. The Conversation screen opens.

Just type the new message, click the **Trackpad**, or press the **Menu** key and select **Send**.

To add an emoticon to your message, press the **Emoticon** soft key button
and swipe your finger or use the **Trackpad** to move to the emoticon you wish to use (see Figure 13-2).

Just click the desired emoticon and it will appear in the message.

> **TIP:** You can also type the characters shown to get the emoticon you want—e.g., ":)" = Smile and "<3<3" = Love Struck.

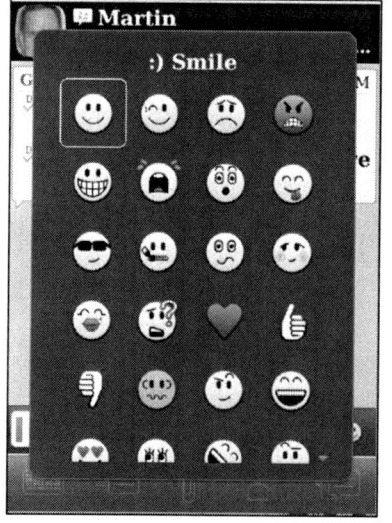

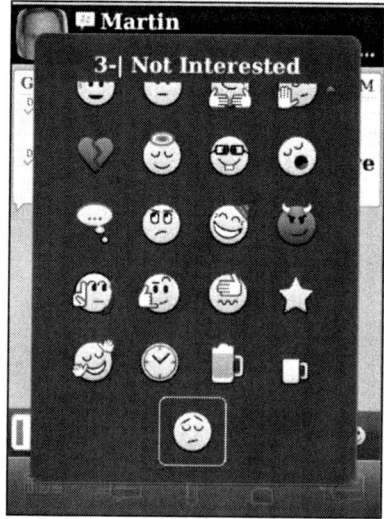

Figure 13-2. Emoticon options in BlackBerry Messenger

334

CHAPTER 13: BlackBerry Messenger, PIN Messaging and More

Sending Files to a Message Buddy

In the midst of your conversation, you can send a file very easily (at the time of publication, you were limited to sending only files that are images, photos, or sound files—ring tones and music).

1. Click the contact in your Conversation screen and open the dialogue with that individual.

2. Press the **Menu** key and select **Send**. Choose whether you wish to send a **Picture**, **Voice Note**, **File**, **Location**, **Messenger** or **BlackBerry Contact**.

3. Navigate to where the image or audio file is stored on your BlackBerry, and select it.

4. Selecting the file will automatically send it to your message buddy.

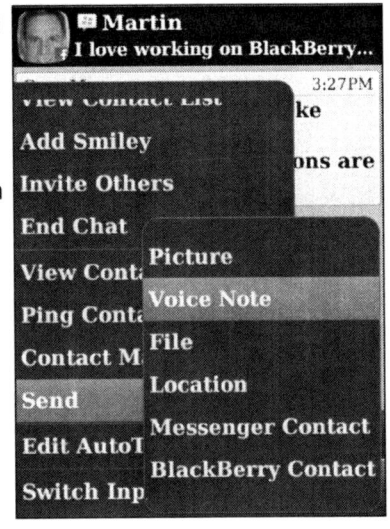

NOTE: When you select **Voice Note**, the **Voice Note Recorder** will start. Just record a message and then choose **Send**.

Some BlackBerry providers (phone companies) may limit the size of file you can send via Bluetooth to a small size, such as 15 kilobytes (kb). Most pictures maybe 300kb or more, and full songs might be 500kb or more.

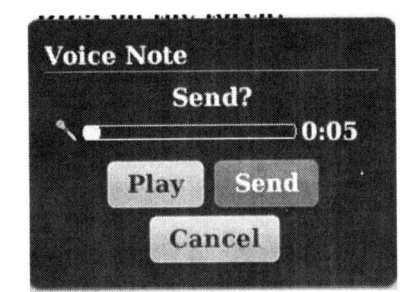

Pinging a Contact

Let's say that you wanted to reach a BlackBerry Messenger contact quickly. One option available to you is to ping that contact. When you ping a BlackBerry user, their device will vibrate once to let them know that they are wanted/needed immediately.

> **TIP:** You can set your BlackBerry to vibrate or not vibrate when you receive a ping in your BlackBerry Messenger Options screen.

1. Open a conversation with a contact from the Contact screen.
2. Press the **Menu** key and scroll down to **Ping Contact**.
3. The dialogue screen will reflect the ping by showing PING!!! in capital red letters.
4. The ping recipient will notice that his or her BlackBerry vibrates and indicates that you have pinged it.

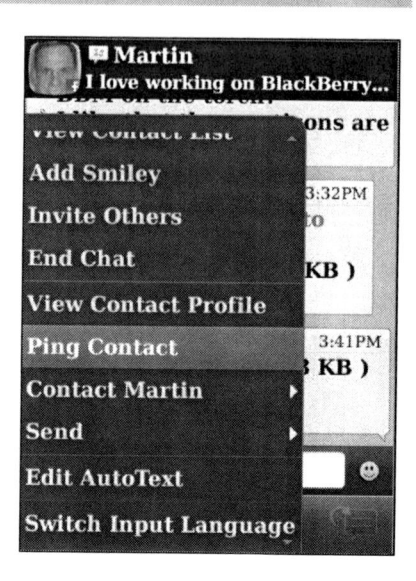

Set Availability to Others (My Status)

Sometimes, you might not want to be disrupted with instant messages. You can change your status to unavailable, and you won't be disturbed. Conversely, one of your contacts might be offline, and you want to know when he or she becomes available. You can set an alert to notify you.

1. Navigate to your main Messaging screen and highlight your name at the top and then press the **Trackpad**.
2. You can also press the **Menu** key and scroll to **View My Profile** and click.

> **TIP:** You can even show what music you are listening to by checking the box next to **Show What I'm Listening To**.

CHAPTER 13: BlackBerry Messenger, PIN Messaging and More

3. Scroll to My Status and choose either **Available** or **Busy**.

4. You can set a personalized status by clicking on **Personalize Status**.

5. Type in your status and that will be available to all your BBM contacts.

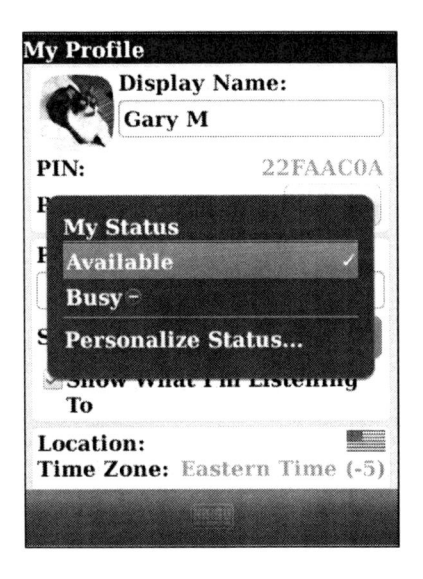

Multi-person Chat with BlackBerry Messenger

One very cool feature on BlackBerry Messenger is the ability to have a multi-person chat with two or more of your Messenger contacts.

To start a multi-person chat, just go to the main screen of the BlackBerry Messenger app, press the Menu key and select New Multi-person chat.

Start up a conversation as you did before. Now, let's say we wanted to invite another person (Martin) to join the conversation; we would press the **Menu** key and select **Invite Others** (see Figure 13-3).

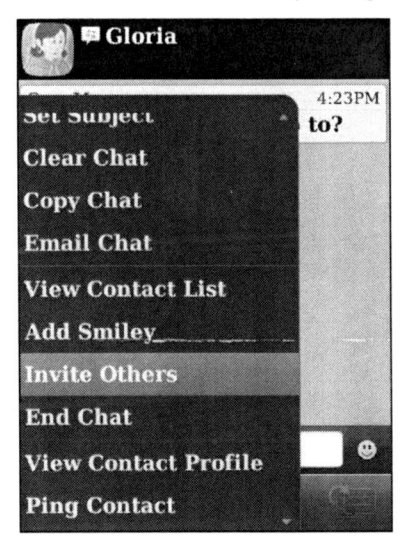

Figure 13-3. Inviting a third party to a BlackBerry Messenger conference

I then find the contact from my Messenger list I want to invite—in this case, Martin.

An invitation is sent for him to join us. When she accepts, it will be noted on the Messenger screen and all three of us can have our conversation.

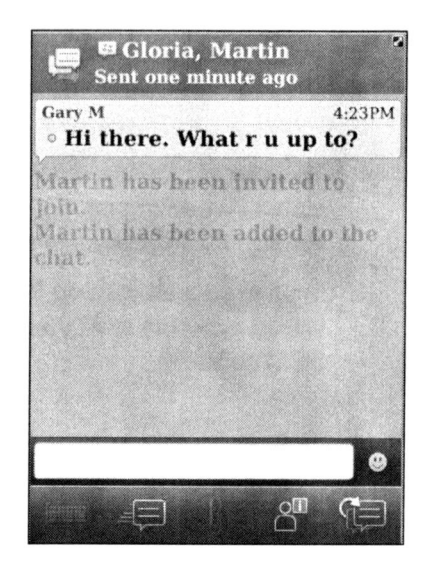

Using Groups

Another great feature of BlackBerry Messenger is the ability to create and use groups. Groups can be very useful for the following things:

- Chat with a private group at once
- Share pictures, and allow group members to comment on them (like Facebook comments)
- Create To-Do Lists and assign items to group members
- Share a group calendar

Create a New BBM Group

1. From your main Messenger screen, just click the **BlackBerry Groups** tab.

2. If you don't have any groups, you will see only the **Create a New Group** option.

3. Give your group a name and description, and even choose a new **Group** icon if you wish.

4. From your new group, click **Members** and then click **Invite a new member** to begin to choose members for your group. Follow the same steps as you did before to invite people.

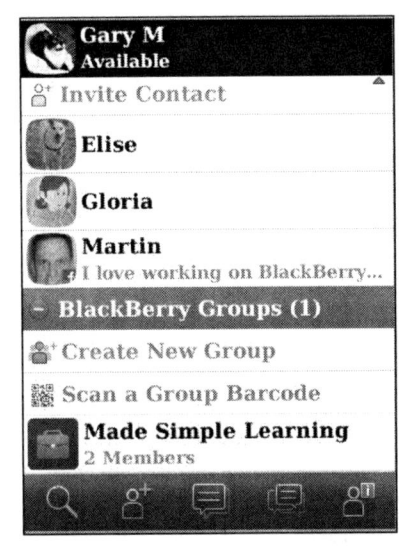

Group Chats, Pictures, Lists and Calendar

Click on the group you just formed to see this Group main screen.

From here you can select:

Members to add or remove members.

Chat to talk with everyone in the group at once.

Pictures to upload pictures or comment on pictures others have uploaded.

Lists to create or manage to-do lists created for the group.

Calendar to view or add to the calendar shared among all group members.

Picture Comments:

Here is an example of a comment placed on a shared group picture. You get the idea – using BBM to share and comment on pictures can be very useful. This can give you and the entire group instant feedback and reactions to any picture – saving on countless phone calls and meetings!

Shared To-Do Lists:

You can also share task lists among all group members and even talk about the tasks with the **Discuss List** feature you see here.

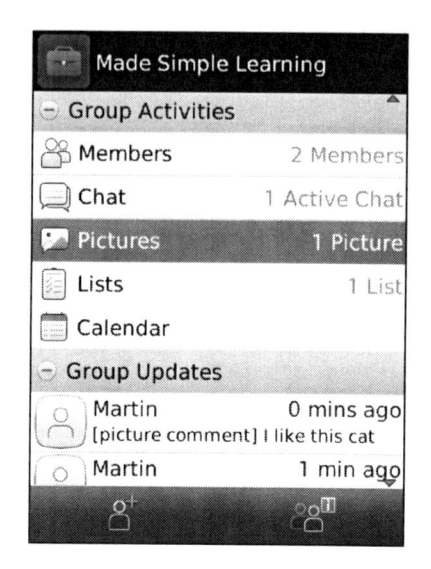

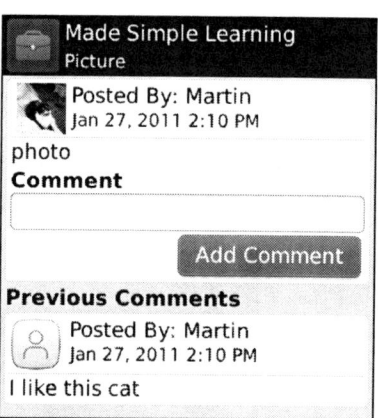

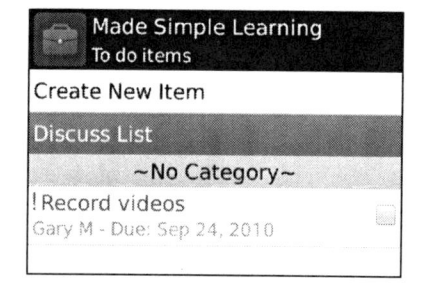

# PIN Messaging

BlackBerry handhelds have a unique feature called PIN-to-PIN, also known as PIN Messaging or Peer-to-Peer Messaging. This allows one BlackBerry user to communicate directly with another BlackBerry user as long as you know that user's BlackBerry PIN number. We'll show you an easy way to find your PIN (Figure 13-4), send an email message to your colleague, and a few good tips and tricks.

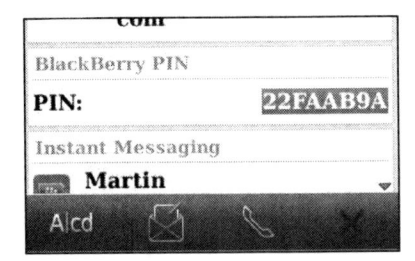

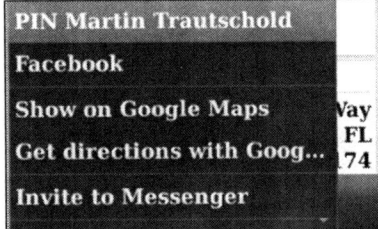

Figure 13-4. PIN messaging on your BlackBerry

Compose an email message to your colleague. In the body type of the email message, type the code letters **mypin** and hit the **space** key as shown in Figure 13-5 — you will then see your pin number in the following format "pin:XXXXXXX," where the XXXXXXX is replaced by your actual PIN. Press the **Menu** key and select **Send**.

*Figure 13-5. Typing **Mypin** to display your PIN*

Just press the **Menu** key and select **Send** to deliver an email message with your PIN number.

You can also use the **Send** soft key at the bottom of the screen.

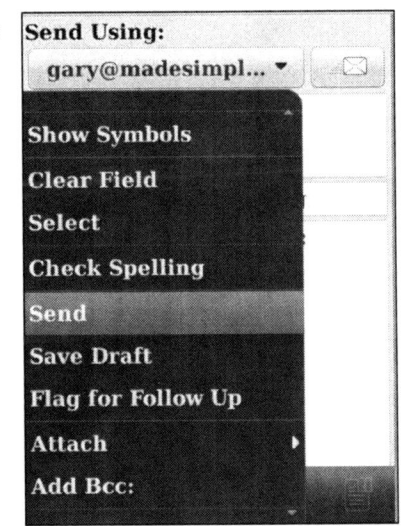

Replying to a PIN Message

Once you receive your PIN message, you will see that it is highlighted in red text in your inbox.

To reply to a PIN message, simply click the message to open it, click the **Menu** key, and select **Reply**, just like with email and other messaging.

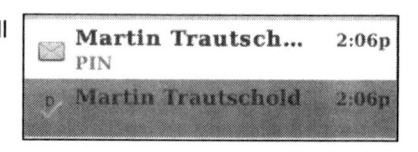

Adding Someone's PIN to Your Address Book

Once you receive an email message containing a PIN number from your colleague or family member, you should put this PIN number into your contact list.

If you don't already have this person in your contact list, then do the following:

1. Highlight the **PIN** number
2. Press the **Menu** key.
3. Select **Add to Contacts** from the menu, and then enter the person's information.
4. Be sure to save your new entry.

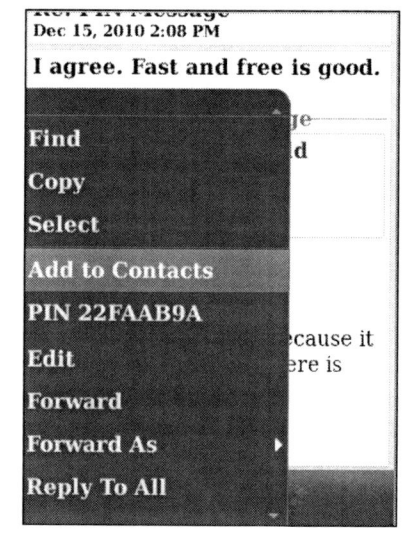

If you already have this person in your contact list, then you should do the following:

5. Select **Copy** from the menu and paste the PIN into the contact record.

CHAPTER 13: BlackBerry Messenger, PIN Messaging and More

6. Then press and hold the **Menu** key until you see the Switch Application pop-up window.

7. Select **Contacts** if you see it.

8. Select **Home Screen** if you don't see **Contacts**—then click **Contacts** to start it from the Home screen.

9. Type a few letters of the person's first, last, or company name to find them, press the **Menu** key, and select **Edit**.

10. Scroll down to put the cursor in the **PIN** field, press the **Menu** key, and select **Paste**.

11. Press the **Menu** key and select **Save**.

Now, next time you search through your contacts, you will have the new option of sending a PIN message in addition to the other email, Text, MMS, and phone options.

Using AIM, Yahoo, and Google Talk Messaging

After you get used to BlackBerry Messenger you will begin to see that it is a powerful way of quickly keeping in touch with friends, family, and colleagues. Realizing that many are still not in the BlackBerry world, you can also access and use popular IM programs like AOL Instant Messenger (AIM), Yahoo Messenger, and Google Talk Messenger right out of the box on the BlackBerry. Individual carriers do have some restrictions, however, and you will need to check your carrier web sites to see which services are supported.

View Video Tutorials and Free Tips at www.MadeSimpleLearning.com   343

Even More Instant Messenger Apps

It is likely you already have a number of Instant Messaging apps installed and we show you how to find them here. If you cannot find the one you want, visit **BlackBerry App World** and look for it there. It is likely that any popular IM application will be available for the BlackBerry.

1. First see if your carrier has placed an **IM** or **Instant Messaging** folder icon in your Applications directory (see Figure 13-6). This may already have the icons installed that you want to use.

2. Navigate to **Applications** and scroll for an icon that simply is called **IM**.

Figure 13-6. Other instant messaging icons on your BlackBerry

3. If you have an **IM Folder** icon, click it, and follow the on-screen prompts.

4. If there is no **IM** icon, click on BlackBerry App World and do a search for Instant Messenger apps or view the Instant Messaging category.

For more information to help you download and install third-party software, please check out Chapter 20: "BlackBerry App World" which is devoted to adding and removing software.

Chapter 14: Social Networking

BlackBerry smartphones aren't just for business executives anymore, but you already know that. Your BlackBerry can keep you in touch in ways beyond the messaging features shown earlier in this book.

Some of the most popular places to connect these days are social networking sites—places that allow you to create your own page and connect with friends and family to see what is going in their lives. Some of the most popular websites for social networking are Facebook, Twitter, and LinkedIn.

In this chapter, we will show you how to access these sites. You will learn how to update your status, "tweet," and keep track of those who are important or simply of interest to you.

We will also show you the new Social Feeds app which puts all your social networking and RSS feeds into one, easy to access place to quickly check up and check in.

Downloading Social Networking Apps

Check in your Applications folder or Social Networking folder for these apps. If you do not already see these apps on your Torch, you can easily download them from **App World**.

1. Start **App World** by clicking on it.
2. Use the **Search** feature in App World and simply type the name of the app—**Facebook**, **Twitter**, **foursquare**, **Gowalla**, **MySpace**, **LinkedIn**, or **Flickr**—to quickly find each of these apps.

3. Download and install any app you want.

> **TIP:** Follow the steps shown in Chapter 20 "BlackBerry App World" to get each app downloaded and installed.

As of publishing time, BlackBerry has its own very capable **Twitter** client. There are also more than half a dozen pretty good Twitter clients, many of them free. You might want to try out **UberTwitter** or one of the others.

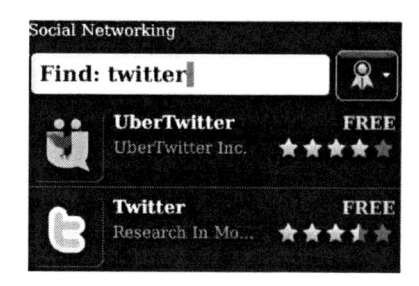

Logging into the Apps

In order to connect to your account on **Facebook**, **Twitter**, and **MySpace**, you will need to locate the icon you just installed and click on it. We use the example of **Facebook** here, but the process is very similar for the rest of the apps.

There is a lot to like about having **Facebook** and these other social networking apps on your Blackberry. You can always stay in touch. Just log in as you do on your computer and you are ready to go anywhere, anytime from your Torch.

> **NOTE:** You might find your **Facebook** icon in your **Social Networking** folder, depending on your carrier.

Once **Facebook** is successfully downloaded, the icon in your **Downloads** folder should look something like this.

1. Touch or click on the **Facebook** icon.

CHAPTER 14: Social Networking

2. Enter your login information and click the **Login** link.

3. You are logged in. Usually, you will only have to log in one time to each of these apps.

Facebook

Facebook was founded in February of 2004. Since that time, it has served as the premier site for users to connect, re-connect and share information with friends, co-workers, and family. Today, over 400 million people use Facebook as their primary source of catching up with the people who matter most to them.

If you have the **Facebook** icon on your BlackBerry already, it will look like this:

1. Just touch or click on the **Facebook** icon to start it.

2. Login if requested as shown above.

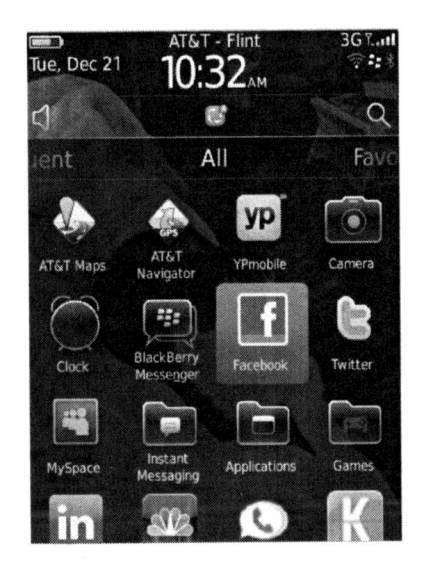

There is a lot to like about having Facebook on your Blackberry. All of your messages will be pushed to your device, just like your email, so you can always stay in touch. You also navigate the site in a very similar way to that on your computer.

View Video Tutorials and Free Tips at www.MadeSimpleLearning.com   347

Facebook Setup Wizard

1. After you login the first time, you will be taken through the **Setup Wizard** screens.

2. By default, all the boxes are checked. We recommend leaving them checked to have the best integration with your BlackBerry apps.

3. Tap **Save** at the bottom to continue.

Some of the integration you will get with Facebook includes:

- See **Facebook Profile** pictures in your **Contacts** list.

- See Facebook calendar events on your BlackBerry **Calendar**.

- Receive a **Facebook** message in your **Messages** app when you or your friends poke you or do a status update.

- Receive a **Facebook** messages when anyone comments on photos you have uploaded.

- Replace existing **Contact** photos with **Facebook friend** photos.

Status Update and News Feed

Once you log on to **Facebook** for BlackBerry, you can to write you own status update.

1. Click on **What's on your mind?**
2. Type your status update.
3. You can also see your **News Feed** from your friends.

> **News Feed**
>
> **Anna Doody Arico** The problem with flannel sheets is that you are so warm and cozy in the morning, your toes won't venture out onto the bedroom floor.
>
> 30 minutes ago Comment · Like
>
> BlackBerry for AT&T

Top Bar Icons

Along the top bar are icons for some of the most use features of Facebook. You will see icons for the **News Feed, Notifications, Messages, Upload a Photo, Friends, Search** or **Places**..

Notification/Email in Facebook:

Click on the Notifications icon (second from left) to see all your notifications in Facebook. The great thing is that these are fully integrated into your Messages Inbox icon. So every notification you receive in Facebook is also shown in Messages.

This saves you the step of going into Facebook to check for notifications.

> **NOTE**: If you choose to integrate your Facebook account into your Social Feeds, (see the end of the chapter) you will also see your notifications there.

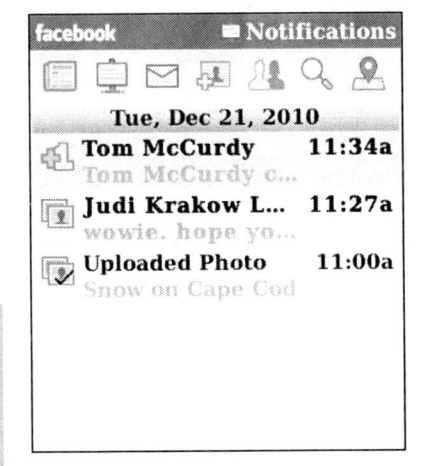

Notifications:
- Tue, Dec 21, 2010
- **Tom McCurdy** 11:34a
  Tom McCurdy c...
- **Judi Krakow L...** 11:27a
  wowie. hope yo...
- **Uploaded Photo** 11:00a
  Snow on Cape Cod

Communicating with Facebook Friends

It is easy to communicate with your friends.

1. Click on the **Friends** icon to see a list of friends displayed.

2. If you have contact information or phone numbers in your **Contacts** app, the icons next to your friend's names will show that.

3. Click on the friend you wish to communicate with to see all the options available to you.

4. In this example, we clicked on Martin's name. Now we can:
   - Send him a message
   - Write on his wall
   - Request his phone number
   - Poke him
   - View his profile
   - View his BlackBerry profile
   - Refresh contact info
   - Disconnect his information from the BlackBerry profile

NOTE: I can also just highlight his contact in my list and press the Menu key to see these same options.

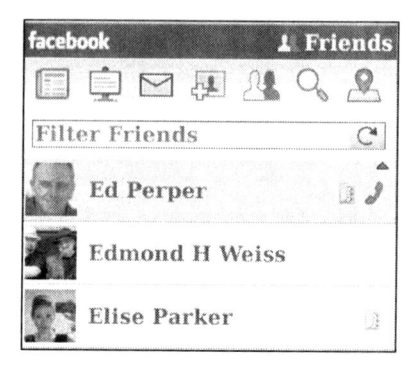

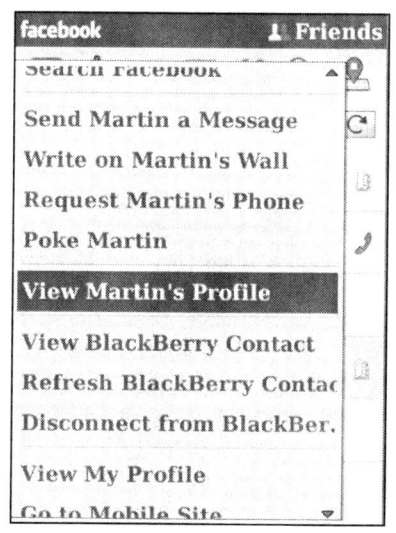

Uploading Pictures to Facebook

An easy and fun thing to do with Facebook is to upload pictures. You can upload pictures from your Torch to Facebook in using these methods:

- From the **Facebook** app - click on the small **Upload a Photo** icon.

- From your **Camera** app – snap a picture and upload it using **Send/Share** menu item or the **Send Picture** icon, then **Send to Facebook**.

- From your **Pictures** app – select a picture, press the **Menu** key, select **Send/Share**, and select **Send to Facebook**.

Once you select a photo to upload, you can specify a caption and a folder. You can also Tag the photo with a Facebook friend's name. We show you the steps from the **Facebook** app, but they are identical when you upload from any of the above options:

1. Click the **Upload a Photo** icon at the top of **Facebook**.

2. Navigate through the folders on your Torch to locate the picture you wish to upload or choose **Camera** at the top to take a new picture.

3. Click on the picture to select it for upload.

4. Now you can write a **Caption**, select an Album for the photo (the default is **Mobile Uploads**), and **Tag This Photo** (left image in Figure 14-1).

5. Click on **Tag This Photo** to identify one or more people from the photo as your Facebook friends. You will be given a small cross-hair to identify the faces of your friends. After the photo is uploaded, your friends will be notified that they have been tagged (right image in Figure 14-1).

6. When you are done tagging, you can click **Upload** to send the picture.

Figure 14-1. *Uploading pictures to Facebook*

> **TIP:** Immediately after you upload the photo from your Torch, you can see it on both on your computer and on your BlackBerry Facebook page.

Twitter

Twitter was started in 2006. Twitter is essentially an SMS (text message)-based social networking site. It is often referred to as a "micro-blogging" site where the famous and not-so-famous share what's on their mind. The catch is that you only have 140 characters to get your point across.

With Twitter, you subscribe to "follow" someone who "Tweets" messages. You might also find that people will start to follow you. If you want to follow us, we are: @garymadesimple or @mtrautschold on Twitter.

Create a Twitter Account

Making a Twitter account is very easy. We do recommend that you first establish your Twitter account on the Twitter web site.

1. On your computer, go to www.twitter.com.
2. Click on the **Join Today** button.

When you establish your account, you will be asked to choose a unique user name and choose a password.

3. You will then be sent an email confirmation. Click on the link in your email and you will be taken back to the Twitter web site. You can choose people to follow, or post your own Tweets on the web site. You can also read Tweets from your friends.

Using Twitter for BlackBerry

To start using Twitter, follow these steps:

1. Click on the **Twitter** icon (it may be located in your **Downloads** folder).

2. The first time you start the app, log in with your Twitter **username** and **password**.

3. Click the checkbox next to **Remember me** to avoid re-entering your information.

4. Click **Sign In**.

5. The initial view in **Twitter** has a small box at the top of the screen that says "What's happening" where you can tweet (send messages) to those who are following you.

6. Under that are the tweets from those you are following.

7. Just scroll through the list of tweets to catch up on all the important news of the day!

Sending Out Tweets

1. Tap in the box that says **What's happening?**

2. Type your Tweet about what is important to you (in 140 characters.)

3. Type your message and then click on the **Update** button to post your Tweet.

Twitter Icons

Along the top of the Twitter screen are eight icons; these are the quick links to the basic **Twitter** functions.

Figure 14-2. Twitter icons at top of screen

Mentions

A "mention" in Twitter speak is a tweet that mentions your @username somewhere in the body of the tweet. These are collected and stored in your Mentions section of your Twitter account.

My Lists

You can create a custom list so that your followers cannot only follow everything you say, they can choose to follow a specific topic (a list) that you create. In this example, I created a list entitle "Made Simple Announcements" so that users can choose to only follow our special announcements if they choose.

CHAPTER 14: Social Networking

My Profile

Just as it sounds, the My Profile icon will take you to your Twitter profile. You can see your followers, your tweets, and your bio. Just click on any field in your profile and a short menu comes up. To edit your information, just choose **Edit Profile** from the menu.

Direct Messages

Direct Messages are messages between you and another user on Twitter.

Click the **Direct Messages** icon to see your direct messages.

To compose a direct message click on the **Compose Message** icon.

Fill in the **To** field with the Twitter user name and then type your 140 character message in the box. When you are done, just click on the **Send** button.

Find People

The **Find People** icon takes you to a search window where you can type in a user name, business name, or last name to search for someone who might be on Twitter.

1. Type in the name and your query will be processed. Press the trackpad to search.

2. Scroll through the results to see if you find what you are looking for. If so, just click on the entry that matches your search and scroll down.

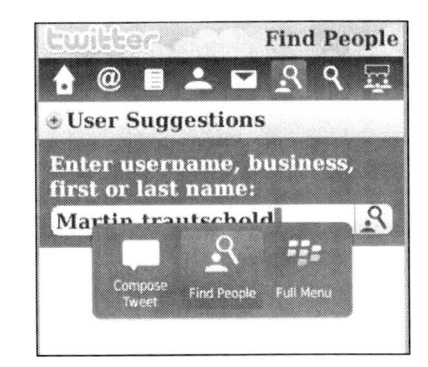

View Video Tutorials and Free Tips at www.MadeSimpleLearning.com   **355**

You will be able to see the statistics of how many followers there are and if they have particular Lists to follow. You can also choose the select **Follow** or **Block** to either follow this contact on Twitter or block all tweets.

Search

The **Search** icon takes you to another search window. Here you can type in any keyword to find opportunities to follow a topic or individual. For example, if you were to type in the word **golf**, you might find popular golf courses, golfers, or driving ranges that all use Twitter. You might also find thousands of individuals who just want to brag about their golf score.

TIP: Be very specific in your search to help narrow down the search results.

Popular Topics

The **Popular Topics** icons will simply list those topics that are currently popular on Twitter. So, if you wanted to scroll through, you might find everything from the New England Patriots to the President's visit overseas. On Twitter, someone always has something to say about everything.

Twitter Options

To get to the Twitter options, follow these steps:

1. From your Twitter home screen, press the **Menu** key.
2. Scroll down to **Options**.
3. You will notice several boxes in which you can click to place or remove a check mark.

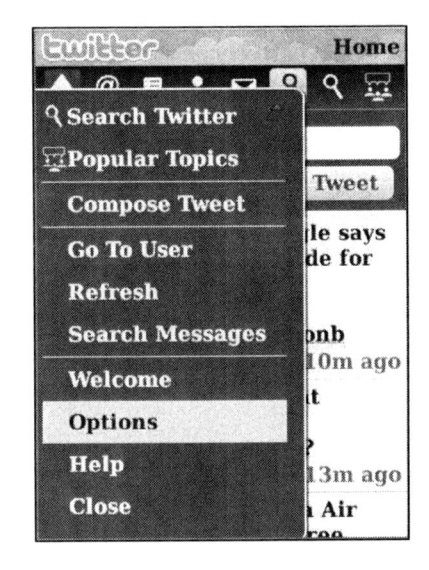

The available options are:

- Tweet refresh in background
- Automatic Tweet Refresh
- Notify on new tweets
- Notify on new replies and mentions
- Specify number of tweets per refresh
- Integrate in Messages app.
- Spell check before sending
- Show navigation bar
- Show tweet box on home screen
- Use system font settings
- Check tweets for personal information before sending
- Connect directly to Twitter when Wi-Fi is enabled
- Upload photos

Just put a check box in the corresponding box next to the option you would like to enable.

LinkedIn

LinkedIn has very similar core functionality to **Facebook** but tends to be more business and career focused, whereas **Facebook** is more focused on personal friends and games. With LinkedIn, you can connect and re-connect with past business associates, send messages, see what people are up to, have discussions, and more.

NOTE: The **LinkedIn** app is available as a free download from the BlackBerry App World.

Download and Sign In to LinkedIn

Locate and download **Linkedin** from BlackBerry **App World**.

Tap the icon, enter your **Email**, **Password** and tap **Sign In**.

Navigating around the LinkedIn App

LinkedIn has a similar icon-based navigation as Facebook. The top of the home screen has a box for posting your profile update (similar to a status update on Facebook.)

Just type your message and then click on the **Post** button.

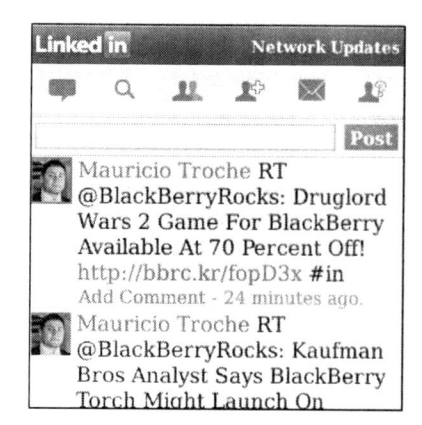

CHAPTER 14: Social Networking

Icons

At the top of the LinkedIn home screen are six icons. The first one is the Network Updates icon that will return you to the home page with profile updates and the ability to update your own LinkedIn status.

Search

Click in the search box and LinkedIn will try to first match your search with contacts in your LinkedIn directory. Next, the search field will show individuals who match the search criteria who may be part of your "Network."

Note: LinkedIn uses a "Network" system. Someone who is a direct contact is considered a 1st degree connection in your network. Someone who is connected to one of your contacts (but not you) is considered a 2nd degree contact and so on.

Connections

The next icon along the top is the **Connections** icon, which shows you a listing of your LinkedIn connections. Scroll down and find the particular connection you are looking for.

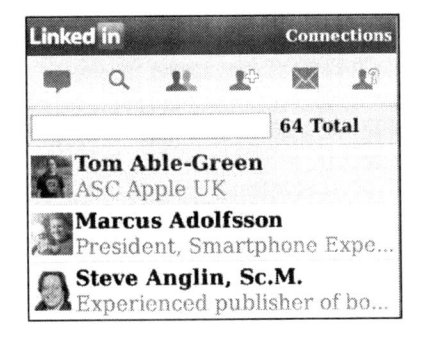

View Video Tutorials and Free Tips at www.MadeSimpleLearning.com   359

> **Tip**: Press the **Menu** key on your BlackBerry when a connection is highlighted and scroll and click on **Link to BlackBerry Contact** in the menu. This will attach all contact information (including the LinkedIn picture) to your BlackBerry contact for that individual.

Invitation

Click on the Invitations icon to see if you have any pending invitations for connections. You can also press the **Menu** key and select **Compose New Invitation**.

Type in the email address of the individual you would like to add to your LinkedIn network.

Edit the text below the email field

Press the **Menu** key and select **Send**.

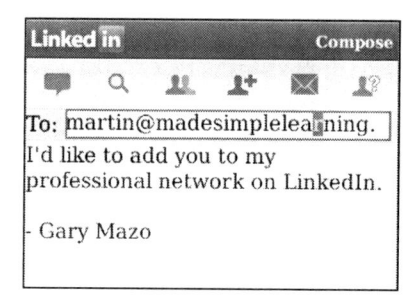

> **Note**: Unlike the **LinkedIn** web based application, you don't have to say how you know the individual with whom you would like to connect. This makes it much easier to send an invitation from the mobile application.

Messages

The Messages icon is where you go for direct messages to your contacts. You can read messages that are sent to you here. You can also press the **Menu** key and select **Compose Message** to send a direct message to any of your **LinkedIn** contacts.

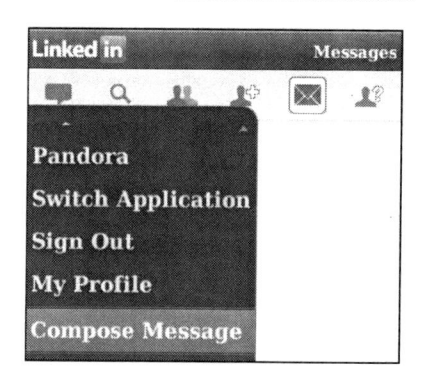

Reconnect

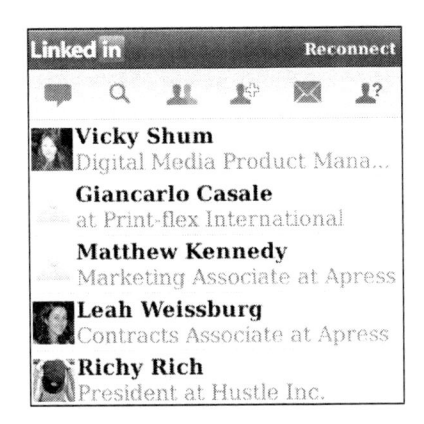

The Reconnect icon  is where you click to see people you might know, people who might be in groups you belong to or people who might have common connections to you. It is a good way to expand your network and see interesting people you might have forgotten about.

Vicky Shum
Digital Media Product Mana...

Giancarlo Casale
at Print-flex International

Matthew Kennedy
Marketing Associate at Apress

Leah Weissburg
Contracts Associate at Apress

Richy Rich
President at Hustle Inc.

LinkedIn Options

Press the Menu key from any section of the **LinkedIn** app and scroll to Options and click. You will notice a selection of radio boxes that you can check or uncheck depending on your needs and desires.

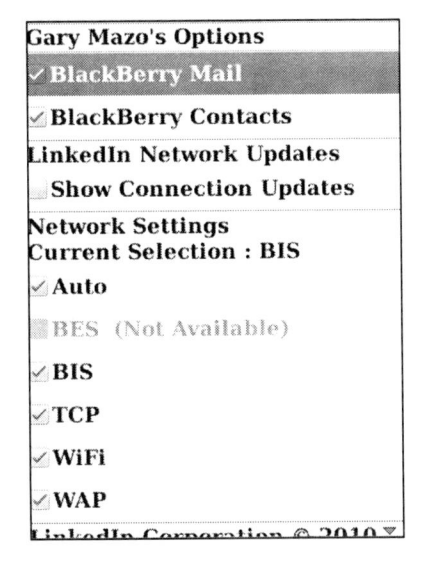

Select **BlackBerry Mail** to have LinkedIn messages displayed in your BlackBerry message list.

Select **BlackBerry Contacts** to have your LinkedIn contacts displayed as part of your Contact directory.

The bottom boxes let you specify your type of network and if you want BES integration of your LinkedIn account.

> **Note**: Some options like BES, WiFi and WAP are only available if your BlackBerry is connected to those networks.

Gary Mazo's Options
- BlackBerry Mail
- BlackBerry Contacts
- LinkedIn Network Updates
- Show Connection Updates

Network Settings
Current Selection : BIS
- Auto
- BES (Not Available)
- BIS
- TCP
- WiFi
- WAP

YouTube

A very fun site, **YouTube** lets you view short video clips on just about everything. Your new Torch is able to view most of the **YouTube** videos without doing anything special. There are two ways to access YouTube on the Torch, through the **YouTube** app or through the **YouTube** web site.

If the **YouTube** app is on your Torch, it will usually be in the Media home screen. You can also go to BlackBerry App World and search for YouTube or go to mobile.blackberry.com and search for the YouTube app download.

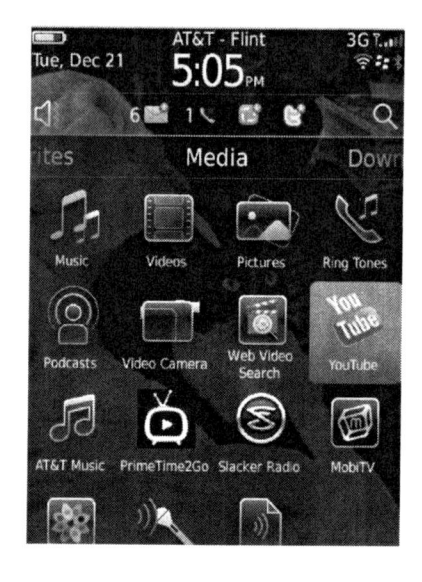

The YouTube App

Navigate to the **YouTube** icon (usually in the **Media** home screen) and touch or click. You will be prompted to either **Go to YouTube** or **Upload a Video**.

In this example, we will choose **Go to YouTube**. What you see is the standard mobile web based version of YouTube.

To access your **YouTube** account, sign in by clicking the **Sign In** button at the top right of the screen.

YouTube on the Web

If you don't see the **YouTube** app, you can still find it using your **Web Browser.** See Chapter 19 for more details on the **Browser.**

1. Click on your **Browser** icon.
2. Click on the top address bar and type in m.youtube.com.
3. Press the **Go** button on the on screen keyboard or **Enter** key on the slide-out keyboard.
4. You should now see a list of the most popular videos (Figure 14-3).
5. Touch or click on the video image itself to start playing the video.
6. Tap on the title of the video to learn more about it.

> **TIP:** Set a web bookmark in your Browser for YouTube by hitting the **Menu** key and selecting **Add Bookmark**. Go to Chapter 19 to learn how to set bookmarks

7. Once the video starts playing, tap the screen to show the controls. See the Chapter 18: "Fun with Videos" for more help about how to enjoy videos on your Torch.

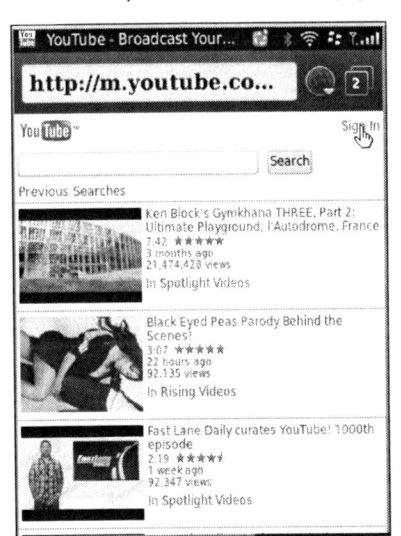

Figure 14-3. YouTube Mobile on the Torch

To search for any video on YouTube, just type anything in the search field. Type in your search and click the **Trackpad** and the search results will display for you.

YouTube Favorites

If you have a YouTube account, you can log in using the BlackBerry **YouTube** app or on the YouTube mobile web site.

Login with your YouTube user name and password. On the next screen, you will see your Videos, Favorites, Subscriptions and Inbox.

Just click on whichever of your items you wish to view. Your favorites are listed just like the popular videos on the YouTube home screen. Just touch or click a video to start.

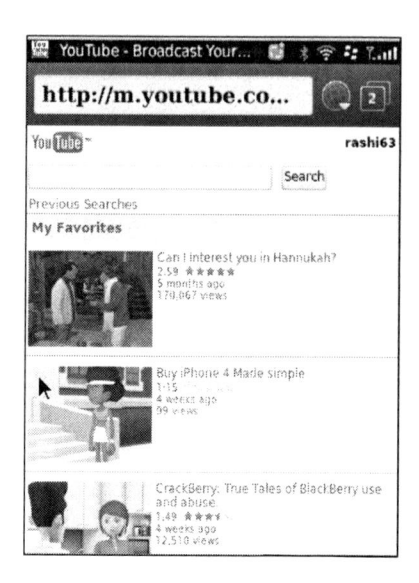

Social Feeds

With all the new Social Networking sites, you can find yourself inundated with messages, updates, feeds and status updates.

New to the BlackBerry in OS 6 is the ability to combine all your important Social Messaging apps into one Social Feed that is easy to quickly view. You don't "have" to use the Social Feed app, but it does make quick viewing of your feeds much more streamlined.

Getting Started with Social Feeds

Touch or click the **Social Feeds** app on your BlackBerry. The app should be listed under the **All** Home Screen.

CHAPTER 14: Social Networking

Usually, when you start the app for the first time you will see that you have no feeds associated with the app, so you will need to set them up.

You should see a Getting Started message under the Social filter. If you slide the screen to the left, you will see another filter for RSS feeds.

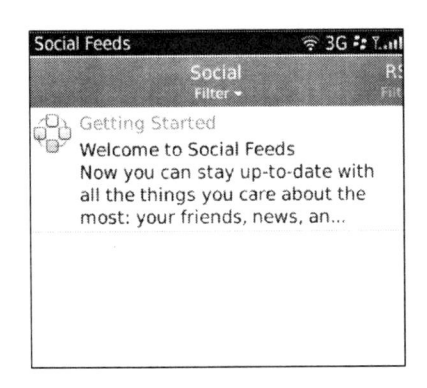

Adding Feeds

Touch the **Getting Started** button or just touch the Social filter heading and you will see all the options for the Social Feeds app. Initially you will see that you are not logged in to your Social Networking sites.

To add feeds to the Social Feeds app:

Touch or click the **Social Filter** header.

Touch or click on the account you want to add to the **Social Feed** app.

Log in to the appropriate app – either **Facebook**, **Twitter**, **BlackBerry Messenger**, **MySpace**, **Google Talk** or one of the other apps listed.

Adding RSS Feeds

RSS stands for "Really Simple Syndication." These are very short news feeds from popular web sites; snippets of full stories and a great way to quickly follow your favorite sites.

RSS feeds are usually viewed in an RSS reader like Google Reader or any other RSS app.

Once your Social feeds are set, just slide the top status bar to the left and you will see the RSS filter. To add RSS feeds:

Touch or click on the header for the **RSS Filter**.

If you have no RSS feeds, you will be prompted to search for a feed or to enter a URL of a web site you frequent. Most web sties have an RSS feed that you can "subscribe" to in order to get snippets of news.

You can also press the **Menu** key and select **New RSS Feed**.

Type in the web address to search for an accompanying feed.

Choose which of the feeds you wish to add to the **RSS Filter**.

In this example, we searched for MSNBC.com and found two available RSS feeds to which we could subscribe.

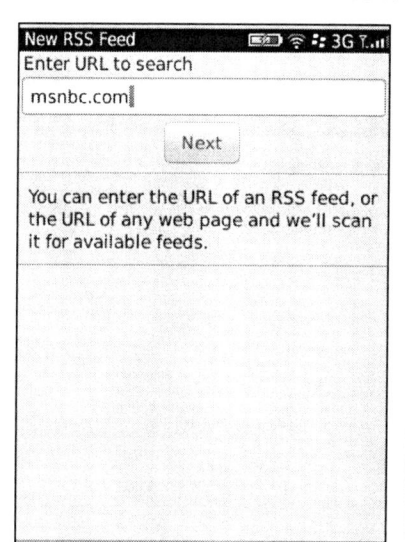

Figure 14-4. Adding RSS Feeds to the **Social Feeds** app.

You can continue to add as many RSS feeds as you would like in this manner.

Using the Social Feeds App

Once the Social and RSS feeds are set up, using the **Social Feeds** app is quite simple. To use your Social Feeds app, just:

Touch and click on the Social Feeds

To view your Social feeds, scroll through those listed under **Social Filter**.

CHAPTER 14: Social Networking

To view your RSS feeds, slide the top status bar to the left and scroll through the feeds listed under the **RSS Filter** header.

Figure 14-5 Social Filter and RSS Filter in the **Social Feeds** app.

Posting to All Social Apps at Once

One of the Great things about the Social Feeds app is that you can make one post and put in on BlackBerry Messenger, Facebook and Twitter at the same time. This certainly saves a bit of time if there is a status update you want to post on all three accounts.

To do this:

1. Make sure you are in the Social Filter.

2. Press the **Menu** key and select **New Post**.

3. Place check marks in the app to which you would like to post. In the example shows, we choose **BlackBerry Messenger**, **Facebook** and **Twitter**.

4. Type your post in the box provided and tap the **Post** button.

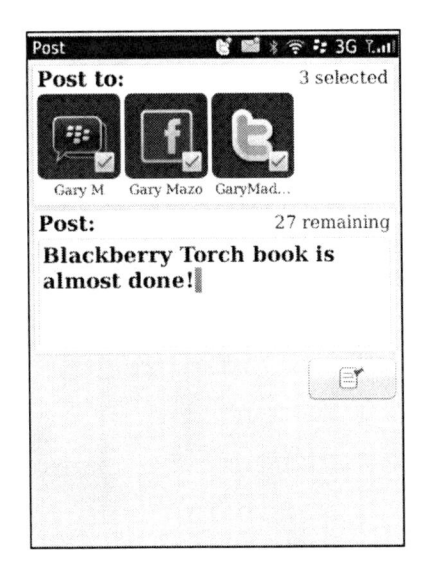

View Video Tutorials and Free Tips at www.MadeSimpleLearning.com   367

Managing RSS Feeds

While in one of the two filters, press the Menu key. In this example, we want to manage our RSS feeds.

Press the Menu key and select **Manage RSS Feeds**. The list of subscribed feeds will now be listed. Touch of click on one of them. You can now edit the name of the feed or uncheck the Subscribe box and remove it from your filter.

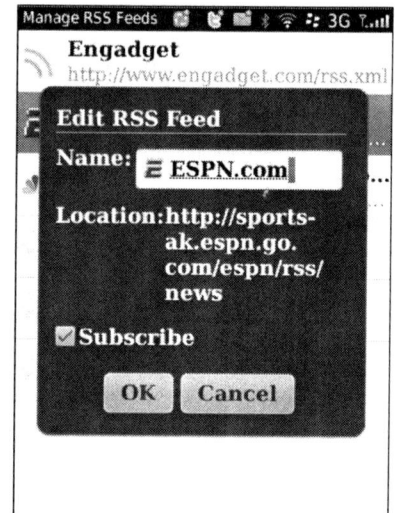

Figure 14-6 Managing RSS Feeds in the **Social Feeds** app.

Chapter 15: Tasks and Notes

The Task Icon

Like your contacts, calendar, and MemoPad, your task list becomes more powerful when you share or synchronize it with your computer. Since the BlackBerry is so easy to carry around, you can update, check off, and even create new tasks anytime, anywhere they come to mind. Gone are the days of writing down a task on a sticky note and hoping to find it later when you need it.

How Do You Get Your Tasks from Your Computer to Your BlackBerry?

You can mass load or sync your computer's task list with your BlackBerry **Task** app.

If your BlackBerry is tied to a BlackBerry Enterprise Server, the synchronization is wireless and automatic. Otherwise, you will use either a USB cable or Bluetooth wireless to connect your BlackBerry to your computer to keep it up to date. For Windows PC users, see Chapter 2; Apple Mac computer users, see Chapter 4.

Viewing Tasks

1. Tap the **Search** icon and type the word **Tasks** to find your **Tasks** icon. You could also tap on the **Applications** folder to see the **Task** app within the folder as shown here.

2. The first time you start tasks on your BlackBerry, you may see an empty task list if you have not yet synchronized with your computer.

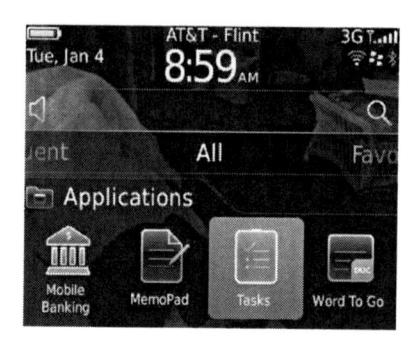

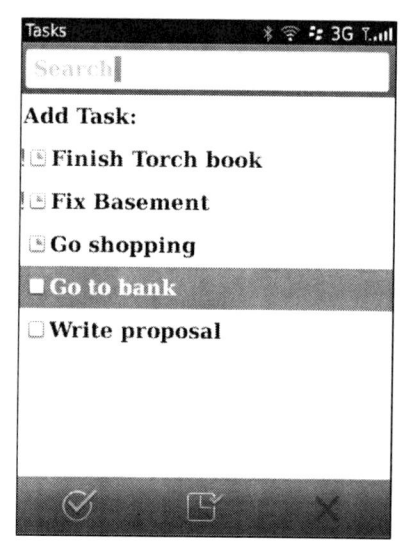

Adding a New Task

Press the **Menu** key and select **New** or just touch or click the **Add Task** bar. Then you can enter information for your new task in the screen in Figure 15-1.

> **TIP:** Keep in mind the way the Find feature works as you name your task. For example, all tasks for a particular "Proposal" should have "Proposal" in the name for easy retrieval.

CHAPTER 15: Tasks and Notes

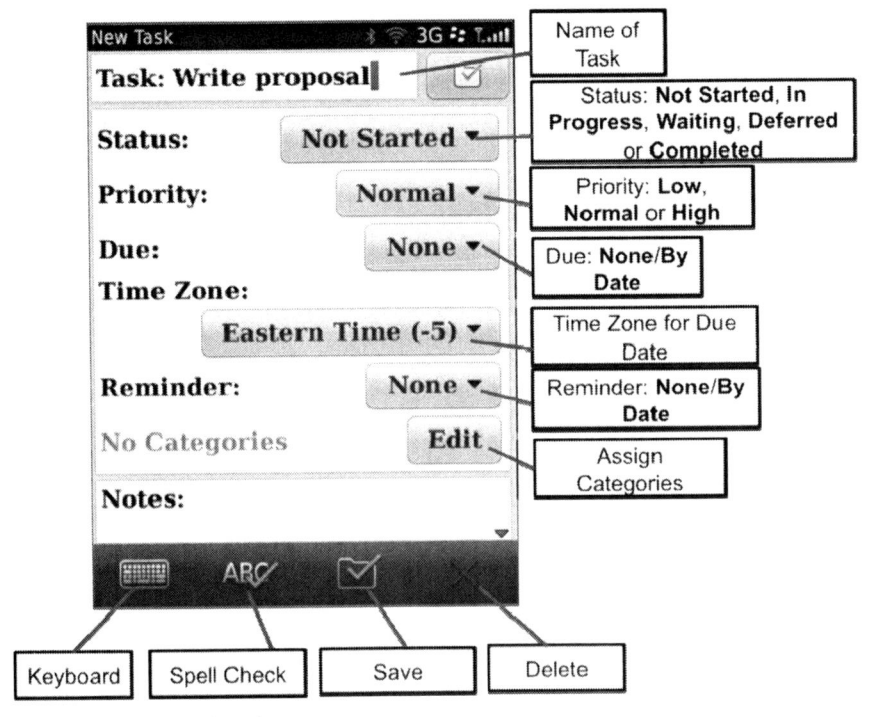

Figure 15-1. New task options menu

Categorizing Your Tasks

Like Address Book entries, you can group your tasks into categories. And you can also share or synchronize these categories with your computer.

Assigning a Task to a Category

1. Highlight the task, click the screen, and open it (see Figure 15-2).

2. You can set the status of your task, as well as the priority and the due date, from this screen.

3. To categorize, press the **Menu** key and select **Categories** or touch the **Edit** button next to the **Categories** heading.

4. Select as many categories as you would like by checking them by clicking the screen or pressing the **Space** key.

View Video Tutorials and Free Tips at www.MadeSimpleLearning.com **371**

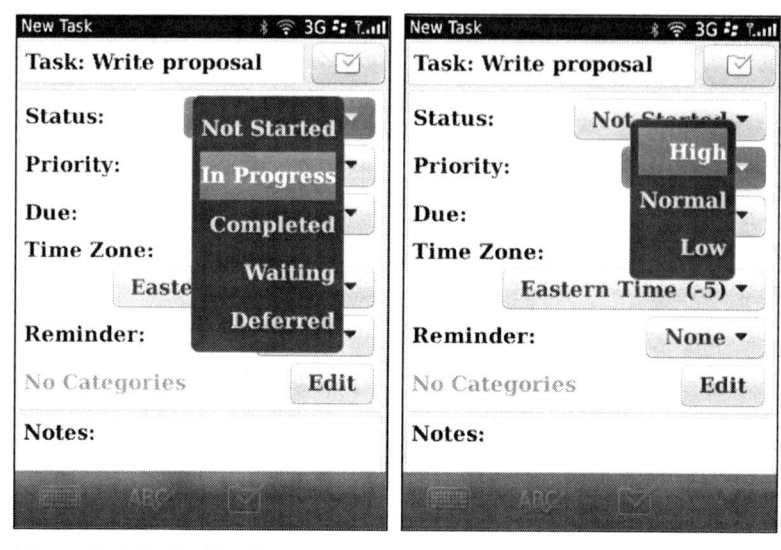

Figure 15-2. Setting New Task status and priority

5. You may even add new categories by pressing the **Menu** key from the **Select Categories** screen and selecting **New**.

6. Once you're done, press the **Menu** key and select **Save** to save your category settings.

7. Press the **Menu** key and select **Save** again to save your task.

Finding Tasks

Once you have a few tasks in your task list, you will want to know how to quickly locate them. One of the fastest ways is with the Find feature. The same Find feature from the address book works in Tasks. Just start typing a few letters to view only those that contain those letters.

In the example here, if we wanted to quickly find all tasks with **Go** in the name Figure 15-3, then we type the letters **go** to quickly see them.

> **NOTE:** If you have the **Universal Search** feature enabled (as it is by default) when you type in the search bar, your BlackBerry searches for matches outside of the current Task app as well.

CHAPTER 15: Tasks and Notes

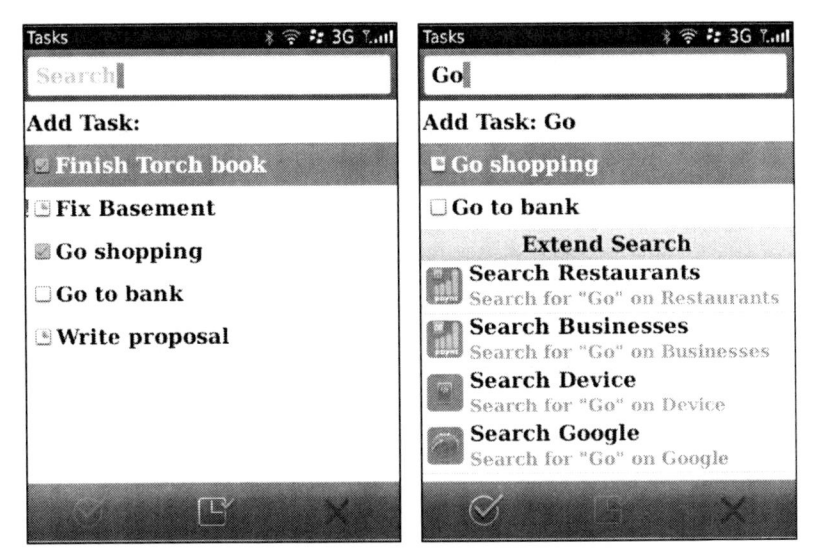

Figure 15-3. Finding exactly the task you are looking for

Managing or Checking Off Your Tasks

1. Scroll down to a task and touch it or use the **Trackpad** to highlight it. (Don't click it, unless you want to open it and make changes.)

2. Once a task is highlighted, you can use the soft keys at the bottom to manage the task.

    - Mark it **Completed**.
    - Mark it **In-Progress**.
    - Or you can **Delete** it.

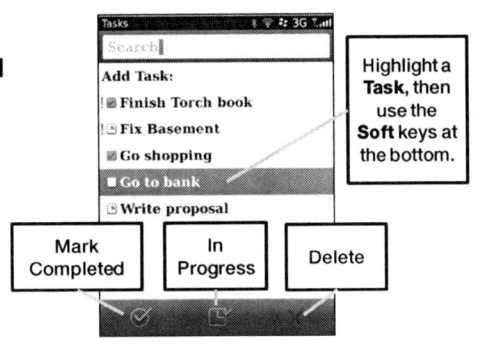

View Video Tutorials and Free Tips at www.MadeSimpleLearning.com   373

Sorting Your Tasks and Task Options

You may sort your tasks by the following methods in your Task Options screen: **Subject (default)**, **Priority**, **Due Date**, or **Status**.

You may also change the **Confirm Delete** field by simply placing a check in the radio box. You may also change the **Snooze** field from **None** to **30 Min**.

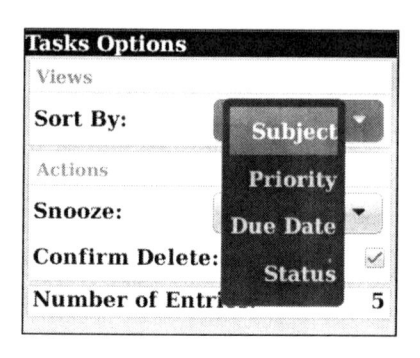

The MemoPad – Virtual Sticky Notes

One of the simplest and most useful programs on your BlackBerry is MemoPad. Its uses are truly limitless. There is nothing flashy about this program—just type your memo or your notes, and keep them with you at all times.

Using MemoPad is very easy and very intuitive. The following steps guide you through the basic process of inputting a memo and saving it on your BlackBerry. There are two basic ways of setting up memos on the BlackBerry: either compose the note on your computer organizer application and then synchronize (or transfer) that note to the BlackBerry, or compose the memo on the BlackBerry itself.

How to Sync Your MemoPad

You can also sync your computer's memo pad or notes list with your BlackBerry **MemoPad** app.

If your BlackBerry is tied to a BlackBerry Enterprise Server the synchronization is wireless and automatic. Otherwise, you will use either a USB cable or Bluetooth wireless to connect your BlackBerry to your computer to keep it up to date. For Windows PC users, see Chapter 2; Apple Mac computer users, see Chapter 4.

The sync works both ways, which extends the power of your desktop computer to your BlackBerry—add or edit notes anywhere and anytime on your BlackBerry—and rest assured they will be back on your computer (and backed up) after the next sync.

1,001 Uses for the MemoPad (Notes) Feature

OK, maybe we won't list 1,001 uses here, but we could. Anything that occupies space on a sticky note on your desk, in your calendar, or on your refrigerator could be written neatly and organized simply using the **MemoPad** app.

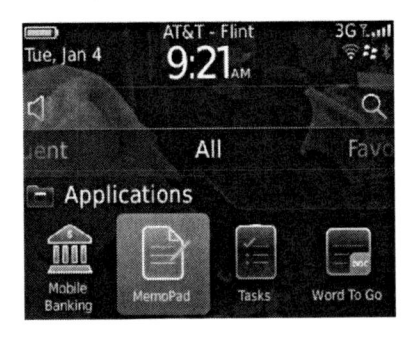

Common Uses for the MemoPad

- Grocery list
- Hardware store list
- Any store list for shopping
- Meeting agenda
- Packing list
- Made Simple Learning videos you want to watch
- Movies you want to rent next time at the video store
- Your parking space at the airport, mall or theme park

Adding or Editing Memos on the BlackBerry

Locate the **MemoPad** icon (your icon may look different, but look for MemoPad to be shown when you highlight it).

NOTE: You may need to first click the **Applications** folder to find it.

To add new memos to the list, simply click **New Memo** at the top of the list.

To open an existing memo, just scroll to it and click it.

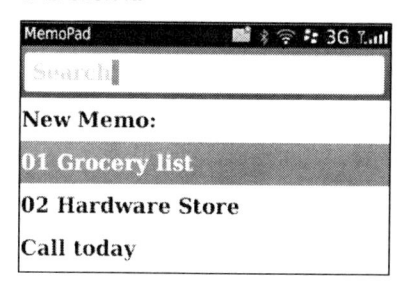

When typing a new memo, you will want to enter a title that will be easy to find later by typing a few matching letters.

1. Type your memo in the body section.
2. Press the **Enter** key to go down to the next line.
3. When you're done, press **Trackpad** and select **Save** or press the **Menu** key and select **Save** or just touch or click the **Save** soft key at the bottom.

Notice, if you have copied something from another app, like **Email**, you could paste it into the memo. (See the Quick Start Guide to learn more about Copy & Paste.)

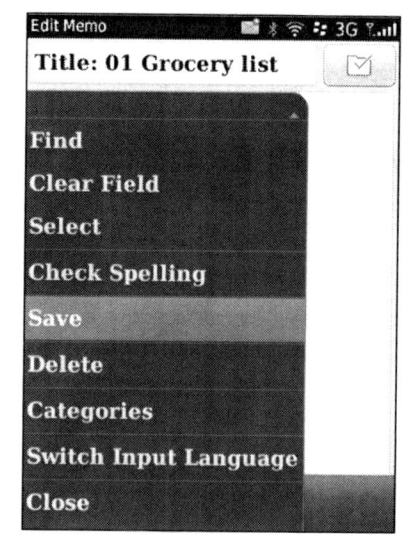

CHAPTER 15: Tasks and Notes

Quickly Locating or Finding Memos

MemoPad has a Find feature to help you locate memos quickly by typing the first few letters of words that match the title of your memos. Example, typing **Gr** would immediately show you only memos matching those three letters in the first part of any word, like "grocery."

> **NOTE**: Because of the Universal Search feature of the BlackBerry, other items might show up on the search. The memos that mach the search criteria will appear first.

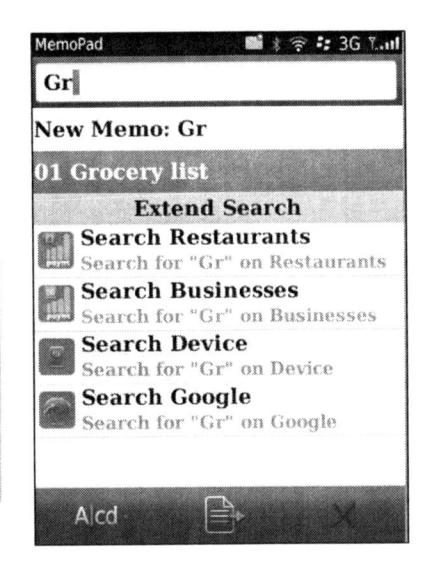

Ordering Frequently Used Memos

For frequently used memos, type numbers (01, 02, 03, etc.) at the beginning of the title to force those memos to be listed in order at the very top of the list. (The reason we started with zero is to keep memos in order after the tenth one.)

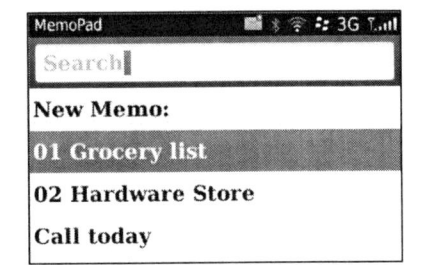

Add a Separate Memo for Each Store

This can help eliminate the forgetting of one particular item you were supposed to get at the hardware or grocery store (and save you time and gas money)!

Viewing Your Memos

1. Scroll down by gently touching the screen and scrolling to a memo, or click in the **Search** field at the top to type a few letters to find the memo you want to view.

2. Click the highlighted memo to instantly view its contents.

View Video Tutorials and Free Tips at www.MadeSimpleLearning.com

377

> **TIP:** Typing **ld** (stands for "Long Date") and pressing the **space** key will insert the date "Tue, 28 Aug 2007," and "**lt**" ("Long Time") will enter the time: "8:51:40 PM" (in the local date/time format you have set on your BlackBerry).

Organizing Your Memos with Categories

Similar to your address book and task list, **MemoPad** allows you to organize and filter memos using categories.

> **TIP:** Categories are shared between your address book, task list, and **MemoPad**. They are even synchronized or shared with your desktop computer.

Similar to your **Contacts** and **Tasks** app, **MemoPad** allows you to organize and filter memos using categories.

First, you must assign your memos to categories before they can be filtered.

One way to be extra organized with **MemoPad** is to utilize categories so all your memos are filed neatly away.

The two default categories are **Personal** and **Business**, but you can easily change or add to these.

> **TIP:** Once you assign Memos to **Categories**, you can use the **Filter** menu command to view only Memos assigned to a specific category.

Switching Applications / Multitasking

From almost every icon on your BlackBerry, MemoPad included, pressing the **Menu** key and selecting **Switch Application** allows you to multitask -- leave your current icon open and jump to any other icon on your BlackBerry. This is especially useful when you want to copy and paste information between apps.

> **TIP:** Pres and hold the **Menu** key to multitask.
> Instead of using the **Switch Application** menu item, try pressing and holding the
>
> **Menu** key. It's a good short cut!

Here's how you jump or switch applications:

CHAPTER 15: Tasks and Notes

1. Press and hold the **Menu** key until you see a little pop-up window with icons appear in the middle of the screen. This is called the Switch Applications window. You can also get to this window by tapping the **Menu** key and then selecting **Switch Application** from the menu (see Figure 15-4).

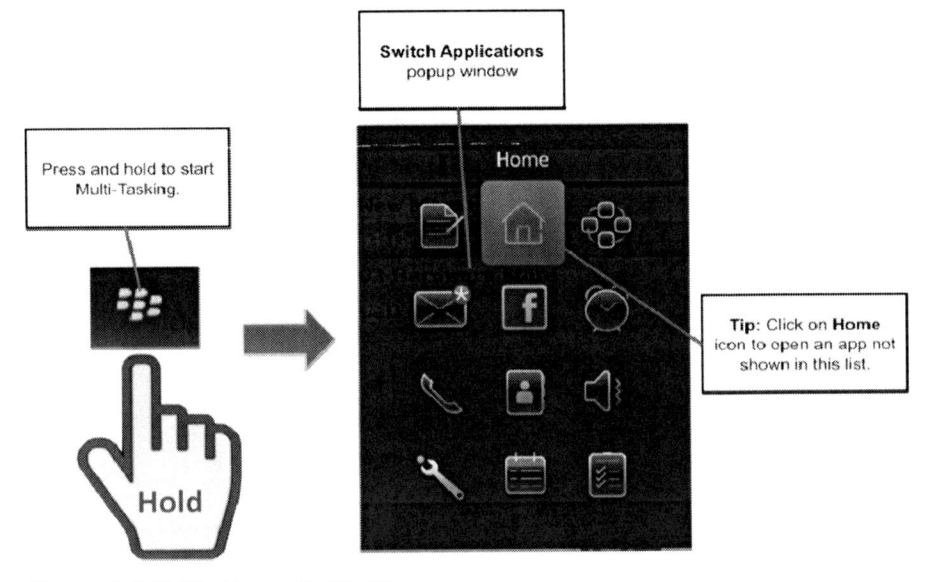

Figure 15-4. Multitasking on the BlackBerry

2. You will now see the Switch Applications pop-up window, which shows you every icon that is currently running.

3. If you see the app you want to switch to, just click it.

4. If you don't see the app you want, then click the **Home Screen** icon. Then you can locate and click the right app.

5. You can then jump back to **MemoPad** or the application you just left by selecting the **Switch Application** menu item from the icon you jumped to.

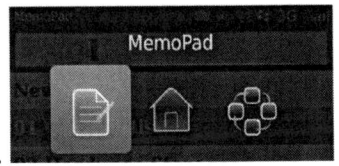

Press and hold the **Menu** key to jump back.

View Video Tutorials and Free Tips at www.MadeSimpleLearning.com 379

Forwarding Memos via Email, SMS, or BlackBerry Messenger

You might want to send a memo item via email, BlackBerry PIN message, or SMS text message to others. If so, you can use the **Forward As** command from the menu. Alternatively, if the keyboard is hidden, you will see the **Forward As** soft key on the bottom of the screen.

1. Highlight the memo you want to send and press the **Menu** key.

2. Select **Forward As**, and then select whether you want **Email**, **Text Message, Group Message, Messenger Contact, PIN, Social Feeds**, or **Twitter** (see Figure 15-5).

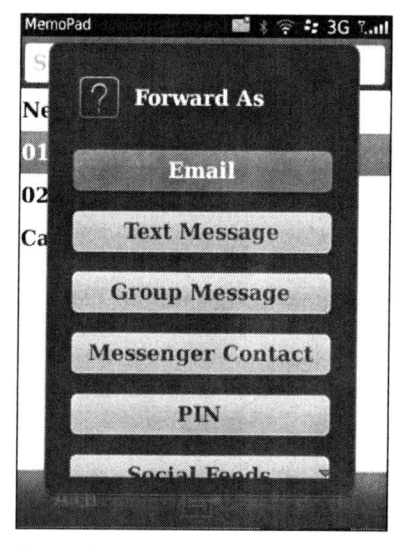

Figure 15-5. Forwarding a memo

3. Finally finish composing your message, press **Send** soft key at the bottom of the screen or the **Menu** key and select **Send**.

Other Memo Menu Commands

There may be a few other things you want to do with your memos. These can be found in the more advanced menu commands.

Start the **MemoPad** icon.

CHAPTER 15: Tasks and Notes

Some of these advanced menu items can be seen only when you are either writing a new memo or editing an existing one.

So you select either **New** and begin working on a new memo, or select **Edit** and edit an existing memo.

From the Editing screen, press the **Menu** key and the following options become available to you:

Switch Application – This is explained above.

Hide Keyboard – Use this to see more of the screen and temporarily hide the keyboard.

Find – If you are in a memo item, then this will allow you to find any text inside the memo.

Clear Field – This will erase the current contents of the memo you are working on - **USE WITH CAUTION!**

Paste – Suppose you have copied text from another program and want to paste it into a memo. Select and copy the text (from the calendar, address book, or another application) and select **Paste** from this menu. The text is now in your memo.

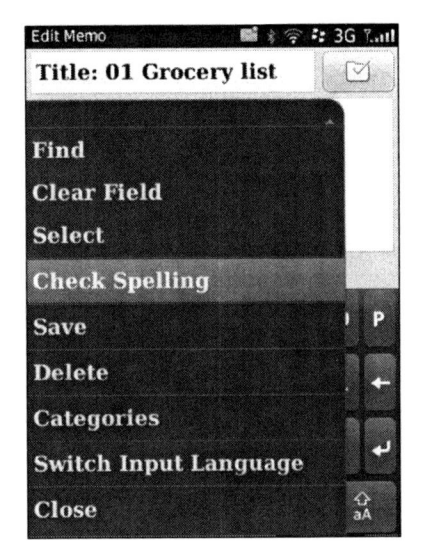

Select – This allows you to do just the reverse—click here and select text from the memo, press the **Menu** key again, and select **Copy**. Now, use the **Switch Applications** menu item to navigate to another application, press the **Menu** key, and select **Paste** to put the text in that application.

Check Spelling – This will run the BlackBerry spelling checker on the currently open memo item.

Save – This saves the changes in the memo.

Delete – This deletes the current memo.

Categories – This allows you to file this memo into either the Business or Personal categories. After selecting **Categories**, you can press the **Menu** key again and select **New** to create yet another category for this memo.

Switch Input Language – Changes to another installed language on the BlackBerry.

Close – This is similar to pressing the **Escape** key.

View Video Tutorials and Free Tips at www.MadeSimpleLearning.com   381

Chapter 16:
Your Music Player

Your BlackBerry is more than your phone and personal digital assistant, your email machine, and your address book. Today's BlackBerry smartphones are also very capable media players. With your BlackBerry, it is easy to carry your music, playlists, podcasts, and more. You can transfer playlists from iTunes and other media programs as well.

Listening to Your Music

One of the things that sets your BlackBerry apart from many of the earlier BlackBerry smartphones is the inclusion of multi-media capabilities and the ability to expand memory with the use of a media card. While some of the most popular formats of digital media are supported, you may have to take some steps to get all your music and videos working on the BlackBerry.

With a good-sized media card (e.g. 32 GB – gigabytes), the media capabilities of the BlackBerry, one might even suggest: **Why do I need an iPhone or iPod? I've got a BlackBerry!**

Syncing Your Music and Playlists

The BlackBerry comes with internal memory, but the operating system (OS) and other pre-loaded programs take up some of that space. What's left over is usually not enough to store all your music.

STEP 1: Buy and insert a media card (if you don't already have one) to boost the memory available to store your favorite music—see the "Boosting Your Memory" section of Chapter 1 for help.

STEP 2: Transfer your music from your computer. If you are Windows user, please refer to Chapter 3. If you are an Apple Mac user, please refer to the information found in Chapter 5.

Playing Your Music

Once your music is in the right place, you are ready to start enjoying the benefit of having your music with you at all times on your BlackBerry.

The fastest way to get to your music is by swiping to the **Media** Home Screen and then touching or clicking the **Music** icon.

You are now presented with various preset options to find and play your favorite music.

All Songs: This shows you every song on your BlackBerry.

Artists: This shows you all artists. Then you can click an artist to see all of his or her songs.

Albums: This shows a list of all albums.

Genres: This shows a list of all genres on your BlackBerry (Pop, Rock, Jazz, etc.).

Playlists: This shows all playlists or allows you to create new ones.

Wi-Fi Music Sync: You can sync your music and playlists wirelessly if this is activated. See Chapter 3 for PC users and Chapter 5 for Mac users.

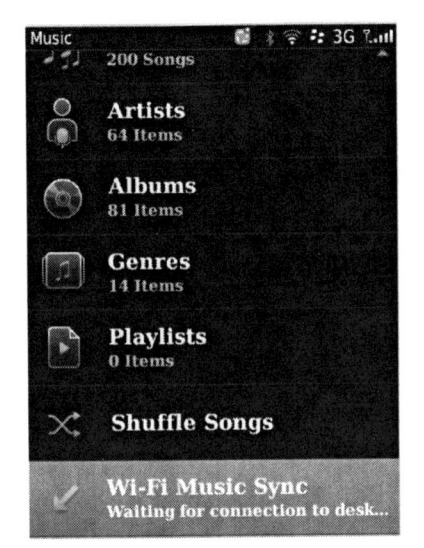

Shuffle Songs: Plays all your music in a shuffle mode or random order.

Finding and Playing an Individual Song

1. Bring up the keyboard (either by pressing the **Menu** key and selecting **Show Keyboard** or using your Convenience key if you have mapped it).

2. If you know the name of the song, then just type a few letters of any word in the song's name in the **Find** field at the top to instantly locate all matching songs.

3. In this case, we type **Love** and see all matching songs. To narrow the list, press the **space** key and type a few more letters of another word in the name of the song.

4. Once you click a song, the music player will open and your song will begin to play.

5. Press the **Pause**, **Stop**, **Next Track**, and **Previous Track** keys (see Figure 16-1).

6. Adjust your volume using the volume keys on the side of the BlackBerry.

TIP: Pressing the **Mute** key on the top of your BlackBerry will also pause or resume playback.

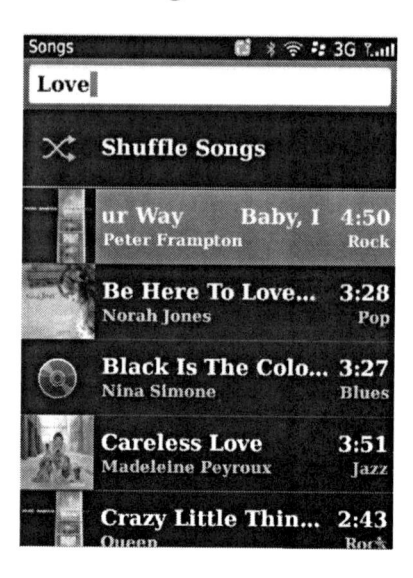

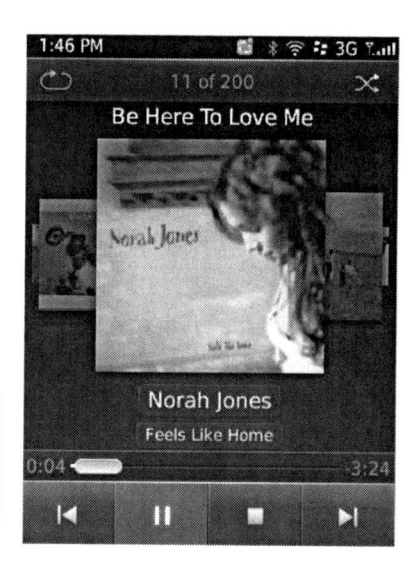

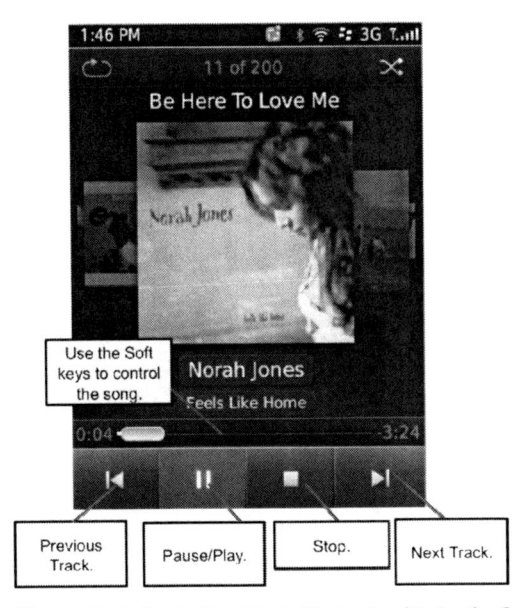

Figure 16-1. Controlling Music Playback with the On-Screen Controls.

Doing Other Things While Listening to Music

You can keep your music playing while you are checking email, browsing the web, or doing just about anything else on your BlackBerry.

Get your music started, as shown previously. Then press the **Red Phone** key to jump out to your Home screen. Start up any icon you want: **Messages**, **Browser**, **Calendar**, **Facebook**, anything. Your music continues playing.

To quickly return to your music player and select another song or playlist, press the **Menu** key and select an item near the top called **Now Playing…** This will instantly jump you right back to the music player (see Figure 16-2).

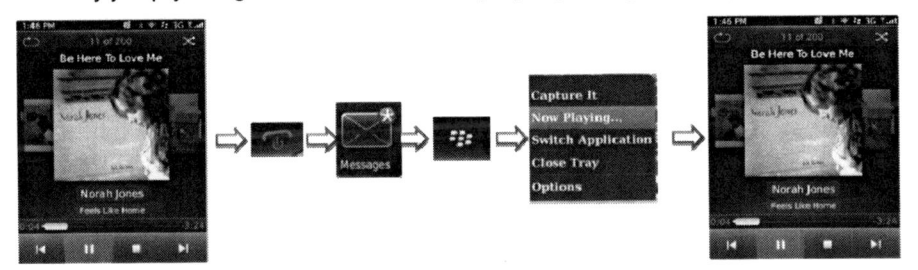

Figure 16-2. Jumping from any app to the Now Playing screen

When a phone call comes in, the music automatically pauses for your call. When you hang up, the music automatically starts up again, picking up where it left off.

The fastest way to pause and silence the music is to tap the **Mute** key on the top-right edge of your BlackBerry. Tap it again to re-start the music.

Using a Song as your Phone Ring Tone

1. Navigate to and play the song you want to use as a ring tone, as described previously.
2. Press the **Menu** key and select **Set as Ring Tone**.
3. Now, the next time you receive a call, your favorite song will be played. You can still set individual ring tones for your contacts – see Chapter 10.

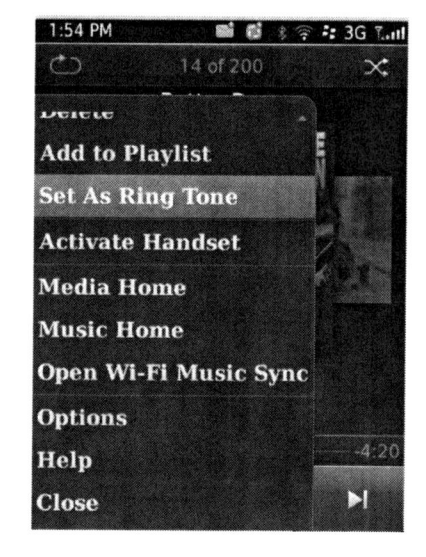

Playing All Your Music

Navigate to your music as just shown, and highlight the first song you wish to play. If you have not set up individual playlists, just highlight the first song, and then press the **Menu** key.

Scroll down to either **Play** or **Shuffle**, and the music player will begin to play all the songs listed.

If you select **Shuffle** (see Figure 16-3), then all the songs listed will be played in a random order.

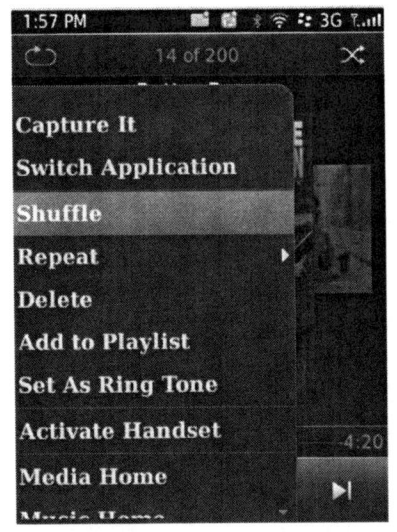

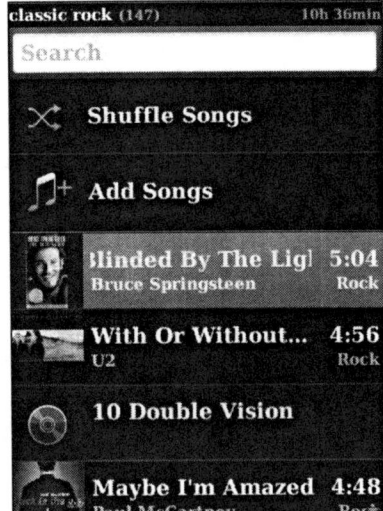

Figure 16-3. Shuffling all of your music

Playing Your Playlists

1. To see the playlists on the BlackBerry, just scroll over to the **Music Player** icon, and click it.

2. Find the **Media** icon or Home Screen and touch or click the **Music** icon from the home screen.

3. Then scroll down to click **Playlists**.

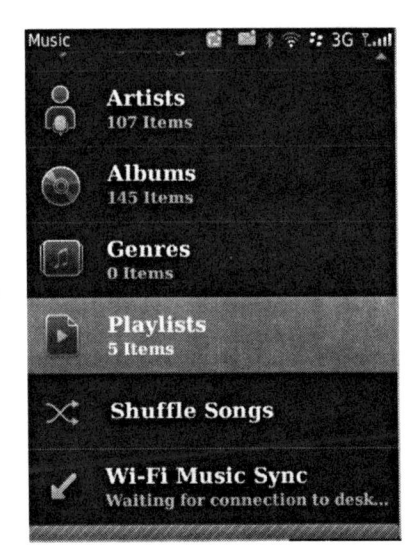

4. You can see that the two playlists synced from iTunes are now listed right on the BlackBerry.

5. Simply click any playlist to begin playing the songs. In this example, we click the **Classic Rock** playlist.

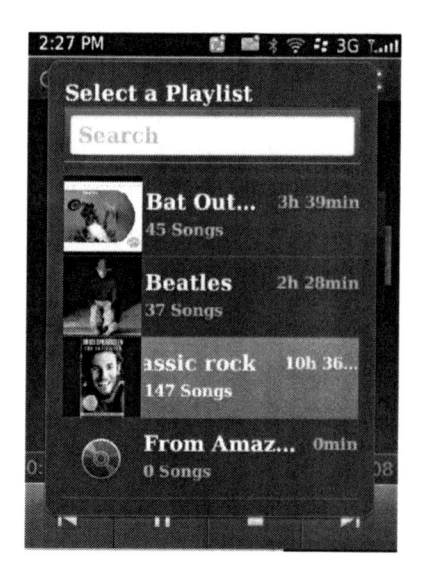

Creating Playlists from Your Computer

Use the supported computer software (for example **iTunes** or **Windows Media Player**) on your computer to create playlists. Use the steps shown in Chapter 3 (PC users) or Chapter 5 (Mac users) to learn how to sync these playlists to your BlackBerry.

Creating Playlists on Your BlackBerry

1. From your Music app, click **Playlists** to get into the Playlists section.

2. Touch or click **New Playlist** at the top of the screen.

> **TIP:** You can also add any song that is playing to one of the existing playlists or to a **New Playlist**. While a song is playing, press the **Menu** key and select **Add to Playlist** form the menu.

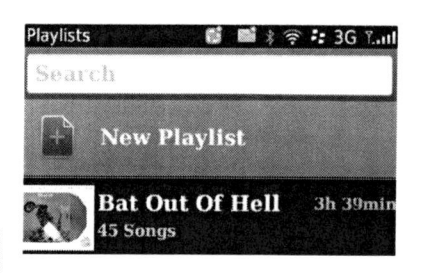

CHAPTER 16: Your Music Player

3. Now, you need to select **Standard Playlist** or **Automatic Playlist**.

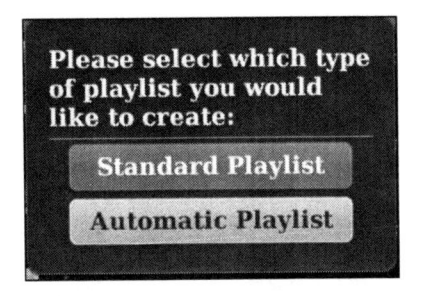

4. Select **Standard Playlist** to select and add any songs already stored on your BlackBerry.

5. Then type your playlist name in the **Name** field at the top, press the **Menu** key, and select **Add Songs** to add new songs or just touch the **Add Songs** button at the top of main screen.

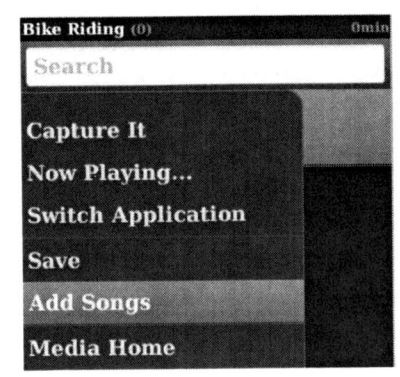

TIP: To find your songs, you can just scroll up or down the list or type a few letters you know are in the title of the song, like and instantly see all matching songs.

6. When you find the song you want, just click it to add it to your playlist.

TIP: You can add songs to any existing playlist by just selecting the playlist and touching or clicking **Add Songs** at the top of the **Playlist** screen.

TIP: The **Automatic Playlist** feature allows you to create some general parameters for your playlists based on artists, songs, or genres. See figure 16-3 below.

View Video Tutorials and Free Tips at www.MadeSimpleLearning.com

Figure 16-3. Generating automatic playlists on the BlackBerry

You can also add music to playlists by using the **Menu** key. This works when you are in the **Add Song** mode (Figure 16-4), or even when you just have a song playing and decide to add it to one of your existing playlists.

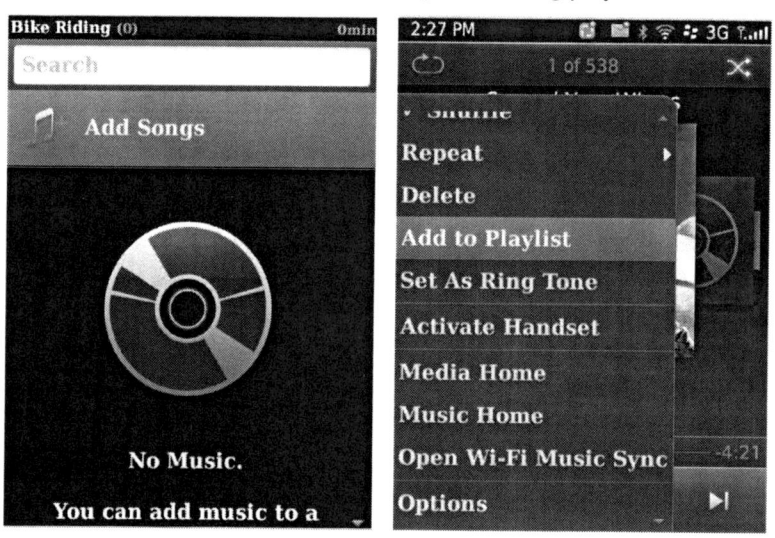

Figure 16-4. Adding individual songs to your playlist

Supported Music Types

The BlackBerry will play most types of music files. If you are an iPod user, all music except the music that you purchased with DRM (Digital rights Management) on iTunes should be able to play on the BlackBerry. However, if you burn your iTunes tracks to a CD (make a new playlist in iTunes, copy your

CHAPTER 16: Your Music Player

iTunes tracks, and then burn that playlist.) The most common audio/music formats supported are the following:

- ACC - audio compression formats AAC
- AAC+, and EAAC+ AMR - Adaptive Multi Rate-Narrow Band (AMR-NB) speech coder standard
- MIDI - Polyphonic MIDI
- MP3 - encoded using MPEG
- WAV - supports sample rates of 8 kHz, 16 kHz, 22.05 kHz, 32 kHz, 44.1 kHz, and 48 kHz with 8-bit and 16-bit depths in mono or stereo. Other WAV file formats may not be supported.

TIP: You can pause (and instantly silence) any song or video playing on your BlackBerry by pressing the **Mute** key on the top of your BlackBerry. Press **Mute** again to resume playback.

Music Player Tips and Tricks

- To pause a song or video, press the **Mute** key.
- To resume playing, press the **Mute** key again.
- To move to the next item, press the **Next** button at the bottom of the screen.
- To move to a previous item (in your playlist or video library,) press the **Previous** button at the bottom left of the screen.

TIP: See all the Music and Media Player shortcuts and hot keys in Part 4 of this book.

Streaming Internet Radio

If you ever get tired of your music, try free internet radio. You can listen to personalized radio stations that play only the music you like. Select from hundreds of pre-set stations, or make your own new station based on your favorite artist.

There are several applications that allow you to set up and listen to streaming internet radio. There are now several great free applications for the BlackBerry. All work well. The two most popular are **Pandora** and **Slacker Radio**, both found in BlackBerry App World. Use the steps in Chapter 20: "BlackBerry App World" to locate and download these free apps.

After you set up your free account, specify the type of music you like and you should be streaming internet radio in minutes.

CAUTION: All internet radio programs are very data intensive. In other words, you should use them only if you have purchased an **unlimited data plan** from your wireless carrier. Without an unlimited data plan, you may be surprised with an extremely high wireless data charge on your next bill.

Pandora Internet Radio

Pandora is an outgrowth of the Music Genome project. Essentially, Pandora is streaming internet radio where you control the radio stations. Essentially you build stations based around your favorite artists.

Installing Pandora

The easiest way to download Pandora is to go to **BlackBerry App World.** If the **Pandora** app is not part of the featured apps, you can find it in the Music section or by doing a search for Pandora.

Starting Pandora for the First Time

Once loaded, you will see the **Pandora** icon, usually in your **Downloads** folder.

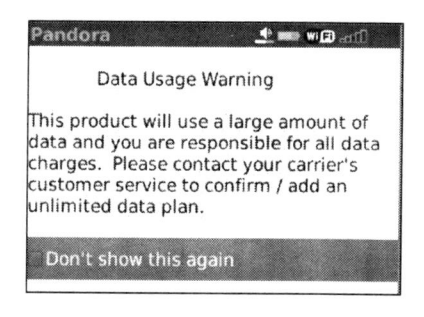

CHAPTER 16: Your Music Player

When **Pandora** starts for the first time, you will be asked if you already have an existing account or if you would like to create a new account

Just type your email address and password (for an existing account) or input your email address and password to create a new account. The best thing about **Pandora** is that it is free!

If this is your first time using Pandora, you will be asked to create a new station. If you have an existing Pandora account, you will see all your stations on the front page along with the **Create a New Station** option.

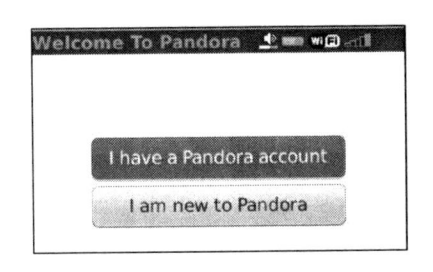

NOTE: Pandora also works on your PC or Mac. You can create your stations, listen on your computer, and then just log in on the BlackBerry. All the stations created on the computer will appear.

Pandora Controls and Options

The interface of Pandora is very clean and intuitive. To play or pause just press the **Play** button or **Pause** button.

To skip to the next song on the station, just click the **Next** button.

Using "Thumbs Up" and "Thumbs Down"

Two buttons that you can click are the **Thumbs Up** and **Thumbs Down** buttons.

If you click the **Thumbs Up** button, a check mark is placed on the song letting Pandora know that you like this song and it can be used again.

Consequently, if you click the **Thumbs Down** button, a very quick check mark appears, and Pandora will skip to the next song and remember to not ever play that song again.

Slacker Radio

Conceptually, **Slacker** is very similar to **Pandora** internet radio. You build stations around your favorite artists or choose from hundreds of existing stations for your listening pleasure. **Slacker** seems to have a few more pre-programmed stations than **Pandora** and has a cache feature, which allows you to stream music to your computer, and then download the music to your BlackBerry media card. This allows you to listen to music without needing an internet connection—perfect for a long airplane trip or whenever you are out of radio coverage for a while.

Like **Pandora**, You can listen to **Slacker** Radio on your PC and Mac, as well as your BlackBerry.

Downloading and Installing Slacker Radio

As you did with **Pandora**, you have two options for downloading **Slacker Radio**. You can find Slacker in BlackBerry App World, or you can also go to its website at www.slacker.com and download the latest build for BlackBerry. Follow the on-screen instructions for download and installation. Accept the license agreement and you will be Slacking in no time!

NOTE: Slacker might ask you for permission to both personal information and connectivity resources. You will need to Save the permissions to continue.

Creating or Logging in to a Slacker Account

As with Pandora, your first screen will ask if you have an existing Slacker account or if you wish to create one. Either log in or create a new account by clicking the button at the bottom of the screen. As with Pandora, you can have Slacker running on your computer also!

If you have an existing account, you will see a screen with your favorite stations displayed.

If you are signing on for the first time, you have two options:

- Scroll down to check out stations by genre.
- Click **Search** at the top and build a new station with music created from your favorite artist (as you did with Pandora).

In this example, we are creating a new station based on music by **John Legend**.

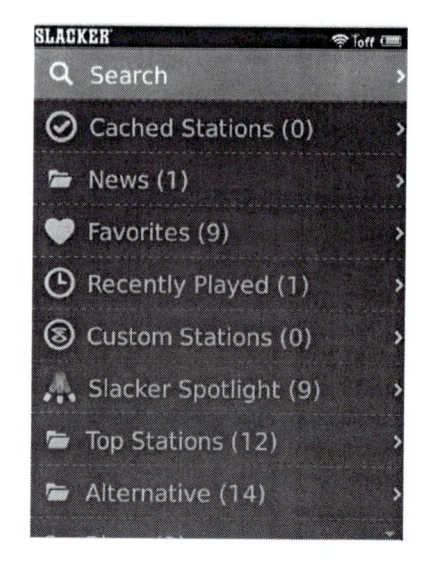

CHAPTER 16: Your Music Player

Choosing Your Station

Once you have some created stations, each time you log on to Slacker, you will be able to choose whether to listen to one of your **Favorites** or listen to other spotlight stations or genres in the Slacker library of music stations.

Slacker Controls

Slacker offers a few more controls and options than does Pandora. You can access the controls via the dock of buttons at the bottom. The Home button takes you back to the Slacker Home screen, where you can choose your station.

The **Next** or **Skip** button takes you to the next song in the station line-up. Notice that Slacker displays a picture of the next song or artist to the right of the album cover of the current song.

NOTE: in the free version of Slacker, you only get 6 skip in a day. The number of skips remaining is listed net to the **Skip** arrow.

If you like the artist or song, just click the **Heart** button for Slacker to remember that.

If you don't want to hear this particular artist or song again, click the **Do Not Play** or **Ban** button.

Slacker Menu Commands and Shortcut Keys

Slacker offers some very cool commands in the menu. Just press the **Menu** button on your BlackBerry and you can:

Upgrade to Slacker Plus

See Song Info

Cache the Station

Share the Station

Remove the Station

Search

See Slacker Settings

Shut Down Slacker

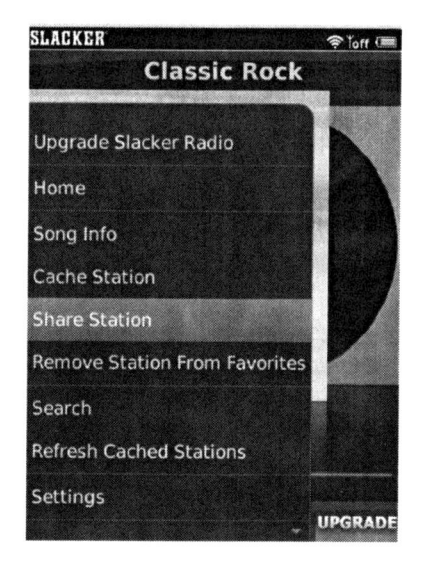

One-Key Shortcuts for Slacker

- **N** = Skip to Next Song (with free Slacker you are limited to six skips per hour)
- **space** = Play / Pause Song
- **H** = Heart (you like the song playing, play more songs like it)
- **B** = Ban Song (you don't like the song, don't ever play it again)

Chapter 17: Snapping Pictures

Many BlackBerry smartphones sold today come with a built in camera—perfect for snapping a quick picture. Some BlackBerry smartphones issued by companies might have the camera disabled or removed for security reasons. Your camera and the photos you take can easily be shared with others via email, MMS messaging, or even uploaded to Facebook and other social media programs.

Using the Camera

Your BlackBerry includes a feature-rich 5.0 mega-pixel camera. This gives you the option to snap a picture anywhere you are. You can then send the picture to friends and family, and share the moment.

Tips for Taking Great BlackBerry Pictures

Tip #1: Use Two Hands. Try to hold the BlackBerry with two hands to make it as stable as possible – shaky hands are one of the top reasons for poor pictures with smartphones.

Tip #2: Control the Flash. Make sure to turn on the flash when you need it. (We show you later how to do this). Sometimes the auto-flash sensor does not work if there is a very bright background. This is especially true when someone has the sun behind them, you should turn on the flash to make sure you see their faces without dark shadows.

Tip #3: Avoid the Zoom. Try to avoid zooming, instead try to move closer to the subject of the photo. The zoom is not a physical zoom, it is a software zoom so the image of a greatly zoomed in photo will look grainy or pixilated.

Tip #4: Scene Modes. Use the scene modes when required. We show you the various scene modes for the camera later in this chapter. For example, if you are out at the beach, then use the **Beach** scene mode. There is even a scene mode for taking pictures of text.

Tip #5: Work the delay. Usually, there is a ½ second or so delay between when you touch the soft key to take the picture and it is actually taken. Make sure to hold the BlackBerry very still until you hear that the picture has been taken.

Tip #6: Remember Landscape. Whenever something is wider than tall, such as an outdoor scene, it is usually better to rotate your BlackBerry sideways and take a picture that is wider than it is tall.

Tip #7: Maximize the Size. Use the camera options to set the picture resolution (number of pixels) to the maximum possible. This will give you the highest resolution images possible with your BlackBerry camera.

Camera Features and Buttons

You can get as involved as you want in your picture-taking with your BlackBerry. Every feature of your photo is configurable. Before we do that, however, let's get familiar with the main buttons and features.

Starting the Camera Application

The camera can be started in one of two ways:

Option #1: The Convenience key

Unless you have re-programmed your Convenience key (see Chapter 7 for details), then pressing the **Convenience** key will start your camera—the one directly below the volume control buttons.

Push this button once, and the camera should start.

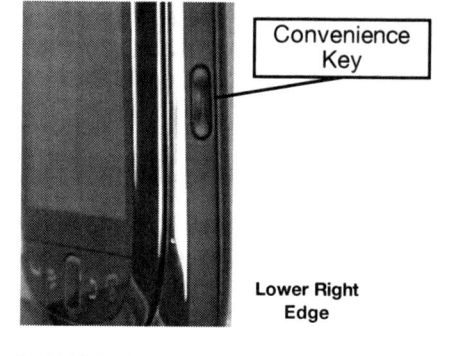

Convenience Key

Lower Right Edge

Option #2: The Camera Icon

Press the **Menu** key to see all your icons. Scroll to the **Camera** icon, and click.

CHAPTER 17: Snapping Pictures

Icons in the Camera Screen

Usually, when you open the **Camera** application, either the last picture you took is in the window or the camera is active. Underneath the picture window are five icons, which are described in Table 21-1.

Table 21-1: The Five Icons on the Camera Screen

Icon	Name	Description
	Last Picture Taken	Use this icon to view the last picture taken by the camera. You can then use this to scroll through pictures taken and upload to Facebook, Flickr, or another site if you have these apps installed (see Chapter 14 "Social Networking").
	Geotagging Options	You can use this icon to either **Show the Overlay** of the coordinates for Geotagging (giving your picture coordinates so others can see the location) or you can **Turn Off** Geotagging.
	Take-a-Picture	Click this icon to take another picture.
	Flash Mode	Touch or click to cycle through the Flash settings – either **Automatic**, **On** (uses more battery) or **Off**.

View Video Tutorials and Free Tips at www.MadeSimpleLearning.com   **401**

BlackBerry Torch Made Simple

Scene Modes   The BlackBerry has many presets for Scenes which will automatically adjust white balance and brightness depending on the scene. There are 11 pre-sets, from **Auto**, to **Portrait**, to **Snow** to **Beach** and more.

You can also set a picture as a wallpaper image for your Home screen (like the desktop background image on your computer).

1. Just open a picture from Last Picture Taken.
2. Press the **Menu** key, select **Set as Wallpaper**, and click.

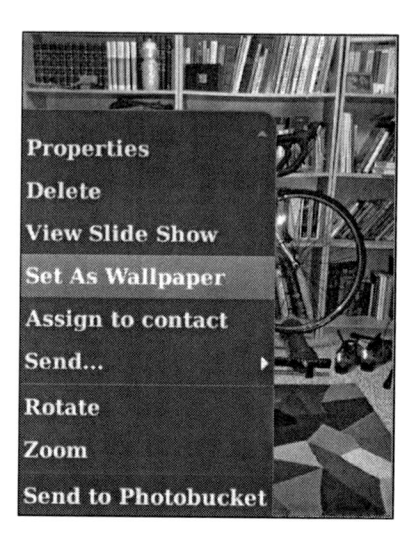

Sending Pictures with the Email Envelope Icon

Touch or click the Last Picture Taken icon. Scroll through the pictures in the Camera Roll and click the **Envelope** icon, which brings up the email dialogue box. You can also press the Menu key and scroll through the same options.

Click one of the six options:

- **Send as Email** (attached to an email message as an image file)
- **Send as Text Message** (a multimedia message as the body of an email—learn all about MMS on Chapter **323**)
- **Send to Messenger Contact** (this option may not be available if you have not yet set up **BlackBerry Messenger**—see Chapter 13 for details)

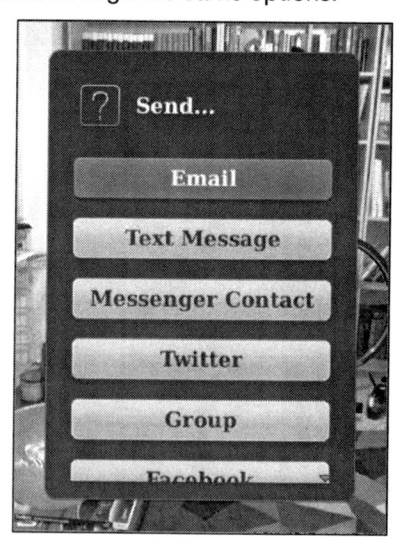

402

You can also send your picture to your **Twitter**, as a **Group** message, or **Facebook** accounts by pressing the appropriate button.

NOTE: These additional options appear only if you have them installed on your device.

Adjusting the Size of the Picture

The size of your pictures corresponds to the number of pixels or dots used to render the image. If you tend to transfer your BlackBerry pictures to your desktop for printing or emailing, you might want a bigger or smaller picture to work with.

1. From the Camera screen, press the **Menu** key and select **Options**.

2. Scroll down to the **Image Size** field and select the size of your picture: small, medium, or large.

3. Press the **Menu** key and save your settings.

TIP: If you email your pictures, then you will be able to send them faster if you set the **Picture Size** field to **Small**.

Geotagging Your Pictures

On your blackberry, you can either enable or disable geotagging. Geotagging means to assign of the current GPS (Global Positioning System) longitude and latitude location to each picture taken with your BlackBerry camera.

TIP: You can learn how to get **Flickr** installed on your BlackBerry in Chapter 20.

Why would you want to enable geotagging?

Some online sites such as Flickr (photo sharing) and Google Earth (mapping) can put your geotagged photo on a map to show exactly where you took the picture (see Figure 17-1 for geotagged pictures on Flikr).

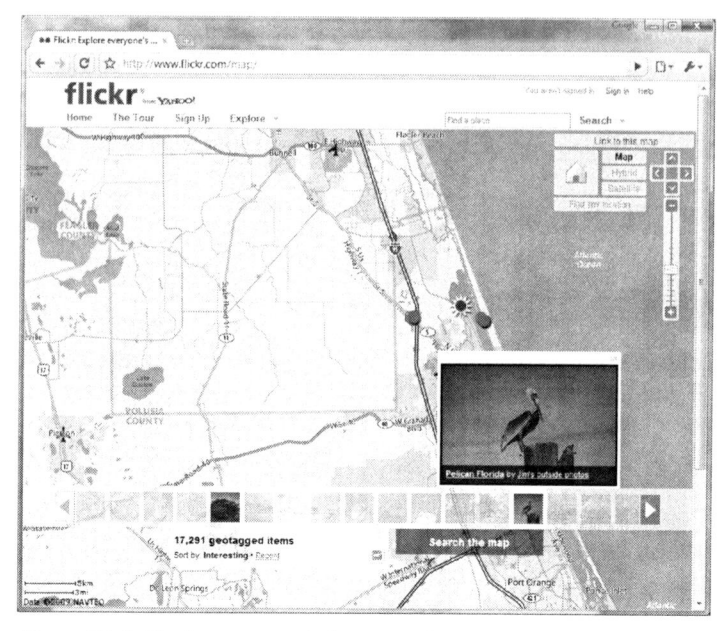

Figure 17-1. Viewing geotagged photos on Flickr

This is a map from www.flickr.com showing what geotagging your photos can accomplish. Essentially it will allow you to see exactly where you snapped your photos and help organize them.

Other programs you can purchase for your computer can organize all your photos by showing their location on a map (to find such software, do a web search for "geotag photo software (Mac or Windows)").

To turn geotagging on or off from the **Camera** screen, touch or click the **Geotagging** icon – second from left at the bottom.

> Make sure your GPS is enabled on your BlackBerry—see Chapter 21.

Select **Hide Overlay** to get rid of the name of your location at the very top of the screen. In this image it says **Barnstable Town, MA, USA**.

You know that geotagging is turned on if you see the red Geotagging symbol, in the lower left of your Camera screen.

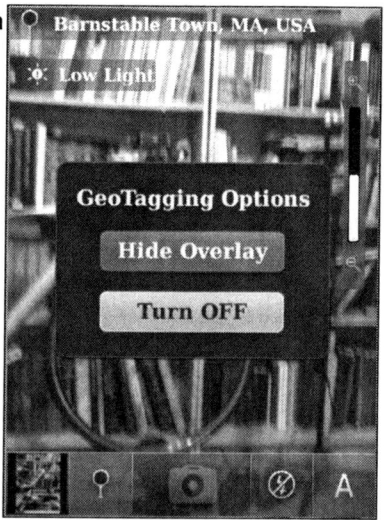

404

CHAPTER 17: Snapping Pictures

Changing the Scene Mode

Usually the automatic white balance works fairly well, however, there may be times when you want to manually control it.

The BlackBerry has a built in **Scene Mode** to make this process easy for you.

Touch or Click the **Scene Mode** icon (furthest to the right) and select form one of the built in eleven options.)

The **Scene Modes** work great. You can always just use the **Auto** setting, but a **Scene Mode** gives you more precise control of the exposure of the image.

The 12 available modes are described well on the BlackBerry itself and are: Auto, Face Detection, Portrait, Sports, Landscape, Party, Close-Up, Snow, Beach, Night and Text.

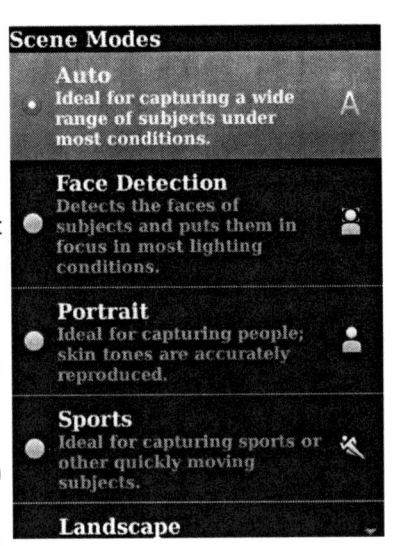

Scene Modes
- **Auto** — Ideal for capturing a wide range of subjects under most conditions.
- **Face Detection** — Detects the faces of subjects and puts them in focus in most lighting conditions.
- **Portrait** — Ideal for capturing people; skin tones are accurately reproduced.
- **Sports** — Ideal for capturing sports or other quickly moving subjects.
- **Landscape**

Adjusting the Picture Size/Quality

While the BlackBerry is not meant to replace a 7- or 8-megapixel camera, it is a very capable photo device. There are times when you might need or desire to change the picture quality. Perhaps you are using your BlackBerry camera for work and need to capture an important image. Fortunately, it is quite easy to adjust the quality of your photos. Realize, however, that **increasing the quality or the size will increase the memory requirements for that particular picture.**

In one "non-scientific" test, changing the picture quality resulted in the following changes to the file size of the picture at a fixed size setting of "Large (2592 1944)":

- Small: Approximately 230k
- Medium: Approximately 2X larger than normal
- Large: Approximately 3X larger than normal

1. Start the **Camera** application.
2. Press the **Menu** key, select **Options**.
3. Scroll to **Image Size**.
4. The choices **Large**, **Medium**, and

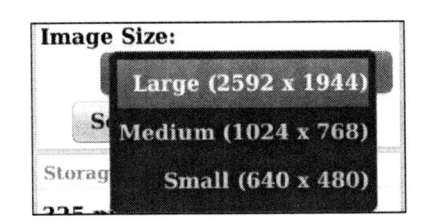

Image Size:
- Large (2592 x 1944)
- Medium (1024 x 768)
- Small (640 x 480)

View Video Tutorials and Free Tips at www.MadeSimpleLearning.com **405**

Small will be available. Just click the desired quality, press the **Menu** key, and save your settings.

Using the Zoom

As with many cameras, the BlackBerry gives you the opportunity to zoom in or out of your subject. Your BlackBerry has a 4x digital zoom. Zooming on the BlackBerry could not be easier. There are a couple of ways to Zoom in and Zoom out:

Frame your picture and gently scroll up (swipe up) on the screen with your finger. The camera will zoom in on the subject.

Use the **Trackpad** to zoom. Scroll up with the trackpad to zoom in and scroll down to zoom out.

TIP: You can also zoom using your **Volume Up** and **Down** keys.

Managing Picture Storage

Your BlackBerry comes with both 4 GB of internal media memory and an option media card slot. If you find that you need to increase your storage capacity, you can purchase a high capacity (e.g. 16 GB or 32 GB) media card. For more information on inserting the media card, please see Chapter 1. See the following for help on storing pictures on the media card.

If you do not have a media card, then you will want to carefully manage the amount of your BlackBerry's main device memory that is used for pictures.

Selecting Where Pictures Are Stored

The default setting is for the BlackBerry to store pictures in the main device memory, but if you have a media card inserted, we recommend selecting that instead.

CHAPTER 17: Snapping Pictures

To confirm the default picture storage location, do the following:

1. Press the **Menu** key from the main Camera screen, scroll to **Options**, and click.

2. Scroll down to the **Store Pictures** field and select **On Media Card** if you have one, or **In Device Memory** if you do not have a media card.

3. Look at the **Pictures Remaining** line under **Storage** to see exactly how many pictures you can store on either the **Media Card** or the **Device**.

Viewing Pictures Stored in Memory

There are two primary ways to view stored pictures.

Option #1: Viewing from the Camera Program

4. Open the **Camera** application and press the **Menu** key.

5. Scroll down to **View Pictures** and navigate to the appropriate image to view (see Figure 17-2).

View Video Tutorials and Free Tips at www.MadeSimpleLearning.com   407

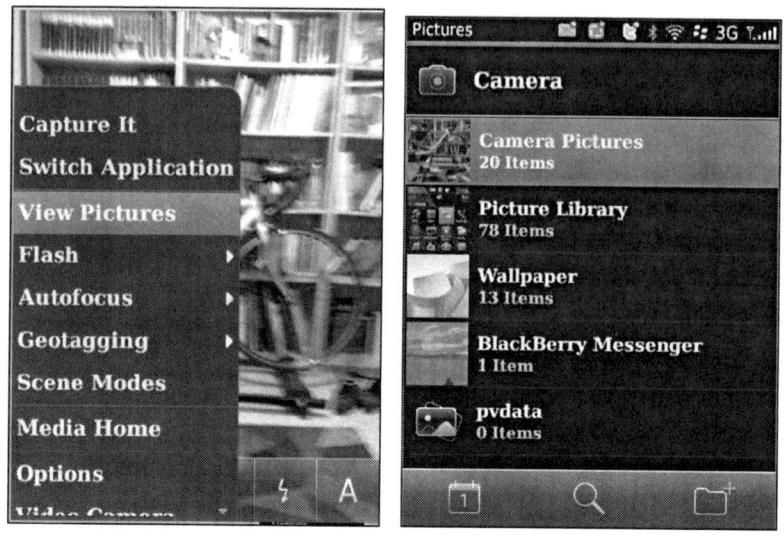

Figure 17-2. Viewing pictures from the Camera app

Option #2: Viewing from the Pictures App

1. Navigate to the **Media** icon and click.
2. Scroll to **Pictures** and click. Your initial options for pictures will be **Camera Pictures** or **Picture Library** (see Figure 17-3).
3. Click the appropriate folder and navigate to your pictures.

Figure 17-3. Viewing pictures from the Pictures app in picture folders

TIP: The bottom left icon is a View by Date icon that will, when clicked, list your pictures by month or year as opposed to by folder.

Picture Viewing Soft Keys

Like many of the other programs we have seen so far, the BlackBerry Torch adds soft keys at the bottom of the picture to put some of the more popular commands right at your finger tips.

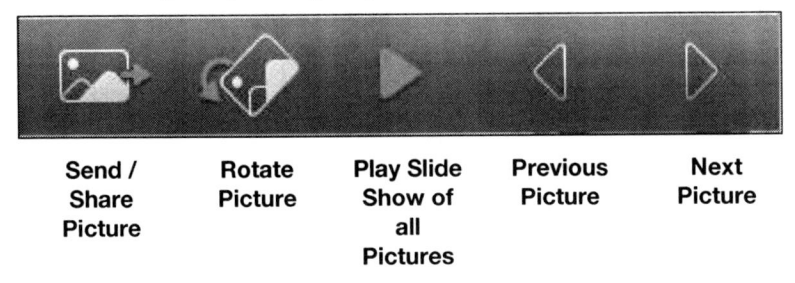

| Send / Share Picture | Rotate Picture | Play Slide Show of all Pictures | Previous Picture | Next Picture |

You can Share, Rotate, rotate the picture, or advance through your pictures by clicking the corresponding soft key.

Viewing a Slide Show

1. Follow the previous steps and press the **Menu** key when you are in your picture directory.
2. Scroll to **View Slideshow** and click.
3. You can also just touch the **Play** soft key to start the slideshow.

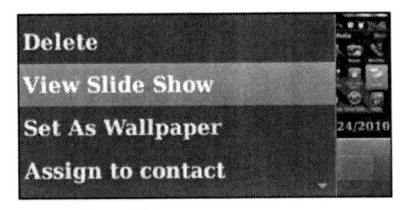

Scrolling Through Pictures

One very cool feature of your BlackBerry Torch is that you can scroll through your pictures just by swiping your finger across the screen.

Just click any picture to bring it into the full-screen mode.

Next, just swipe your finger to the left or right to advance through the pictures in that particular folder.

Adding Pictures to Contacts for Caller ID

As discussed previously, you can assign a picture as a caller ID for your contacts. Please check out our detailed explanation in Chapter 10.

TIP: You can set the **Convenience** key from the middle of the Camera Options screen.

This will allow you to set the **Convenience** key to start your camera.

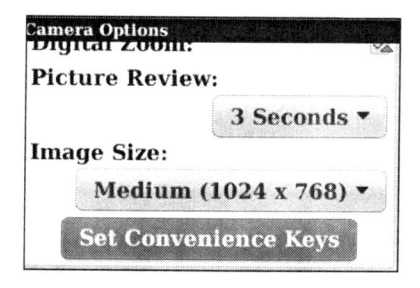

Transferring Pictures To or From Your BlackBerry

There are a few ways to remove pictures you have taken from your BlackBerry and transfer pictures taken elsewhere onto the BlackBerry.

Method 1: Send via email, multimedia messaging, or BlackBerry Messenger. You can email or send pictures immediately after you take them on your camera by clicking the **Envelope** icon. You may also send pictures when you are viewing them in your **Media** application. Click the **Menu** key and look for menu items related to sending pictures. See figure 17-4 below. Or, look for a soft key at the bottom of the screen.

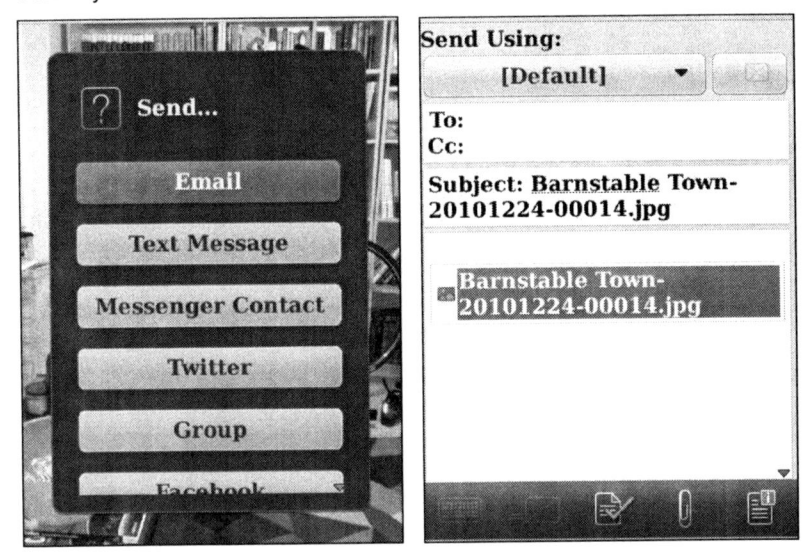

Figure 17-4. Sending options for pictures

Method 2: Transfer using Bluetooth. If you want to transfer pictures to/from your computer (assuming it has Bluetooth capabilities), you can. We show you how to pair or connect Bluetooth devices in the "Bluetooth" section of Chapter 1.

Method 3: Transfer using your computer with special computer software.

Transferring pictures and other media to your computer is handled using the media section of your desktop software. For a Windows computer, see Chapter 3; for an Apple computer, see Chapter 5.

Method 4: Transfer using USB Drive mass storage mode. This assumes you have stored your pictures on a media card. (What's a media card? See Chapter 1)

The first time you connect your BlackBerry to your computer, you will probably see a **USB Drive** question. If you answer yes, then your media card looks just like another hard disk to your computer (just like a USB flash drive). Then you can drag and drop pictures to/from your BlackBerry and your computer. For more details, see Chapter 1.

Chapter 18: Fun with Videos

We have already taught you how to get the most from music and pictures on your BlackBerry; however, your BlackBerry can handle even more media than what we have discussed so far. For example, your BlackBerry can also handle videos ranging from short videos you shoot on the BlackBerry to full-length movies.

Working with Videos on a BlackBerry

In addition to a camera, your BlackBerry also comes with a built-in video recorder that enables you to capture your world in full-motion video and sound when a simple picture will not work. Clicking the **Videos** icon lets you play all videos you record or transfer to your BlackBerry from your computer.

Some BlackBerry providers include a free 4, 8 or 16 GB media card with the devices they sell. All this extra storage and the strong video capabilities make the BlackBerry a very capable point-and-shoot video camera and video player.

Transferring Videos to Your BlackBerry

Check out Chapter 3 (Windows users) and Chapter 5 (Mac users) to learn how to transfer videos and other media (e.g., pictures and songs) to your media card.

Your Video Camera

One of the features of your BlackBerry is the inclusion of a video camera in addition to the still camera. The video camera is perfect for capturing parts of a business presentation or your child's soccer game. Like pictures, videos can be emailed or stored on your PC or Mac for later use.

CHAPTER 18: Fun with Videos

Using the Video Camera

It's a simple matter to start the **Video Camera**; simply follow these steps to do so:

1. Scroll over to the **Media Home** screen and click on the **Video Camera** icon.

2. The BlackBerry should detect that you have a media card installed and ask you if you want to save your videos to the card. We recommend saying **Yes**.

3. Use the screen of the BlackBerry as your viewfinder to frame your picture. Try to have equal amounts of background around your image.

4. When you are ready to record, press the **Record** button. When you're finished recording, press the **Pause** button.

5. You will then see options at the bottom of the screen to **Continue Recording**, **Stop**, **Play**, **Email**, **Send**, **Rename** or **Delete** the video.

View Video Tutorials and Free Tips at www.MadeSimpleLearning.com   413

6. Press the **Menu** button from within the **Video Camera** application after the video is recorded and choose one of the following options:
   - Switch Application
   - View Videos
   - Auto Focus
   - Upload to YouTube
   - Send As Email
   - As Text Message
   - Send to Messenger Contact
   - Media Home
   - Options
   - Camera
   - Help
   - Close

7. Adjust your video options by selecting **Options** from the menu.

8. In this menu, you can adjust the video light. Your choices are to have a constant light from the camera's flash or to set this option to **Off**.

9. You can also adjust the **Scene Mode** options – these are:
   - Auto
   - Portrait
   - Landscape
   - Close-up
   - Beach

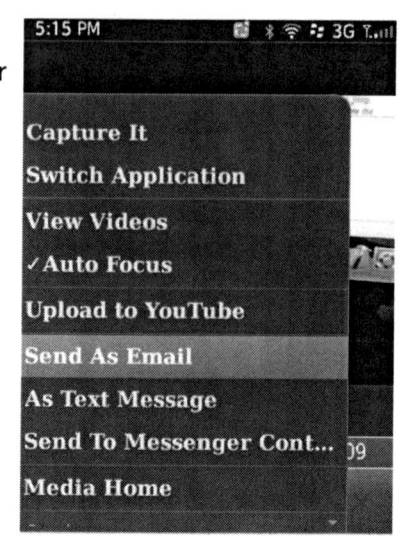

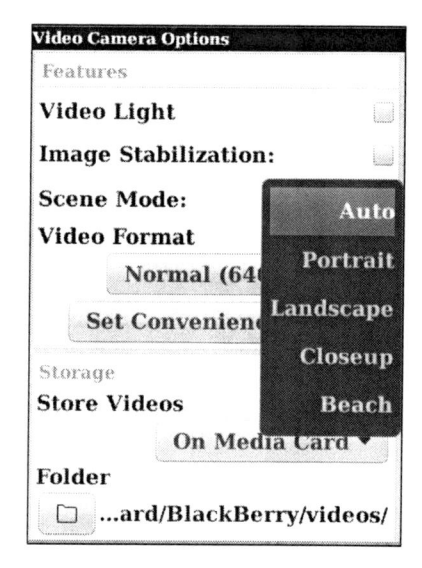

414

10. You can also adjust the resolution of the camera in the **Video Format** option, selecting either **Normal (640 x 480)** or to send as an **MMS (176 x 144 pixels)**. The latter option delivers a lower quality image and a smaller file size (see Figure 18-1).

11. Finally, the **Store Videos** option lets you store your videos **On Media Card** or **On Device**. You should definitely choose the **On Media Card** option if you have a media card because this will give you more space.

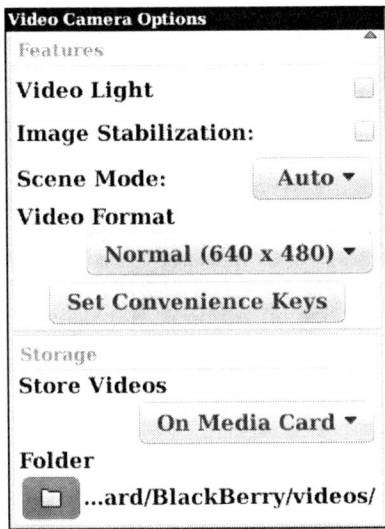

Figure 18-1. Adjusting video options

Converting DVDs and Videos to Play on the BlackBerry

One cool thing about your BlackBerry is that you can use it is as a portable media player to help entertain your kids or yourself when you travel. If your car is equipped with Bluetooth, you can use this technology to send audio from your BlackBerry to your car stereo, so everyone can listen to the movie.

To convert a DVD, you need to follow these steps:

1. Copy the DVD onto your computer.

2. Use a video encoder program to transcode the video (convert it so it is viewable on your BlackBerry).

> **CAUTION:** Please respect copyright laws! As people who make our living from intellectual property (e.g., BlackBerry books, videos, and so on), we strongly encourage you to respect the copyrights of any material you are attempting to copy to your BlackBerry.

For security and copyright reasons, some DVDs cannot be copied onto a device like the BlackBerry. Before you buy a video encoder or converter program, make sure that it supports the file formats that will play on your BlackBerry. Many of the user enthusiast forums such as www.pinstack.com, www.crackberry.com, or www.blackberryforums.com offer tutorials for video conversion. (We did not include details here because the process varies, based on which computer operating system you are running.)

Supported Video Formats on the BlackBerry

Video formats are fast-changing, and there are multiple versions of the BlackBerry operating system which support a variety of video formats. As of the time of the writing of this book, the following video formats were supported;

.mp4

.mov

.3gp

.3gp2

.avi

.asf

.wmv

To see if this changes, Simply do a search for "supported video formats Torch." This link worked at the time of writing:

http://docs.blackberry.com/en/smartphone_users/deliverables/18577/Supported_audio_video_file_formats_60_1018040_11.jsp

Viewing Videos on the BlackBerry

The BlackBerry contains a very sharp screen that is perfect for watching short videos. Another nice feature: The video player is as easy to use as the audio player.

CHAPTER 18: Fun with Videos

Playing a Video

Playing a video is fairly straight forward:

1. Tap the **Video** icon in your **Media** folder.

2. From the video list screen, you can type a few letters to find the video you want to play. In this case, we typed BBM to find the Made Simple Learning video tutorials related to BlackBerry Messenger.

3. Tap any video you wish to play. Tap **Now Playing** to access the last video you were playing.

4. The **Video Player** screen looks very similar to the **Audio Player** screen – just touch the video or click the trackpad to pause or play a video, using the volume controls on the side of the BlackBerry.

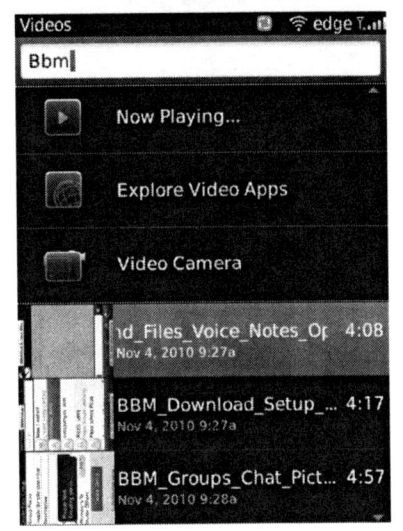

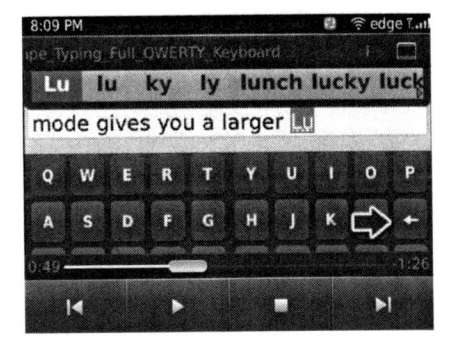

5. Touch the video you want to see with your finger to make it start playing.

View Video Tutorials and Free Tips at www.MadeSimpleLearning.com   **417**

> **NOTE:** Videos only played in **horizontal** (landscape) mode at the time of writing.

Showing or Hiding controls

You can tap the screen to show or hide the controls at the bottom of the screen (see Figure 18-2).

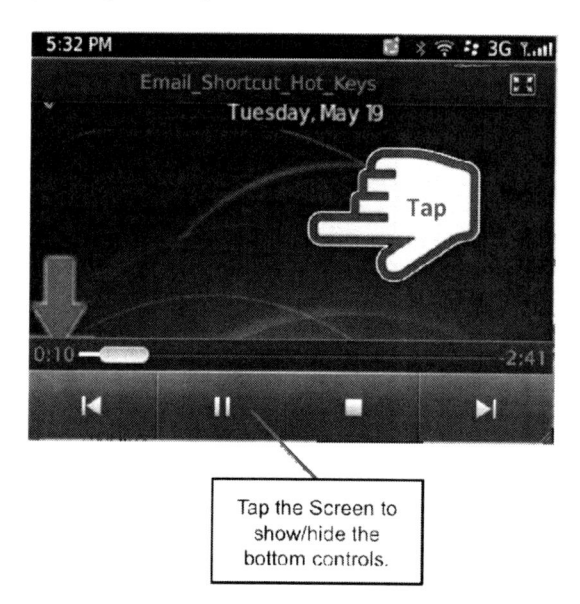

Video still image from the Made Simple Learning BlackBerry Torch video tutorials.

© 2011 Made Simple Learning.

Figure 18-2. Tapping the screen to see video controls

You can use the slider bar at the top of the controls to move to any other section of the video. Just touch it and drag it left or right (see Figure 18-3).

Video still image from The Dark Knight Movie

© 2007 Warner Brothers and Legendary Pictures.

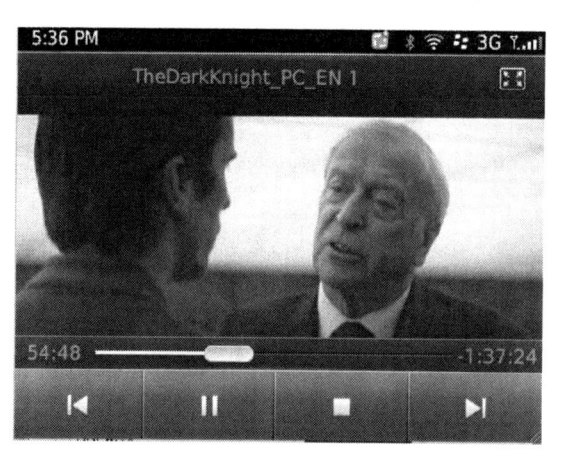

Figure 18-3. Drag the slider to move the video location.

At the top of the controls, you will also see two times. The time on the left is the amount of time already played, while the time on the right, a negative number, counts down how much time remains to be played in the video.

> **NOTE:** The Shuffle and Repeat buttons located between the two times above the top row of controls did not work when playing a video at the time of writing.

This list describes the four buttons along the bottom of the video player, from left to right:

- **Previous Track**: Takes you to the previous video in the list.
- **Play / Pause**: Pauses the video if it's playing and plays the video if it is paused.
- **Stop**: Stops the video and moves you back to the beginning of the video.
- **Next Track**: Takes you to the next video in the list.

The **Video Player** application includes a fairly extensive list of options you can choose from by clicking the **Menu** key. Specifically, pressing the **Menu** button brings up the following list of menu options for the application:

- **Switch Application:** Allows you to go to any other open application.
- **Video Size:** Choose between **Fit Screen**, **Full Screen** or **Actual size**.
- **Delete:** Deletes the current video.
- **Send to:** Sends to any installed photo sharing apps like Photobucket or Flikr.
- **Activate Handset:** Plays the audio of the video through your headset.
- **Media Home:** Go to the Media Home Screen.
- **Videos Home:** Go to the Video directory
- **Options:** Adjust the video options.
- **Help:** Displays contextual help with **Video Player** application.
- **Close:** Closes the application.

Chapter 19:
Web Browser

A requirement for any smartphone today is that you need to be able to get online and browse the Web. While smartphone browsing the Web on your smartphone will never be a substitute for desktop browsing, you might be pleasantly surprised by surfing the Web on your BlackBerry.

BlackBerry OS 6.0, found on the Torch, is the first web kit based browser for a BlackBerry. That means that pages load faster, pinching and zooming is quick and accurate and videos play in the browser as well.

In this chapter, we will show you how to get online, how to set your **Browser** bookmarks, how to manipulate your browsing history, and how to use some *hotkeys* to maneuver around the **browser** program quickly.

Web Browsing on Your BlackBerry

One of the amazing features of smartphones like the BlackBerry is that you can use them browse the Web with ease and speed right from your handheld. Plus, an ever-increasing number of web sites are now supporting mobile browser formatting. These sites sense you are viewing the site from a small mobile browser, and they automatically reconfigure themselves for your BlackBerry so they can load quickly. The specialized mobile version of some sites load more quickly than when using their desktop browser equivalents!

Starting the Web Browser and the Start Page

Tap the **Browser** icon to start browsing.

CHAPTER 19: Web Browser

You will usually be taken to the **Start Page** shown here where you see **Bookmarks**, **History** and can type a web address at the top.

Your wireless carrier may have made some changes, however and you may see your carrier's web page or something else.

Tap the **Pages** icon in the upper right corner to open or view more web pages.

TIP: You can simply type your web search right in the address bar without first going to your favorite search engine (Google, Yahoo, etc.)

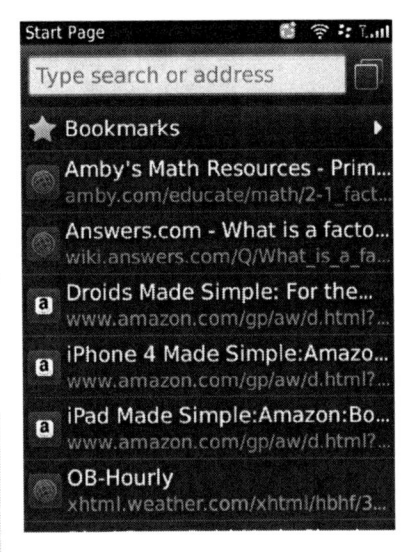

Using Tabs for Browsing

Tabbed browsing has long been available on desktop browsing to make it easier to jump between separate web pages – set as tabs. For the first time, tabbed browsing is now available on the BlackBerry in OS 6.0.

Opening and Using Tabs

Touch the **Tab** icon next to the **Address** bar of the **Browser**. If multiple tabs are already open, you will see a small number in the **Tab** icon.

Touch the **Tab** icon and the open web tabs will be shown at the top of the web page. Just swipe through the tabs to see the open pages.

To jump to one of the open tabs, just touch it and the page will load.

Swipe your finger left or right to view more open screens.

Tap the red **X** under the tab to close it.

Tap any tab to open and view it.

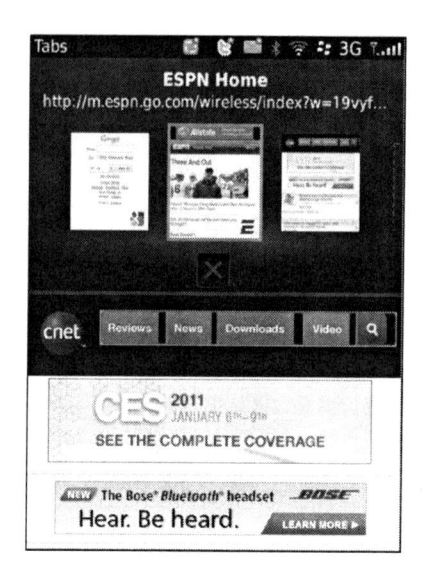

Opening a New Tab

Swipe all the way to the end of the open tabs and touch the **New Tab** icon (box with the plus sign in it) and click to load a new web page into a new tab. Choose from one of the items in your history or a bookmark or type in a new web site. Once the page is loaded, it will be set as a new tab in the browser.

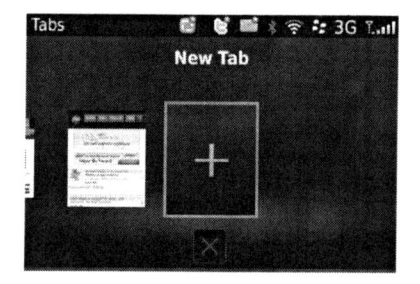

Using Browser Keyboard Shortcuts

You can use various keys on your slide-out keyboard to perform common functions on your Browser. Some of the more useful ones are: **A** = Add Bookmark, **S** = Browser Options, **G** = Go To (Brings you to the Start Page where you can type a web address, view history and bookmarks), **R** = Refresh current page, **I** = Zoom In, **O** = Zoom Out.

> **TIP**: See all the **Browser** keyboard shortcuts listed in Part 4 of this book.

Using the Browser Menu

Like most other applications on the BlackBerry, the heart of the **Browser** program lies in the capabilities you access from the program's **Menu** options. One push, one press, and/or one click can mean you are off to specific sites, bookmarked pages, recent pages, your internet history, and much more.

All the fun begins on the **Browser** program's **Start** page. The **Start** page view includes a **Web Address** box; a **Web Search** box (e.g., Google, Yahoo!, and so on); a **Bookmarks...** section that shows your bookmarks; and a **History...** section that shows, just as you'd expect, a list of recently visited pages (see Figure 19-1).

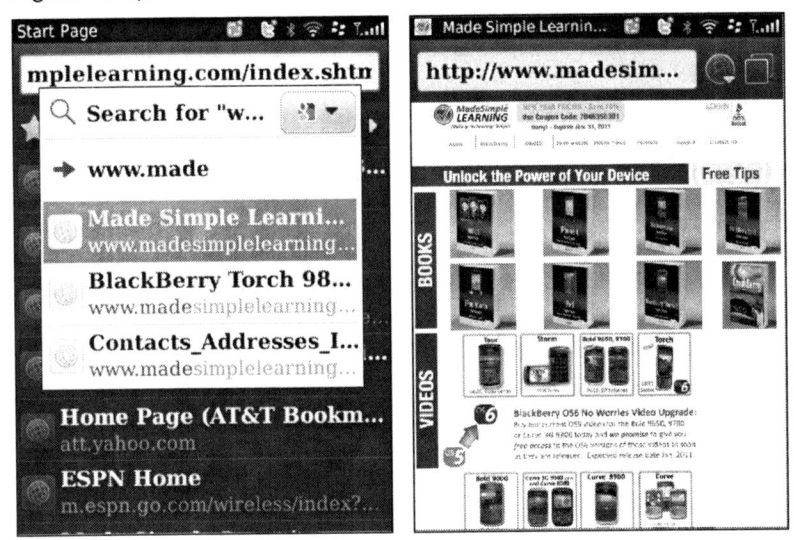

Figure 19-1. Type in a web address from the browser start page

To visit a Web site, you can Tap one of the links in **Bookmarks...** or **History...**; alternatively, you can type in a new URL, and then Tap it. (Remember: You use the **space** key for the "." in the address.)

Exploring the Browser Menu Options

What follows is a list and short explanation of the **Menu** options that become visible when viewing a web page with the BlackBerry's native **Browser** program (see Figure 19-2). You can view these options by pressing the **Menu** key as you view a web page:

- **Switch Application:** Jumps or switches over to other applications, while leaving the current web page open.
- **Show Keyboard:** Brings up the on screen keyboard. This is only available when the Torch is closed.
- **Find on Page:** Searches for text on a web page.

- **Select:** Puts the Browser in Select mode to select text or pictures to copy to the clipboard.
- **Open Link in New Tab:** When the cursor is on a link to another page, you can open up a new tab with the new web page.
- **Add Link to Bookmarks:** Sets the new link as a bookmark in the web browser for future retrieval.
- **Send Link:** Enables you to send the web link to **Email**, **Text Message**, **PIN**, **Messenger Contact**, **Group Message**, **Social Feeds** or **Twitter**.
- **Copy Link:** Copies the web link to then paste into another app.
- **View Image:** Opens up a separate window with the web image highlighted.
- **Save Image:** Saves a web image to your **Pictures** folder.
- **Send Image Link:** Enables you to send the image link to **Email**, **Text Message**, **PIN**, **Messenger Contact**, **Group Message**, **Social Feeds** or **Twitter**.
- **Copy Image Link:** Copies the image link to then paste into another app.
- **Zoom:** Zooms you in on specific part of the page, so you can see that part of the page more clearly. You can zoom in several levels.
- **End Zoom:** Zooms you out, so you can see more of the page.
- **Refresh:** Updates the current web page.
- **Go To...:** Allows you to type in a specific web address.
- **Add to Bookmarks:** Sets the current page as a **Favorite** (i.e., a bookmark). For obvious reasons, this feature is extremely useful.
- **Add to Home Screen:** Creates an icon on the **Home** screen that will automatically take you to this web page in the future.
- **Send/Copy Page Address:** Very similar to **Send Link** described above, but with the address for the web page.

CHAPTER 19: Web Browser

- **Page Properties:** Displays the properties for the current web page.
- **Tabs:** Shows you the open web tabs at the top of the screen. Just move between the tabs and touch the page you wish to go to.
- **Bookmarks:** Lists all your bookmarks.
- **History:** Shows your entire web browsing history.
- **Downloads:** Displays your web downloads.
- **Options:** Sets the **Browser** program's **Configuration**, **Properties**, and **Cache** settings.

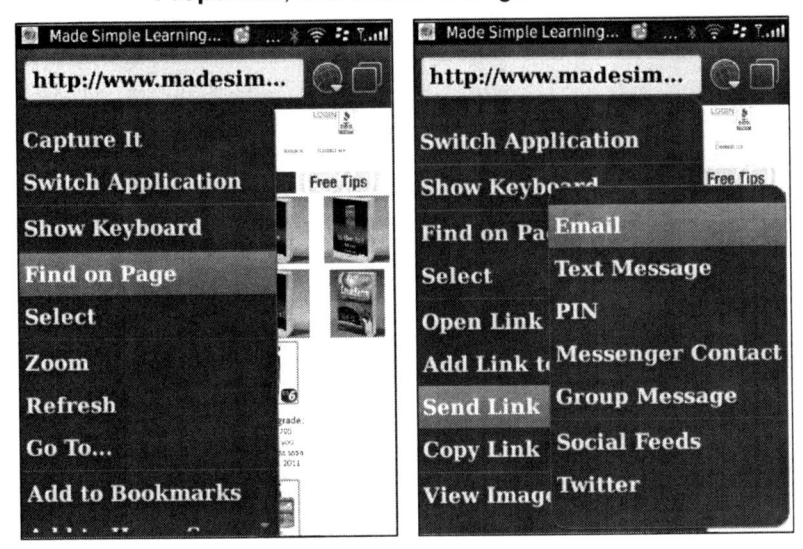

Figure 19-2. Reviewing *menu options in the native Browser program*

- **Help:** Shows on-screen text help for the **Browser** program. This is useful when you forget something, and need quick assistance.
- **Close:** Terminates the current instance of the Browser program and exits to the **Home** screen.

Using Your Address Bar

The first thing you will want to know how to do is get to your favorite web sites. In the case of one of this book's authors, that site is www.google.com. On your desktop computer, you simply type the web address or (URL) into your browser's **Address Bar** box. You will also see an **Address Bar** on the BlackBerry.

BlackBerry Torch Made Simple

You can touch the address bar to highlight it and just type in a new web address or, you can use the **Go To...** menu command to type in your web address. Follow these steps to do so:

Open the **Browser** program by pressing and clicking the **Browser** icon (see Figure 19-3).

Slide open your keyboard and start typing a web address or a web search string.

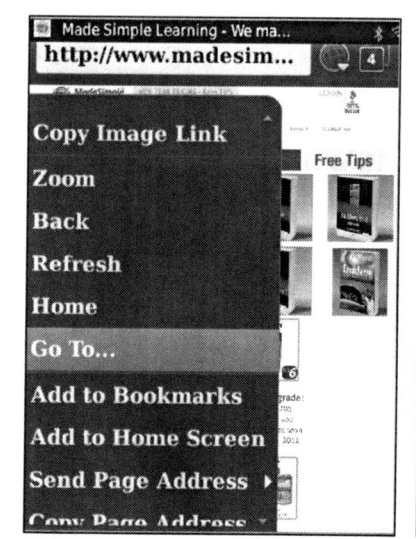

Figure 19-3. Using the Go To... Command from the menu.

> **TIP:** A fast way to get to the top of any web page (and be able to type a new web address or search) is to tap your finger at the very top of the screen. You will instantly see the address bar appear for you to tap and type in it.

CHAPTER 19: Web Browser

The **Address Bar** comes up with the "http://www." part in place, waiting for you to type the rest of the address. Simply type in the web address of the site you wish to visit. Remember: pushing the space key will insert the dot "." automatically.

Or, just touch and highlight the **Browser Bar** and then type in a new web address.

Press the **Enter** key when you are done. The **Browser** program will take you to the entered web page.

> NOTE: As you type in the web address, the BlackBerry will begin searching for matching web sites. If you see the site listed, just scroll to it and click the **Trackpad** to load it.

Once you have type in a few web addresses using the **Go To...** command, you will notice that they appear in a list below the web address bar the next time you select **Go To...** section. You can select any of these sites by scrolling down, and then pressing and clicking one (see Figure 19-4).

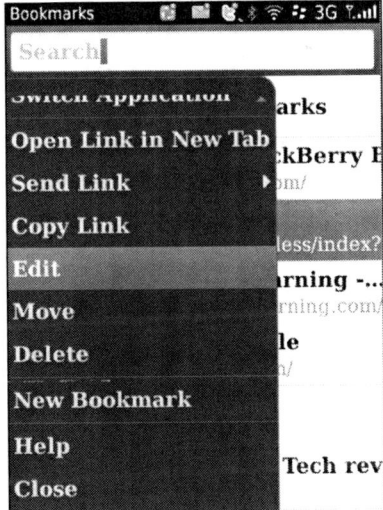

Figure 19-4. Frequently visited web sites in the Bookmarks... (if set) and History... sections

View Video Tutorials and Free Tips at www.MadeSimpleLearning.com 427

> **TIP:** You can save time by editing a bookmark. If you want to enter a web address that is similar to one that you have a bookmark for, you can highlight the previously entered address, press the **Menu** key, and then select the **Edit** option.

Using the Globe Shortcut Menu

Tap the **globe** next to the address bar to see a drop-down menu with quick links to:

Add to Bookmarks (add this page to bookmarks).

Add to Home Screen (add as an icon to your **Home** screen).

Send Page Address (email or otherwise send this web address).

Copy Page Address (copy this address to paste elsewhere).

Bookmarks (view your bookmarks list).

History (view your history).

Zoom In and Out of Web Pages

One of the great features of the new BlackBerry browser is the ability to Pinch to Zoom in or out of a web page.

There are two primary ways of zooming in and out;

1. Double tapping – just double tap on a blank area of a web page to zoom in an double tap again to zoom out.

2. Pinch to zoom – using your thumb and forefinger, spread your finger out to zoom in and pinch them closed to zoom back out (see Figure 19-5).

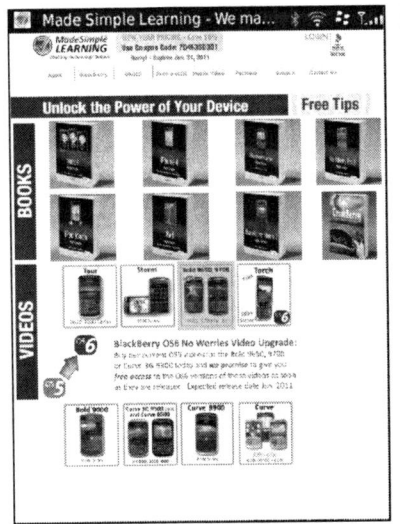

Figure 19-5. Zooming in and out of web pages.

Copying or Sending a Web Page or Link

The BlackBerry's native **Browser** program makes it easy to copy or send a page you might be viewing to someone else. Follow these steps to do so:

1. Press the **Menu** key while viewing a web page in the **Browser** program.

2. Scroll down to the **Send Page Address** option, and then Tap it.

3. You will now see the option to send the page address via **Email**, **Text Message**, **PIN**, **Messenger Contact**, **Group Message**, **Social Feeds** or **Twitter**.

4. Select **Copy Address** to copy the web address to the clipboard; from here, you can easily paste the URL into any other app, email or memo.

5. Alternatively, touch the **globe** next to the address bar and select **Send Page Address** or **Copy Page Address** to do the same thing.

Adding Bookmarks

One of the keys to a great web browsing experience on your BlackBerry is the liberal use of the **Bookmarks...** feature. Your BlackBerry comes with a couple of bookmarks already set. It is very easy to customize your bookmarks to include all your favorite sites for easy browsing.

> **TIP:** You can instantly find bookmarks by typing a few letters of a bookmark's name. You do this in exactly the same way you look up contacts in the **Address Book** program. Remembering this tip can save you a considerable amount of time!

Bookmark Naming Tips

One of the keys to a positive web browsing experience is to name add and name bookmarks in a way that makes them easy to find later. In the next section, we will cover how to set up an example bookmark that lets us look up our local weather instantly.

1. Open the **Browser** program and use the **Go To...** command (or the G keyboard shortcut) to input a favorite web page. In this example, we will type in *www.accuweather.com*.

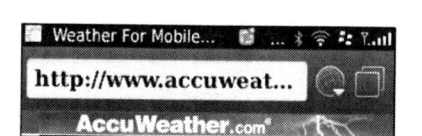

CHAPTER 19: Web Browser

2. Type in your ZIP code or city name to see your current weather (see Figure 19-6).

3. Once the page loads with your local weather, press the shortcut letter **A** on your slide out keyboard to add a new bookmark.

4. The full name of the web address is now displayed. In this case, you will probably see *AccuWeather Quick Look Weather for Marstons Mills, MA*. You may want to rename the URL (see the upcoming section on naming booksmarks well).

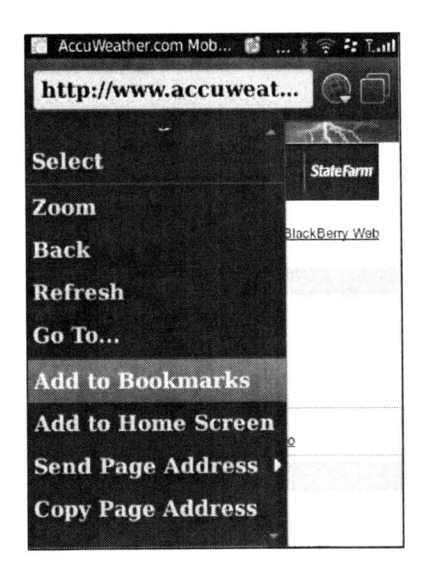

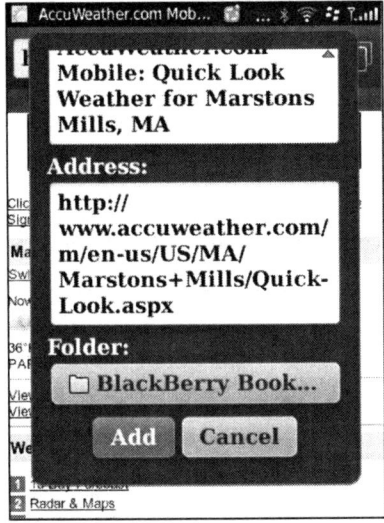

Figure 19-6. Adding a bookmark for your local weather

5. In this case – in most cases, really – we recommend changing the bookmark's name to something short and unique.

NOTE: If you were to bookmark four different weather forecasts, the default bookmark names would all show up as *AccuWeather*. This would be rather useless if your goal were to get straight to the ten-day forecast.

View Video Tutorials and Free Tips at www.MadeSimpleLearning.com   431

Be sure to keep the following tips and trick in mind as you edit and rename your bookmarks:

- **Keep all bookmark names short:** You will only see about the first 10-15 characters of the name in your list because the screen is small, so any characters beyond that won't be displayed.
- **Make all bookmark names similar, but unique**: If you were to add four bookmarks for the weather in New York or your area, you might adhere to this advice by naming them along these lines:
    - NY – Now
    - NY – 10 day
    - NY – 36 hour
    - NY – Hourly

This approach lets you instantly locate all your forecasts by typing the letters *NY* in your bookmark list. Only those bookmarks with the letters *NY* will be displayed.

Viewing the Start Page or Home Web Page

You might prefer to see your list of bookmarks rather than a **Home** page when you open the **Browser** program. The reason is probably obvious: this will allow you to use the **Find** feature in the bookmark list to instantly locate a bookmark, press it, and click it.

This approach speeds up the time it takes to get to favorite bookmarked web pages, whether you want to see the local weather or fire up your favorite search engine.

It might also be that your BlackBerry automatically opens up to your list of bookmarks, but you would prefer to see a particular **Home** page instead. If so, you can follow these instructions to change a given page your **Home** page:

1. Tap the **Browser** icon.
2. Press the **Menu** key and select **Options**.
3. Tap **Start Page** or type in a web address under Home Page(see Figure 19-7).

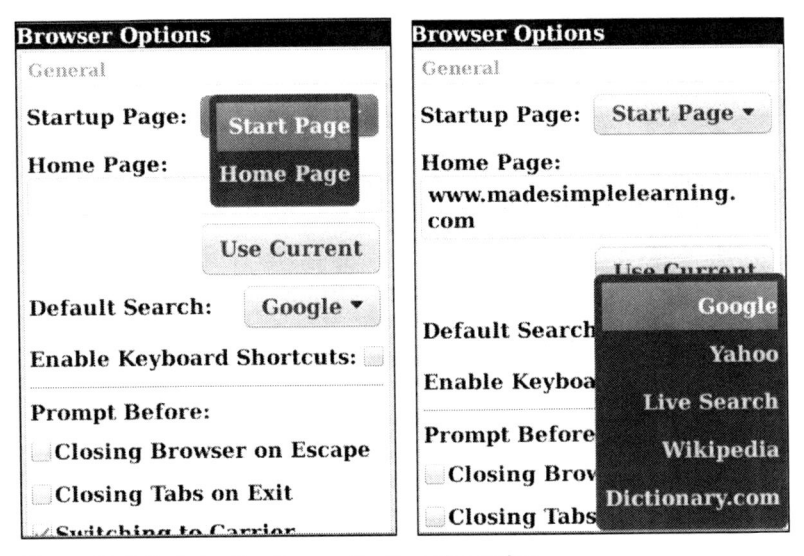

Figure 19-7. Exploring the *Browser Configuration* options

4. Scroll to **Default Search** and select the default search engine for your web browser.

5. under Web Content, we suggest keeping all the boxes checked for images loaded and popups blocked.

Using Your Bookmarks to Browse the Web

At this point, you are ready to begin using your bookmarks to simplify the process of browsing to your favorite sites on the Web. Follow these steps to do so:

1. Tap the **Browser** icon.

2. If you don't see your list of bookmarks automatically when you start the **Browser** program, press the **Menu** key, scroll down to **Bookmarks** option, and Tap it (see Figure 19-8).

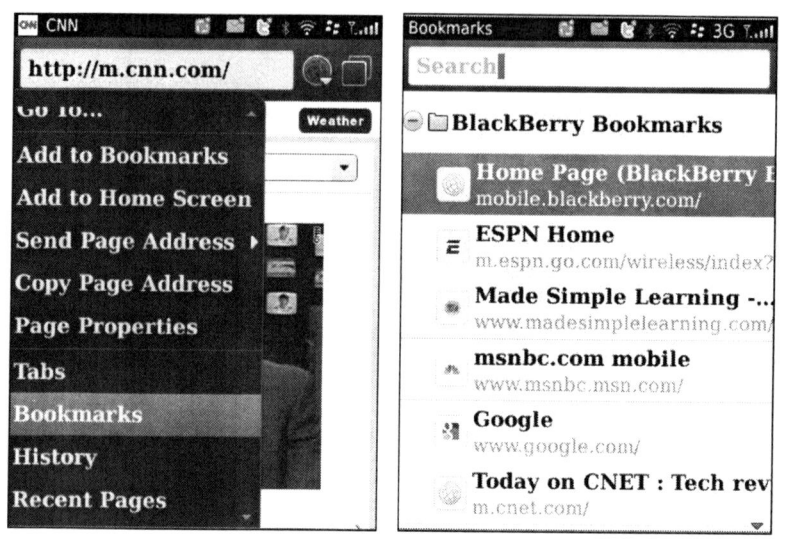

Figure 19-8. Viewing your list of bookmarks

3. All of your bookmarks will be listed, including any default bookmarks that were put there automatically by your phone company.

You might want to Tap a particular folder to open all the bookmarks contained within or to see whether the bookmark you need is located within a given folder. To open a folder of bookmarks, Tap that folder to see its contents.

However, if you have a lot of bookmarks, then you should use the **Search:** feature and type a few letters that match the bookmark you want to find. For example, in Figure 19-9, typing the letters *go* will immediately find all bookmarks with *go* in the bookmark name (e.g., *Google*).

Once you get familiar with your bookmark names, you can type a few letters and find exactly what you need quickly and easily.

Searching with Google

Google also has a mobile version of its site that loads quickly and is quite useful on your BlackBerry.

To get there, just go to www.google.com in your BlackBerry's **Browser** program.

The BlackBerry's native Browser program includes a built-in **Search Address** box at the top of the **Start** page. You can type anything into the **Search Address** box and built-in search engine you configured above will start to search for matches (see Figure 19-9). Pressing the **Search Address** box and clicking the **Dropdown Arrow** icon to the right of the **Search Address** box will show you a list of availabe search engines.

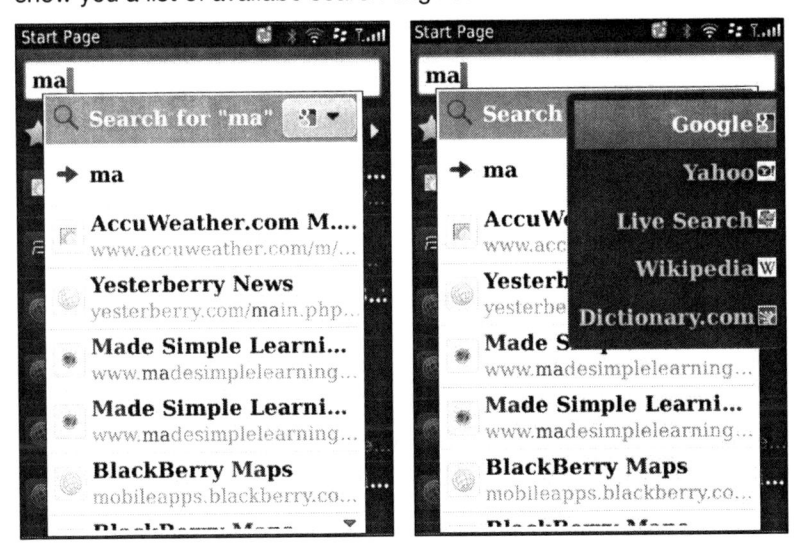

Figure 19-9. Changing the Browser program's default search engine

We highly recommend creating an easy-to-use bookmark for this site and all your favorite web sites. Just as on your computer, you can type in your search string and hit the **Enter** key. This saves you the time it takes to scroll to, press, and click the **Search** button). If you want to find (and even call) pizza restaurants in a certain ZIP code or city, then you would enter in *pizza* and your ZIP code or city (see Figure 19-10).

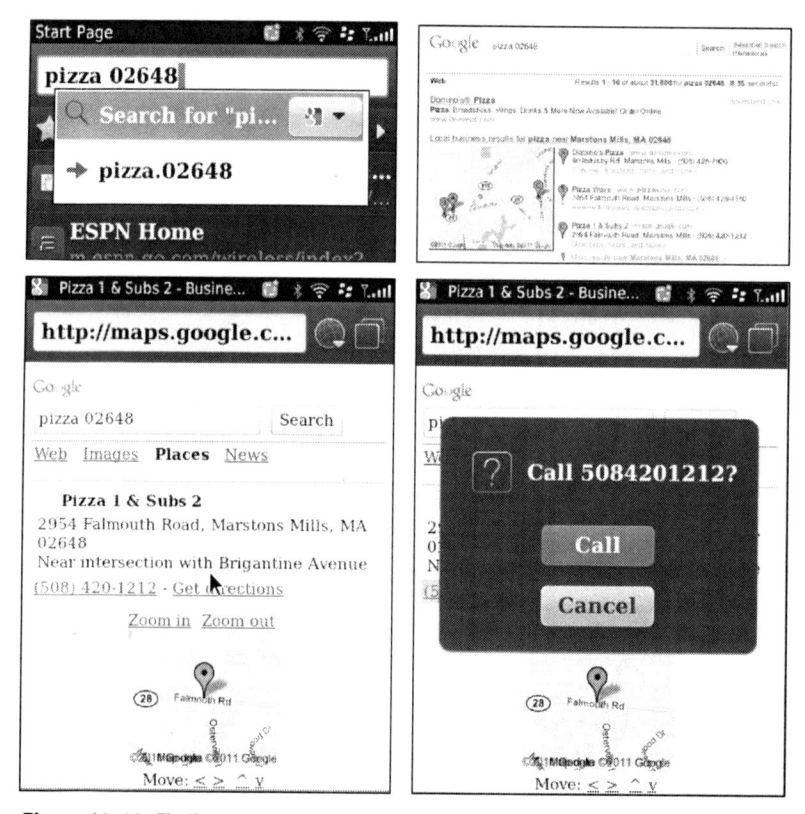

Figure 19-10. Finding (and calling) a pizza place in a given location

Viewing a Google Search Location

It might be that you will want to determine the location of a place you find in a Google search on Google. For example, assume I want to figure out the location of the pizza place I found and called in the preceding example. In this case, I can tap the name of the restaurant from my search results.

Once I Tap the link to the pizza place, Google will display a map that shows the location of the pizza place. If I were to tap the **Get Directions** option, I could find a quick path from my current location to the restaurant – all from my BlackBerry!

Finding Places with Google Maps

After you start using **Google Maps** (shown in Chapter 21), you may not use the web-based Google search very much on your BlackBerry for finding directions and businesses.

Chapter 20: BlackBerry App World

App World is where you go to find thousands of great applications for your new BlackBerry. In this chapter we will show you where to find software for your BlackBerry and how to download and install both from the **App World** and from outside sources.

The App World Concept

Online application stores are all the rage in the world of smartphones these days. One of the great things about a BlackBerry is that you can find applications for it in lots of places – not just the sanctioned **App World**. The BlackBerry **App World** online store is certainly the easiest place to find software as it is all in one, well-organized location. App World also keeps track of your purchases and downloads so you can always re-install or even transfer the purchases to a new BlackBerry.

Downloading App World

You should have **BlackBerry App World** already pre-installed on your Torch, however, if you cannot find it using the Search icon, then you'll need to download it.

1. Tap the **Browser** icon.
2. In the address bar of the browser, enter this address: **mobile.blackberry.com**.
3. Scroll up or down the page to locate **BlackBerry App World** and click the link.
4. Then click **Download Now** and click the **Download** button on the next page.

Starting App World for the First Time

The **App World** icon may be in your **Downloads** folder. If you want to move it to your **Home** screen, just follow the directions in Chapter 7.

Find the **App World** icon and touch or click it. The first time you run the **App World** program, you will have to scroll down a very long license agreement to click the **I Accept** button at the bottom.

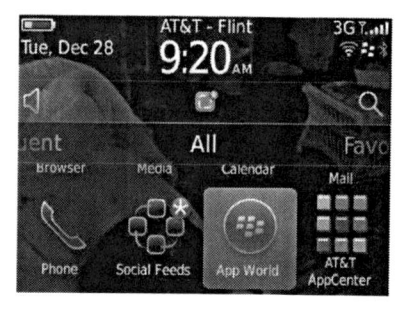

BlackBerry ID, Credit Card, Direct Bill or PayPal Account

NOTE: Depending on your country and carrier, you may be offered different payment options for BlackBerry App World. In the US, you have options of being direct billed by the carrier (App World purchases are added to your phone bill), Credit Card (add a new credit card), or use a PayPal account. You may not see all of these options on your BlackBerry.

The first time you use **App World**, you will be asked to create a **BlackBerry ID**, which you will then link to your payment option.

If you have a PayPal account, all you need to do is input your PayPal user name and password when you purchase an app. If you do not have a PayPal account, go to www.paypal.com from your computer and follow the instructions to setup your account. Setting up a PayPal Personal account is an acceptable requirement for most users.

Downloading Themes from App World

One of the cool parts about your BlackBerry is that you can find many applications that let you tailor the way it looks and behaves to suit your particular tastes. A BlackBerry "theme" changes the look and feel of all the screens, colors and fonts. Just touch the **Categories** soft key as described below, and you can search for the **Theme** category. You can find detailed instructions for doing so in "Personalize Your BlackBerry" in Chapter 7.

Featured Programs

The opening screen of the **App World** program shows large icons for its featured items. Many of these items are free; others need to be purchased (you will learn more about this later in this chapter). Flick left or right through the featured programs to find one that interests you. To learn more about a program or to download it, just touch or click it.

Categories, Top Downloads, and Search

Along the bottom of the opening **App World** screen, you will see four small icons: **Categories**, **Top 25**, **Search**, and **My World** (see Figure 20-1). Each category gives you a different way to look for, download, or manage apps on your BlackBerry. Simply touch any of these soft keys to go into that area.

BlackBerry Torch Made Simple

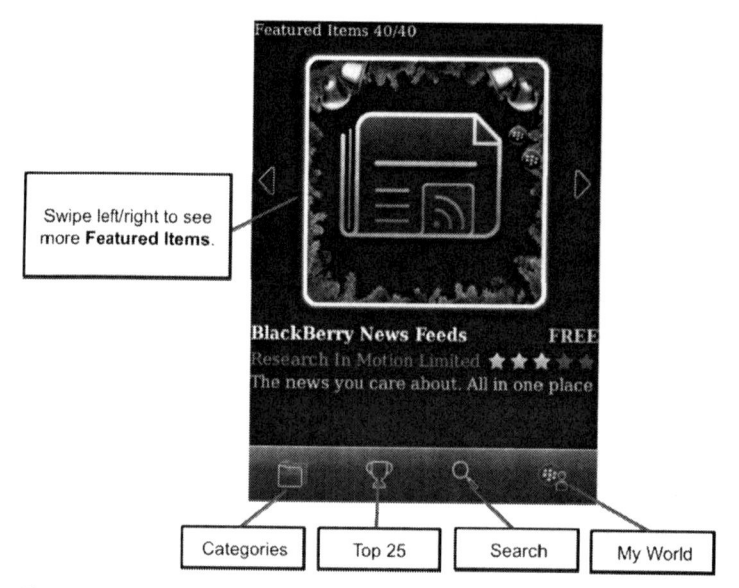

Figure 20-1. The layout of BlackBerry App World

Categories

As its name suggests, clicking the **Categories** icon shows you all the categories of applications available on the new BlackBerry App World site. The number of apps in any particular category is listed in parentheses next to the category name.

At the time of the writing, there were more than twenty categories, ranging from **Games** to **Sports** to **Finance** to various kinds of references and other items.

Categories
IM & Social Networking (150)
Maps & Navigation (148)
Music & Audio (567)
News (417)
Photo & Video (76)
Productivity (407)
Reference & eBooks (3874)
Shopping (62)
Sports & Recreation (286)
Test Center (6)
Themes (1140)

Clicking a category brings up more information about a particular application that may interest you.

Often, a category will have various sub categories listed once you touch or click the category name, For example, in the Category of Games, there are at

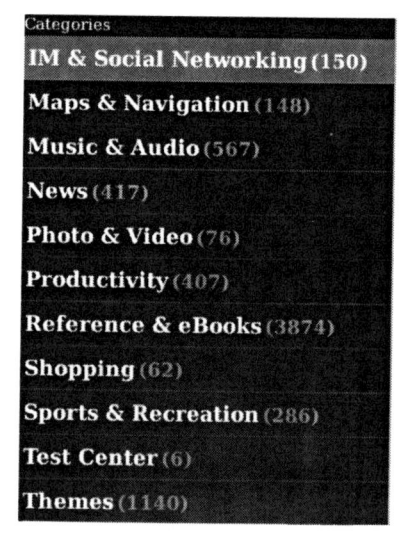

Games
Action (79)
Arcade (99)
Board Games (82)
Cards (73)

440

CHAPTER 20: BlackBerry App World

least ten sub-categories listed.

Once you click a category, the icons for the available programs will show on your screen. Flick up or down to see more apps in the list. You can read reviews or see screen shots for most applications before you decide to download them.

Top 25

As its name implies, clicking the **Top 25** icon will show you the most downloaded free applications from the **App World** site. As before, click any program to see screen shots, read reviews, or download the app to your BlackBerry.

Once you touch or click on **Top 25**, the first list you will see will be Newest at the top.

Slide the **Trackpad** to the right of use your finger and slide the screen from right to left to advance to the other **Top 25** screens. The categories under **Top 25** are; **Newest**, **Free Apps**, **Paid Apps**, **Top Themes** and **Recently Updated**.

View Video Tutorials and Free Tips at www.MadeSimpleLearning.com   441

Top 25 Paid Apps

Clicking the **Top** 25 and then sliding over to **Paid Apps** will show you the most downloaded applications from purchased from the App World site (prices start at $0.99). As you did previously, click any program to see screen shots, read reviews, or download an app to your BlackBerry.

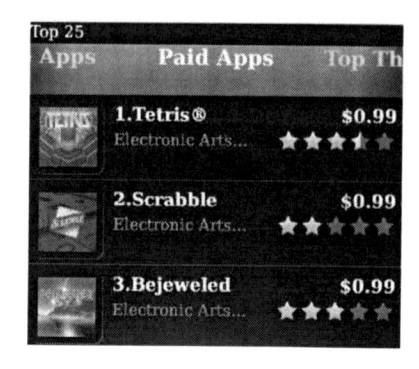

Search App World

The **App World** program has a very good, built-in **Search** tool. Touch the **Search** soft key at the bottom to go into search mode.

If you have an idea of what you might be looking for, but don't know its exact name, just type a word about the general subject into the **Search** bar. In this example, we want to find music applications that are currently available, so we enter *music* into the **Search** bar, which causes all available music-related apps to be displayed (see Figure 20-2).

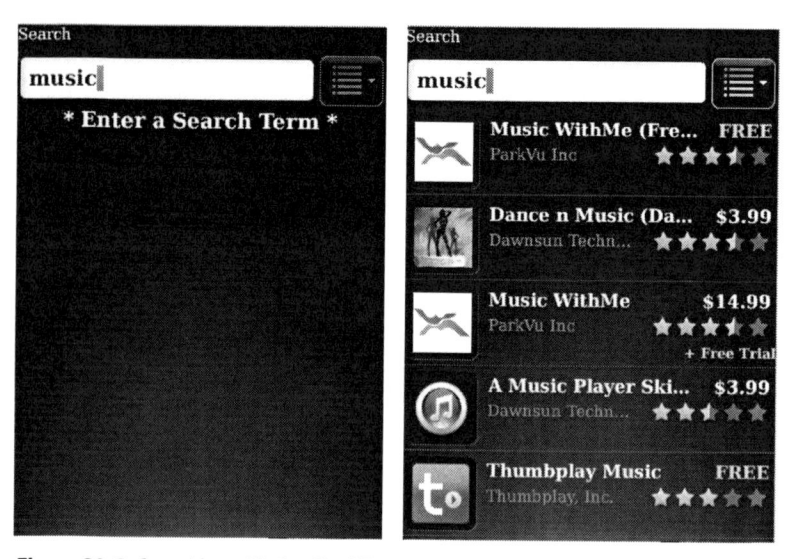

Figure 20-2. *Searching with the BlackBerry App World program*

CHAPTER 20: BlackBerry App World

> **TIP:** After you see your search results, you can sort the search results by clicking the **Sort** icon in the upper right corner of the screen (see Figure 20-3).

Figure 20-3. Sort Search results

Choose the sorting criteria; the choices are to sort by **Relevance**, **Newest**, **Rating**, **Price**, **Vendor** or **App Name**.

Downloading Apps

As the time of writing, the **BlackBerry App World** program accepts the PayPal payment system or credit card or carrier billing. (This may change in the future.)

As we mentioned above, the first time you start up **App World**, you need to create a **BlackBerry ID**, which is linked to your PayPal account.

Changing Payment Options

To set up or change your payment options, touch the **My World** soft key and then press the **Menu** key and select **Payment Options** from the menu. Login with your **BlackBerry ID** (see Figures 20-4 and 20-5).

View Video Tutorials and Free Tips at www.MadeSimpleLearning.com   443

Figure 20-4. Logging in to your BlackBerry ID.

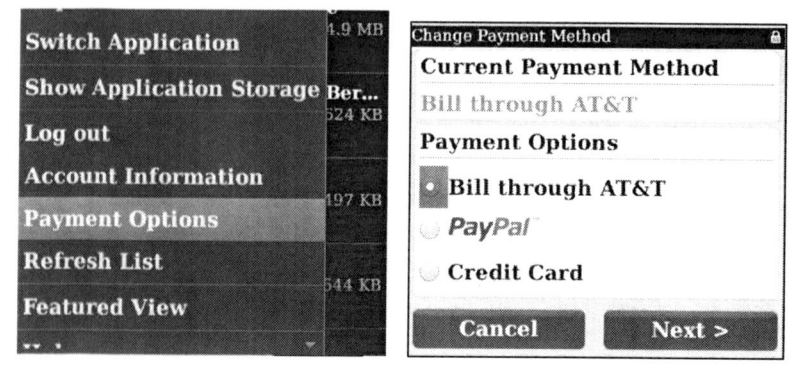

*Figure 20-5. Choosing payment options for **BlackBerry App World**.*

Downloading and Purchasing an App

Scroll through the applications listed, as you did previously. When you find an app that you want to download, just click the **Download** button from the **Details** screen.

CHAPTER 20: BlackBerry App World

If the application is not a free download, the **Download** button will be replaced by a **Purchase** button. Click the **Purchase** button and input your BlackBerry ID information to buy the app.

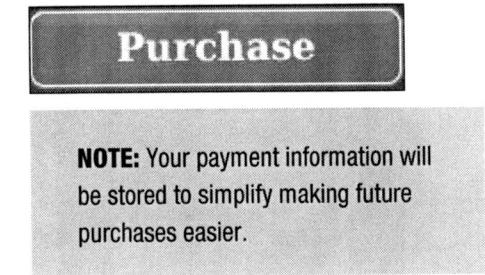

NOTE: Your payment information will be stored to simplify making future purchases easier.

If this is your first time purchasing an app, you will need to log in or set up your BlackBerry ID and link it to a PayPal account, credit card or carrier billing account. PayPal can be tied to any major credit card or your bank account.

You will see a progress bar showing the progress of your application as it downloads, prepares to install, and then installs on your BlackBerry.

You may also be asked to grant the application **Trusted Application** status. We generally recommend saying **Yes** to this.

You can then choose to **Run** the application or select **OK** to go back to browsing for more apps.

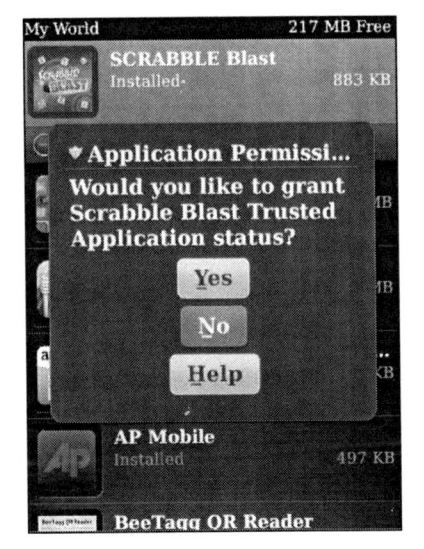

View Video Tutorials and Free Tips at www.MadeSimpleLearning.com   445

Using the My World Area

All the apps that you purchase or download for free are listed in the **My World** area of the **App World** program.

Touch or click the **My World** icon (in the lower right-hand corner of the **App World** program) to see a complete list of your downloaded or purchased programs.

NOTE: If you restart your BlackBerry or do a battery pull, you will be logged out of your account, and you will be asked to re-enter your BlackBerry ID password.

The My World Menu Commands

The menu associated with the **My World** area of the program has several commands associated with it. Follow these steps to access them:

1. Highlight any app in the **My World** area and press the **Menu** key.

2. From the menu that pops up, you can: **Switch Application**, **Show Application Storage Log in** or **Log Out** of your account, view **Account Information, Payment Options, Refresh List**, **View** the application highlighted, **Run** the selected program, **Review** the program, **Recommend** the program to someone else, **Archive** or **Delete** the program.

NOTE: If application updates are available, these will also be indicated in the **My World** context menu.

Archiving or Deleting Programs

After you start to download lots of new apps to your BlackBerry, you may decide that you do not want to use some of them or that you simply want to free up space for new apps.

To delete or uninstall an app, follow these steps from within the **My World** area of the **App World** program.

1. Highlight the app you wish to remove and press the **Menu** key.

2. Select **Archive** to temporarily remove the highlighted program from your BlackBerry. Select **Delete** to permanently remove the app. On the next screen, you will be asked to confirm your selection to remove the program.

3. Most programs will require you to reboot (reset) your BlackBerry to complete the uninstall process.

TIP: If you are removing more than one application, select **Reboot Later** until you have selected and uninstalled all the apps. Then, when you're finished uninstalling apps, select **Reboot Now** to remove all the apps at once. Following this approach will save you a lot of time waiting for your BlackBerry to reboot!

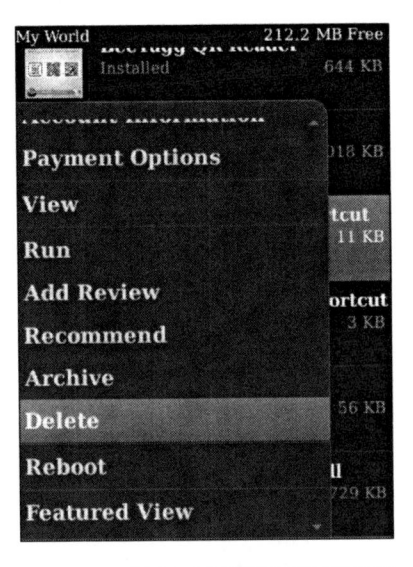

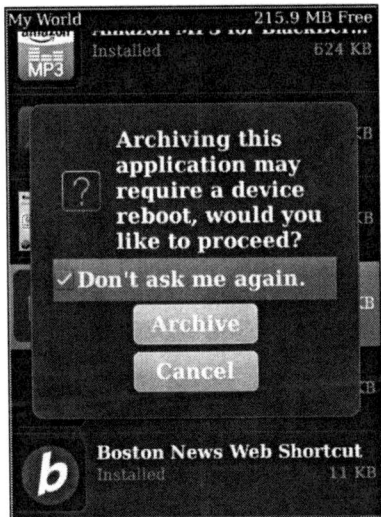

Adding or Removing Apps

Your BlackBerry comes with most of the major apps you will ever need already installed. However, there are literally thousands of third-party apps available in virtually any category you can think of that can help you get the most out of your BlackBerry.

There are apps for productivity, reference, music, and fun – including lots of great games. One nice thing about the BlackBerry platform is that while there is an official **App World**, there are also lots of ways to find and download apps.

Downloading New Software Outside of App World

One of the cool things about your BlackBerry is that, just as on your computer, you can go onto the Web and find software to download. You can download everything from ringtones, to games, to content that is pushed out to your BlackBerry on a regular basis (using the same approach that you *probably* see with email on the BlackBerry – each phone company or carrier uses a slightly different approach).

> **CAUTION:** Web sites change frequently. All of the web site images you see in this book are accurate at the time of writing. And while it's quite possible that a given page or site shown in this book will look exactly the same on your BlackBerry, it is also possible that the page or site will look completely different on your device. It is even possible that the site or page is no longer available. If a site is missing or looks completely different, we recommend searching for a link with approximately the same name.

Several news sites now allow you to put an icon on your **Home** screen that will automatically launch a given news site. For our purposes, let's say we wanted to put the icon for ABC News on our **Home** screen (see Figure 20-6). Follow these steps to do so:

1. Scroll down to the link for ABC News and select **Download** or **Get Web Signal** (depending on the site.)

CHAPTER 20: BlackBerry App World

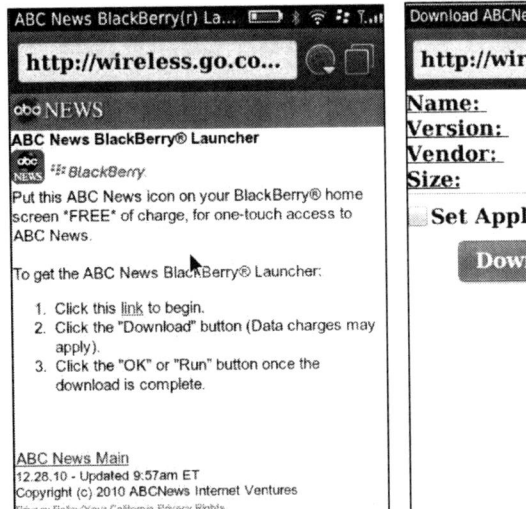

*Figure 20-6. Downloading the **ABC News** app*

2. Tap the **Download** option to place the icon for *ABC* on your BlackBerry.

3. Once the program has successfully installed, you will see a screen that looks like the image to the right.

4. In the **Downloads** home screen, you should see the **ABC News** icon. Tap the icon to go directly to the mobile version of the *ABC* site.

Changing Your Default Downloads Location

By default, your BlackBerry saves the icon of all programs you download to the device's **Downloads** Home Screen. Follow these steps to change the default folder your BlackBerry saves new programs to.

1. Press the **Menu** key from your Home screen and select **Options**.

2. Next, touch or click on the **Downloads Folder** button to change the download default location to a new location.

TIP: Select **Home** if you want your new icons easily accessible!

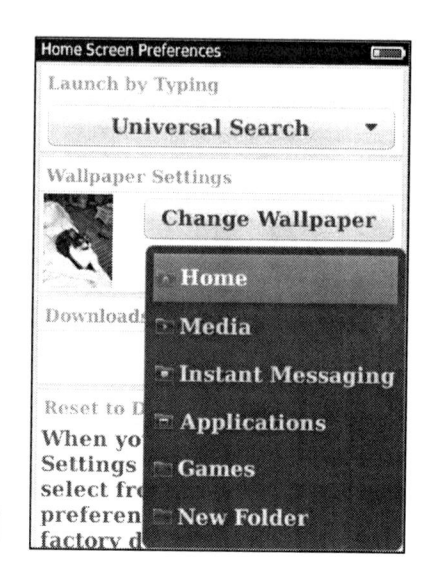

Finding More Software

There are quite a few additional places where you can find software and services for your BlackBerry beyond **BlackBerry App World**.

Web Stores

You can find quite a few online stores that carry software for your Blackberry. You can usually purchase software directly from these stores. Here's a short list of online sites that sell software for the BlackBerry.

- www.crackberry.com
- www.shopcrackbery.com
- www.bberry.com
- www.eaccess.com
- www.handango.com
- www.mobihand.com

Reviews of Software, Services, and More

There are so many programs, services, and other items available for your BlackBerry that it can be difficult to know where to start or what is worth having. Fortunately, you can also find many web sites that provide reviews of all the software, services, ringtones, themes, wallpaper, accessories, and other items and services of interest for your BlackBerry. Some of the sites that provide these reviews include the following:

- www.crackberry.com
- www.bbhub.com
- www.berryreview.com
- www.blackberrycool.com
- www.blackberryforums.com
- www.boygeniusreport.com
- www.howardforums.com (The RIM-Research In Motion Section)
- www.pinstack.com
- www.RIMarkable.com
- www.blackberryrocks.com

Official BlackBerry Sites

In addition to the many web sites dedicated to providing and reviewing programs and services for the BlackBerry, you might also find it worthwhile to check out the official BlackBerry Solutions Catalog, which you can find online at: http://www.blackberrysolutionscatalog.com/

The software and services included in the Solutions Guide are tailored more for business users than individuals; however, you might find items of items on the site. Another official site is the "Built for BlackBerry" software site, which you can find at: http://appworld.blackberry.com/webstore/.

Removing Software from Your BlackBerry

There will be times when you wish to remove a program from you BlackBerry, but you are not connected to your computer at the time. Fortunately, it is an easy and intuitive process to remove programs when using the BlackBerry itself.

You have a couple of ways to remove any software program installed on your BlackBerry, which you'll learn about in the next couple of sections. In addition to removing the software you add to the device, you can also remove some of the pre-installed apps that shipped with your BlackBerry.

Option 1: Deleting Apps from App World

Go into the **My World** section of **App World**, highlight or select the app you wish to delete, press the **Menu** key and select **Delete**.

Option 2: Deleting Apps from the Home Screen

We'll cover the easiest way to delete an app from your BlackBerry first. This approach relies on deleting the app directly from the **Home** screen. Follow these steps to do so:

1. Navigate to the app you want to delete, press and hold on it.
2. Select **Delete** from the pop-up menu. (You could also select Delete from the menu).
3. Your BlackBerry will show a warning screen. Tap the **Delete** button to confirm that you want to delete the application.

NOTE: Not every program can be deleted.

Option 3: Deleting Apps from the Options Icon

You can also delete applications from the **Options** menu. Follow these steps to do so:

1. Tap the **Search** icon and type "Applications."
2. Tap the **Options** icon.
3. Tap **Application Management**.

4. Now type a few letters of the application you want to delete to find it. Or, you can scroll down the list of applications.

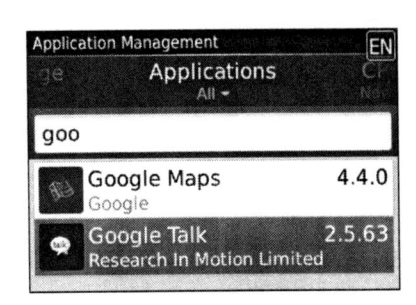

5. Tap on the app you want to delete and tap the **Delete** key as shown.

6. Once you highlight the application you want to delete, you can Tap the **Menu** key and select **Delete** and confirm your choice.

7. Usually, the BlackBerry needs to restart or reboot to complete deleting the application. If so, Tap **OK** or **Yes** to finish the process of removing the application from your BlackBerry.

TIP: If you are removing a number of apps, wait to reboot until you delete them all.

Option 4: Deleting Apps with the Desktop Software

Check out Chapter 3 if you use a Windows computer or Chapter 5, if you are a Mac user to see how to add and remove apps via the BlackBerry Desktop Software.

Chapter 21:
Traveling: Maps & More

Regardless of whether you intend to leave the country with your BlackBerry, there are a few travel-related things you should know how to do. For example, everyone with a BlackBerry can probably benefit from one of the many mapping applications for the BlackBerry that rely the device's built-in GPS (Global Positioning System). Such applications can help you find navigate to your destination with turn-by-turn directions, find nearby stores with services or products you might need, or even mark where you parked your rental car so you can find it when you're ready to leave.

If you to intend to travel to another country, however, there are some definite steps you should take before you, not least so you do not get surprised with a huge data or voice roaming phone bill. We'll cover these topics and many others related to traveling with a BlackBerry in this chapter.

International Travel: Things to Do Before You Go

Some BlackBerry smartphones are well equipped for international travel such as your BlackBerry Torch with its SIM card run on networks called "GSM" that are fairly common throughout the world. So connecting overseas should be easier than phones without a SIM card.

In any case, we always recommend that you call your cell provider well in advance of a trip to see whether there is an **international** feature you can turn on in your BlackBerry.

Avoiding a Shockingly Large Bill

One thing you want to avoid when traveling is returning home to a shockingly large bill with high data or voice roaming charges. For example, we have heard of people who returned home after a trip abroad to find a phone bills over $500 in monthly data and voice roaming charges. You can avoid such a surprise by taking a few easy steps before and during your trip.

Before Your International Trip

There are a few simple steps you can take before you leave to ensure that your BlackBerry will always work – no matter where you might be in the world.

Call Your Phone Company

You should also contact the provider that supplies your phone service before you leave. When you call, you should check with your wireless carrier about any voice and data roaming charges you might incur when traveling. You can also try searching on your phone company's web site for this information, but usually you will have to call the company's Help Desk and specifically ask what the voice roaming and data roaming charges might apply for the country or countries you plan to visit. If you use email, SMS Text, MMS messaging, web browsing, and any other data services, you will also want to specifically ask about whether any of these services are charged separately when traveling abroad.

Some phone companies offer something that might be called an **International or Global Traveler** rate plan that you will need to activate before your trip. In some cases, you must activate such a plan to use your BlackBerry at all; in other cases, activating such a plan will allow you to save some money on the standard data and voice roaming charges. You should check out these plans in advance to see whether they can save you some money while you are on your trip, especially if you need to have access to data while you are away.

TIP: If you decide to purchase an International Rate plan, then you should create a reminder calendar event for a day or two after you return reminding yourself to turn off this extra rate plan. The phone company will be happy to continue charging you the extra money until you contact them to turn off the plan.

Ask About Using a Foreign SIM Card

In some cases, your BlackBerry phone company won't offer special deals on international data roaming plans, or its rates will be unreasonably high. In these cases, you may want to ask your phone company to unlock your BlackBerry, so you can insert a SIM card you purchase in the country you're visiting. In many cases, inserting a local SIM card will eliminate or greatly reduce data and voice roaming charges. However, you should carefully check the cost of placing and receiving international calls on that foreign SIM card. Using a foreign SIM card may save you hundreds of dollars, but it's best to do some web research or try to talk to someone who has recently traveled to the same country for advice before settling on that approach.

Airplane Travel: Getting into Airplane Mode

Some airlines force you to completely power-down your electronic devices during take-off and landing, but then allow the use of "approved electronic devices" (read: you must turn off your BlackBerry's wireless radio) while in flight. This is easy to do – you simply tap the top of your **Home** screen to bring up the Manage Connections screen. Then tap **Turn All Connections Off.**

You know the radios are all off when you see the word **OFF** next to your radio in the upper right portion of the screen.

Tap **Restore Connections** to turn all connections back on, or, in some cases you can turn on **Wi-Fi** while in flight – to do that, tap **Wi-Fi** to selectively turn that radio on.

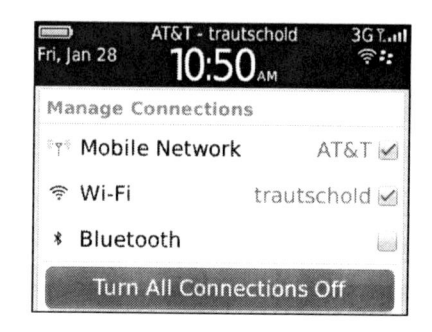

Things to Do When Abroad

Once you've completed the steps described so far to prepare for your trip, you will need to address some additional issues once you get to your destination. The next sections explain the things you need to keep in mind after you arrive at your destination.

Getting Your BlackBerry Ready

Once you arrive at your destination, you will need to do a couple of thing to make sure your BlackBerry is ready to use while you're traveling.

Step 1: Make Sure the Time Zone Is Correct

When you arrive, you will need to make sure your BlackBerry is displaying the correct time. Usually, your BlackBerry will auto-update your time zone when you arrive at a new destination. However, if it doesn't, you can manually adjust the time zone (see Chapter 7 for help with setting the time zone).

Step 2: Turn Off Data Roaming If It's Too Expensive

If you were unable to find out about data roaming charges from your local phone company, try to contact the phone company in the country you're traveling to find out about any data roaming or voice roaming charges.

The worst case: If you are worried about the data roaming charges, and you can do without your email and web browser while you're away, then you should disable data services for the duration of your trip. Follow these steps to do so:

1. Tap the top of your **Home** screen to bring up the **Manage Connections** screen.

2. Tap **Mobile Network Options** near the bottom.

3. Your **Data Services** will most likely say **On**, but there is a secondary tab for **While Roaming**. We recommend setting the switch to **Off** in the **While Roaming** field, so that you can continue to receive data in your **Home** network.

4. Press the **Escape** key; your new settings will be saved automatically.

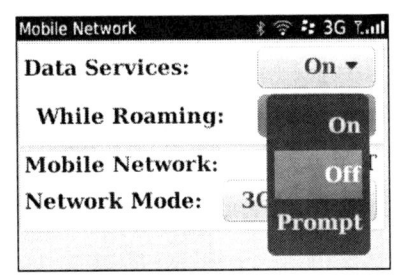

Setting the **Data Services** value to **Off While Roaming** should help you avoid any potentially exorbitant data roaming charges. (You still need to worry about voice roaming charges, but at least you can control those by watching how much you talk on your phone).

The nice part about using the **Off While Roaming** setting: When you return to your **Home** network, all your data services (e.g., email, web browsing, and so forth) will work automatically.

> **NOTE:** If you want to use email, web browsing, GPS mapping and other services that require a data connection, you will have to set the While Roaming to On.

Step 3: Register with the Local Network

If you are having trouble connecting to the local wireless network, you may need to register your BlackBerry on the local network (see the steps shown in "Host Routing Table Register Now" section at the end of Chapter 12 for more information).

Step 4: Look for Free Wi-Fi Networks

In your hotel, international chain restaurants, coffee shops, some other types of restaurants and public libraries, you may find free Wi-Fi networks that will allow you to connect to the Internet from your BlackBerry. If you use Wi-Fi you will not be incurring any wireless data roaming charges.

Returning Home

As you did when you arrived at your travel destination, you will need to make a few changes to your BlackBerry's settings once you arrive back home before your device will work as expected.

Step 1: Check Your Time zone

As mentioned in the section on traveling abroad, your BlackBerry usually will auto-update your time zone when you arrive at a particular location. If the device does not auto-adjust after you return home, you can manually adjust your time zone (see Chapter 7 for help with setting the time zone).

Step 2: Reset Your Data Services to On

If you have turned the **Data Services** field to **Off**, then you will need to make sure to reset it to **On** or **Off When Roaming**, so you can receive data when you return home.

Step 3: Register Your BlackBerry on the Local Network

Depending on the actions you took while preparing for your trip or while traveling, you might need to re-register your BlackBerry on your local phone network (see the steps shown in "Host Routing Table Register Now" section at the end of Chapter 12 for more information).

Step 4: Turn Off Your Special International Plan

The final step is optional. If you have activated some sort of special international roaming rate plan with your BlackBerry phone company, and you do not need it any more, contact the company to turn it off to save yourself some money.

BlackBerry Maps, Google Maps, and Bluetooth GPS

In addition to the myriad ways your BlackBerry to help you manage your life, the device can also take you places – literally. With the aid of software that is either pre-loaded on the BlackBerry or easily downloaded from the Web, you can use your BlackBerry to find just about any location, business or point of attraction.

Enabling GPS on Your BlackBerry

You will need to turn on the GPS location (GPS is the longitude and latitude of your current location based on triangulation from orbiting satellites). GPS is on your BlackBerry is required to enable mapping software to track your precise location. You will also need to turn this feature on if you want to turn on the geotagging (see Chapter 17) feature for your camera. Follow these steps to do so:

1. Tap the **Search** icon from your **Home** screen and type "GPS" into the search bar.
2. Tap the **Options** icon and select **Location Settings**.
3. Make sure that the **Location ON** property is selected under **Location Services** as shown.
4. Leave **Location Data** and **Location Aiding** as **Enabled**.
5. Finally, press the **Menu** key and select **Save**.

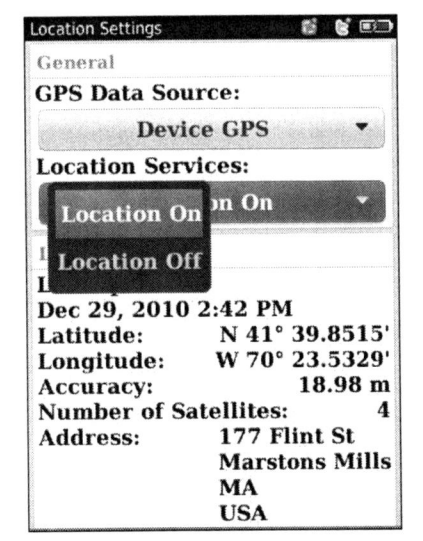

Using BlackBerry Maps

Your BlackBerry usually ships with the **BlackBerry Maps** software, which is a very good application for determining your current location and tracking your progress through your GPS receiver.

> **NOTE**: Some carriers will install their own version of mapping software. If you do not have **BlackBerry Maps** installed, go to go to the mobile.blackberry.com web site from your BlackBerry and download the free **BlackBerry Maps** software.

Follow these steps above to enable GPS use with the **BlackBerry Maps** program.

Viewing a Contact's Map

Sometimes you may want to look up the location of one of the contacts in your device. For example, perhaps you want to meet near the contact's home, and you will use the address information in the contact's profile to look up where exactly the contact lives. Follow these steps to view a map for the contact (see Figure 21-1):

1. In the **Blackberry Maps** program, press the **Menu** key from the main **Map** screen.
2. Scroll down and select **Find an Address**.

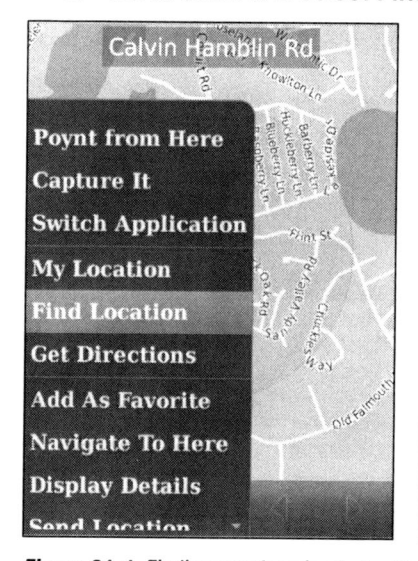

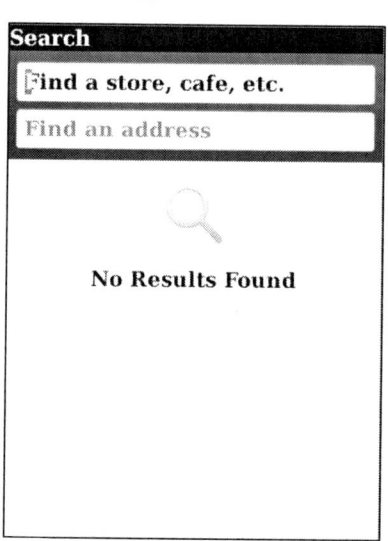

Figure 21-1. Finding your location in the BlackBerry Maps program.

460

CHAPTER 21: Traveling (Maps & More)

3. Just start typing a name or address. If the name is stored in your Contacts app, you will begin to see matches in the window below.

4. Scroll down to tap the name of a contact.

5. A map of that contact's location will appear on your screen.

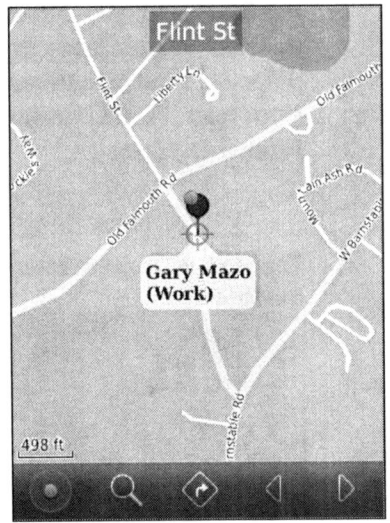

Getting Directions with BlackBerry Maps

After you map a location, it is easy to get directions to a particular location.

1. Press the **Directions** soft key in the middle of the bottom row.

2. If you need to adjust the **Start, End or Route Options** tap those buttons.

3. For **Start/End**, you can customize to select the address you just mapped, **Your Location, Select from map**, **Find Location**, or **From Favorites**.

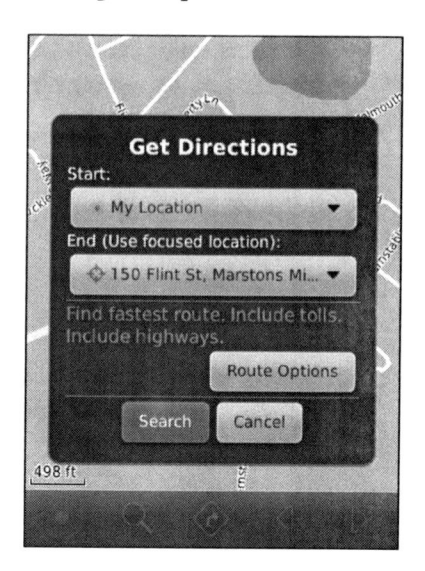

4. Tap **Route Options** to adjust the following:

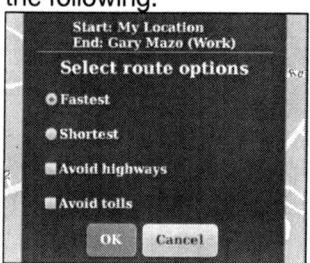

5. After you tap **Search**, your route will be mapped and displayed in summary.

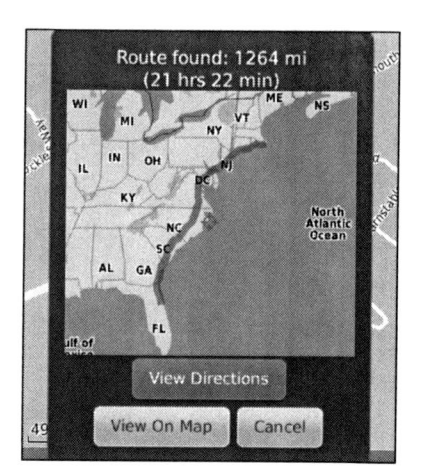

CHAPTER 21: Traveling (Maps & More)

6. Tap **View Directions** to see the list of turn-by-turn directions.

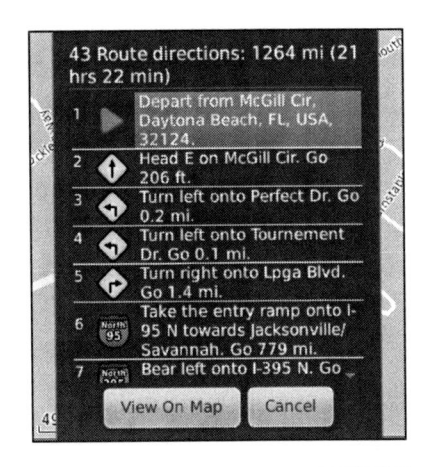

7. Tap **View on Map** to see these directions on the map.

At the time of writing, the GPS will track you along the route, but it will not give you voice prompts or turn-by-turn voice directions.

You can also press **Menu** key and then **Add as Favorite** to mark a location as a **Favorite** on your BlackBerry.

BlackBerry Maps Menu Commands

You can accomplish a great many tasks with the **BlackBerry Maps** application. You can get a sense of what it lets you do by pressing the **Menu** key, which brings up a screen with a ton of options you can choose from simply by scrolling, pressing, and clicking. Specifically, you can perform the following tasks from the screen displayed by pressing **BlackBerry Maps** program's **Menu** key:

- **Switch Application:** Jumps you to any other application while keeping the current application open.
- **My Location** – Zooms to your current location on the map.
- **Find Location:** Jumps you to an address you enter.

View Video Tutorials and Free Tips at www.MadeSimpleLearning.com   463

- **Zoom To Point:** Takes you from street level to the stratosphere (keyboard shortcut keys: **L** = Zoom In, **O** = Zoom Out).

- **Local Search:** Searches for restaurants, stores, or points of interest near your current location.

- **New Directions:** Lets you find directions using your location history or by typing in new addresses.

- **Get Link:** Connects you to a web site so you can find an address.

- **Get/View Directions:** Switches you to **Text** mode if you are in **Map** mode. Otherwise it brings up the **Get Directions** screen shown above.

- **Clear Map:** Erases the current map or route on the screen.

- **Add as Favorite:** Adds your current location as a Favorite for easy retrieval on the map.

- **Navigate to Here:** If the location on the map is not your Current Location, you can immediately get directions to that spot.

- **Display Details:** This will display the longitude and latitude of the selected location.

- **Send Location:** Forwards your map location via Email, Text Message, Group Message, Messenger Contact, PIN, Social Feeds or Twitter.

- **Send Directions:** Sends your directions via Text Message, Group Message, Messenger Contact, PIN, Social Feeds or Twitter (Only available if you are currently in **Direction** mode.)

- **Copy Location –** This will copy the location to the clipboard so you can paste it in another application.

- **Zoom to Point:** Shows the map detail around the currently selected point in your directions; this option is available only when viewing directions.

- **Favorites –** This will display those destinations set as **Favorites** in **BlackBerry Maps**.

- **Track up** or **North Up –** Changes the orientation of the tracking arrow in the app.

- **Options:** Changes **Global Map Options**. For example, this choice lets you disable the backlight timeout settings, change units from metric (kilometers) to imperial (miles), change the font and turn on **Address Recognition**.

- **About:** Shows information about the current provider of the mapping data and software.
- **Help:** Displays a contextual **Help** menu.

Google Maps

If you have ever used **Google Earth**, then you have seen the power of satellite technology in mapping and rendering terrain. **Google Maps Mobile** brings that same technology to handheld devices, including the BlackBerry.

With Google Maps, you can view 3-D rendered satellite shots of any address, anywhere in the world. To get started with this amazing application, you need to first download it onto your BlackBerry. Follow these steps to do so:

1. Tap the **Browser** icon from your **Application** menu.
2. Press the **Menu** key, scroll to the **Go To** command, and Tap it.

> **TIP:** Tap the letter G on your slide-out keyboard to execute the **Go To** command.

3. Enter this address to perform an *over the air* (OTA) download onto the BlackBerry: www.google.com/gmm.
4. Tap **Download Google Maps**, and the installation program will begin (see Figure 21-2).

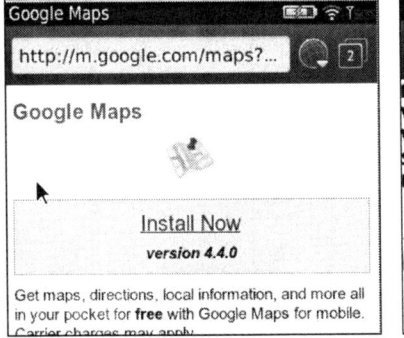

Figure 21-2. Downloading the Google Maps program

5. Tap the **Download** button on the next screen.

6. Finally, you will see a screen indicating that the application was successfully installed. Select OK to close the window or Run to start Google Maps Mobile right away. Or, you can Tap the Google Maps icon on the BlackBerry, which should be in the Downloads folder. You might get prompted to reboot your device – if so, just Tap **Reboot**, and then try to run the program.

The first time you start up **Google Maps Mobile**, you will need to read the terms and conditions for the program, and then click **Accept** to continue.

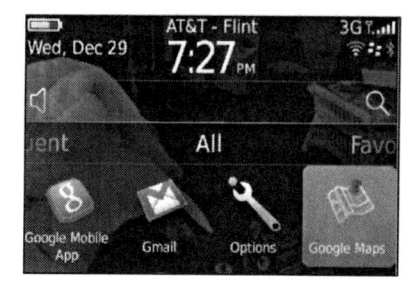

Google Map Commands and On-Screen Icons

Google Maps Mobile is full of great features – most of which you can access right from the program's menu. One cool new feature is **Street View**, which **Google Maps Mobile** tells you about at the first screen.

When you first start the program, it will try to determine your location based on the built-in GPS or the cell tower closest to your phone.

To Zoom In or Out, tap the Zoom icons:

Jump to the next or previous location mapped or person in Google Latitude, tap these arrows:

To see a list of results or the list of turn-by-turn directions, tap this:

To force the map to center on your location,

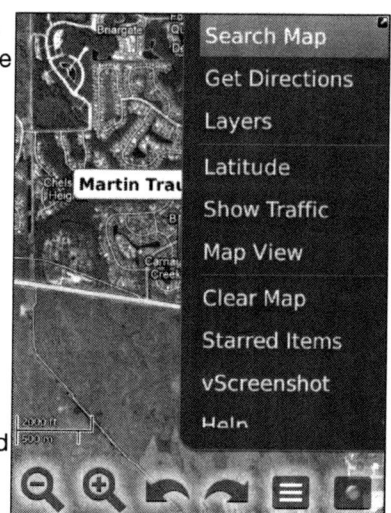

466

tap the blue dot:

Google Maps Search - Finding an Address or Business

It's easy to find an address or business with **Google Maps Mobile**. Follow these steps to do so:

1. Tap the **Google Maps** icon to start the application.
2. Press the **Menu** key and choose **Search Map**.
3. Click on **Popular Categories** to see specific categories such as **Restaurants**, **Shopping**, **Entertainment**, etc.
4. You can type just about anything in the search string: an address, type of business and ZIP code, business name and city/state, and so on. For example, if you wanted to find bike stores in Winter Park, Florida, you could enter *bike stores winter park fl* or *bike stores 32789* (assuming you know the ZIP code).
5. You can also hold the **Green Phone** key and search by voice.
6. Finally, press the **Search** button and select **OK**. Or you can press the **Enter** key on the keyboard to begin the search.
7. Your search results may show a number of matching entries. Just scroll to locate an entry. If you only see one match, touch the List button to see all the potential matches. Touch that entry to select it, and then tap the screen to see details for that entry (see Figure 21-3).

View Video Tutorials and Free Tips at www.MadeSimpleLearning.com   467

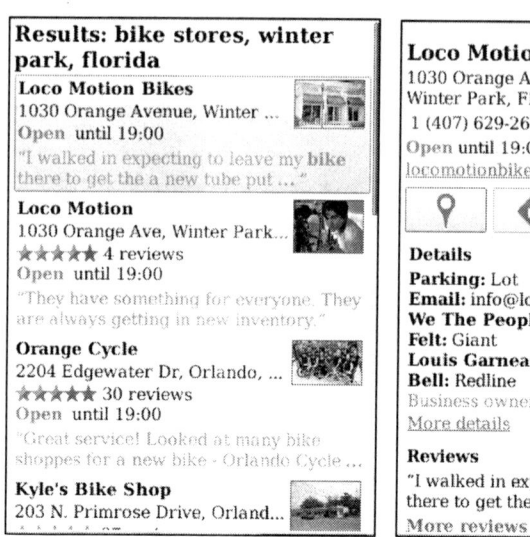

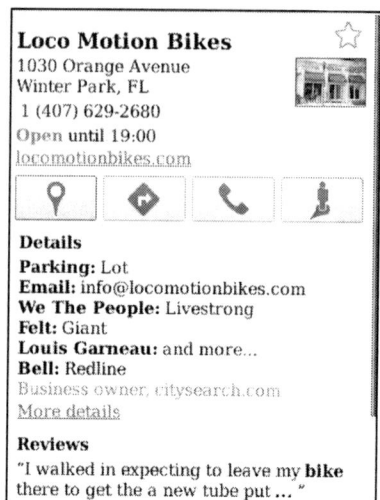

Figure 21-3. Search results in *Google Maps Mobile*

8. Press the **See on Map** button at the top left to see the search results on the map.

9. To see the new street view, Tap the **Street View** button which is the farthest to the right of the four buttons at the top (see Figure 21-4).

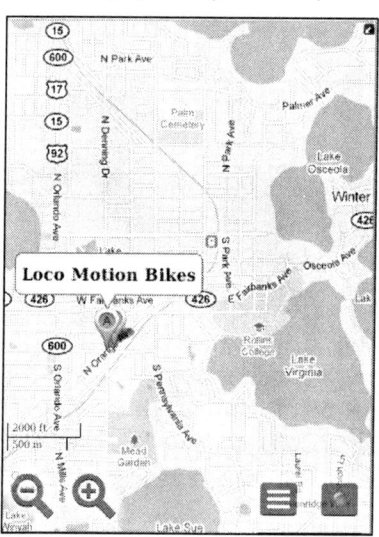

Figure 21-4. Viewing map results with Street View

Keyboard Shortcut Keys

Google Maps includes quite a few keyboard shortcuts that you let you get at the program's core functionality quickly and easily. To use these, slide out your keyboard. Here is list of the available shortcut keys and what they do:

O	Zooms out.
2	Brings up the **Layers** menu.
I / 3	Zooms in.
5	Toggles between **Map/Satellite** view.
7	Show or hide **Traffic** status.
A / *	Shows the **Starred** items.
Q / #	Shows or hides lists (e.g. **Latitude** friends, directions list, etc.).
0	Shows your **My Location**.

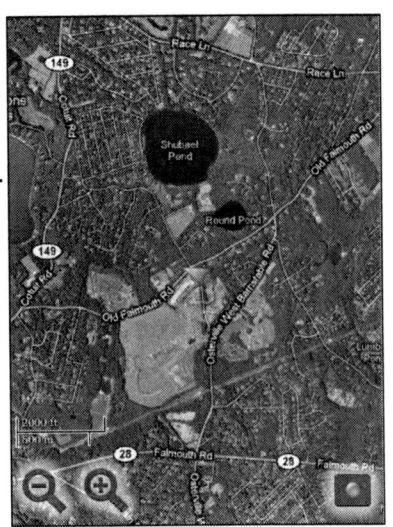

Getting Directions

Obviously, one of the key reasons you might want to use Google Maps Mobile is because it can provide turn-by-turn directions for your BlackBerry that integrate nicely with the device's built-in GPS feature. Follow these steps to get directions on your BlackBerry:

1. Press the **Menu** key and select **Get Directions**.
2. Set your **Start Point**; the default will be **My Location.** and then
3. Set your **End Point**
4. Choose your **Travel by** (mode of transport).
5. Tap **Show Directions** to display turn-by-turn directions to your destination.

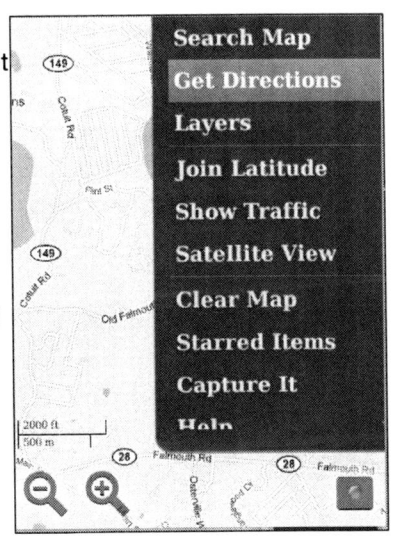

View Video Tutorials and Free Tips at www.MadeSimpleLearning.com

> **TIP:** With the most current version of **Google Maps**, you can actually choose from **Car**, **Public Transit**, **Walking** and **Bicycle** as modes of transport for your directions. The **Bicycle** choice tries to route you on bike paths and bicycle-friendly roads.

Google Latitude

Another new feature in **Google Maps Mobile** is **Google Latitude**, which enables people in your **Google Contacts** to see where you are in real time – you can also see them once you accept their invitations. Follow these steps to enable the **Google Latitude** feature:

1. Press the **Menu** key and select **Join Latitude** to get started.
2. Once you set this feature up, your location on the map is shared with whomever you choose.
3. From your **Friends** list, press the **Menu** key and select **Add friends** to invite others to "share" their location with you. You will then be able to see your friends on the map.

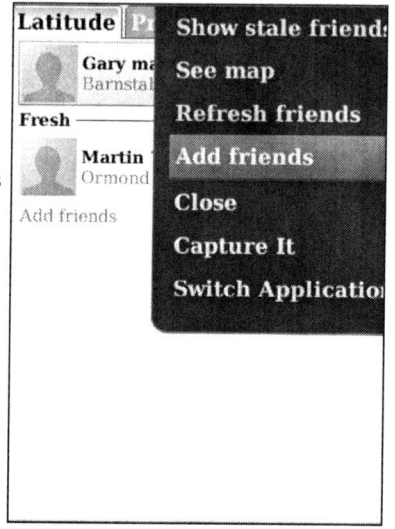

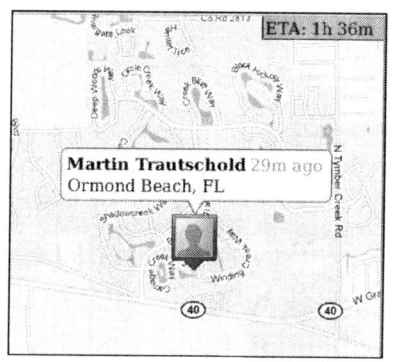

> **NOTE:** While your kids may not like it, this is a great way to check in on them and see that they are where they are supposed to be. One of the authors used this feature to track his son soon after he got his driver's license.

Layers – Finding More Things Nearby

The newest version of **Google Maps Mobile** (version 4.4.0 at the time of writing) includes a new feature called **Layers**. Essentially, this feature will show you

CHAPTER 21: Traveling (Maps & More)

Wikipedia entries for things nearby, including traffic conditions, transit lines, or personal maps. The data will be overlaid on the current map.

In this example, I have used the **Latitude** feature to locate Martin in Florida. I then used the feature to find things nearby (see Figure 21-5).

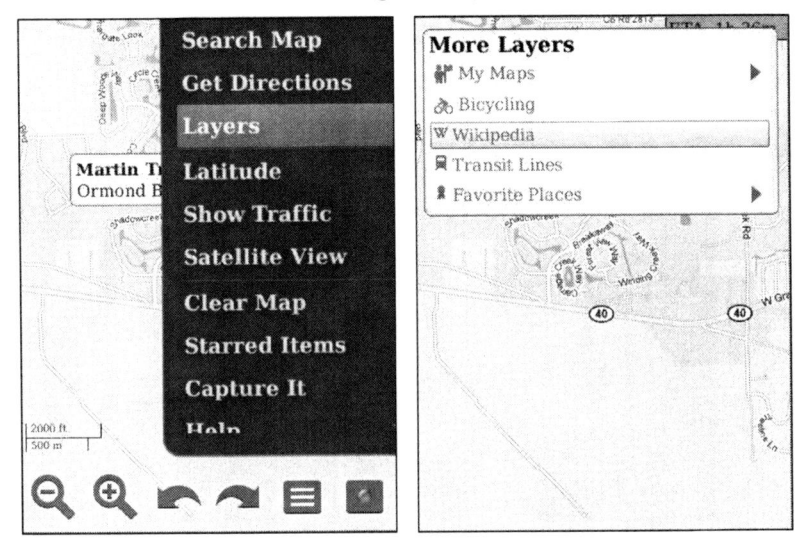

Figure 21-5. Using Layers in Google Maps Mobile

In the Wikipedia entries, I see that the Museum of Arts and Sciences is near him, so I click it. I can now find more details about the museum (see Figure 21-6).

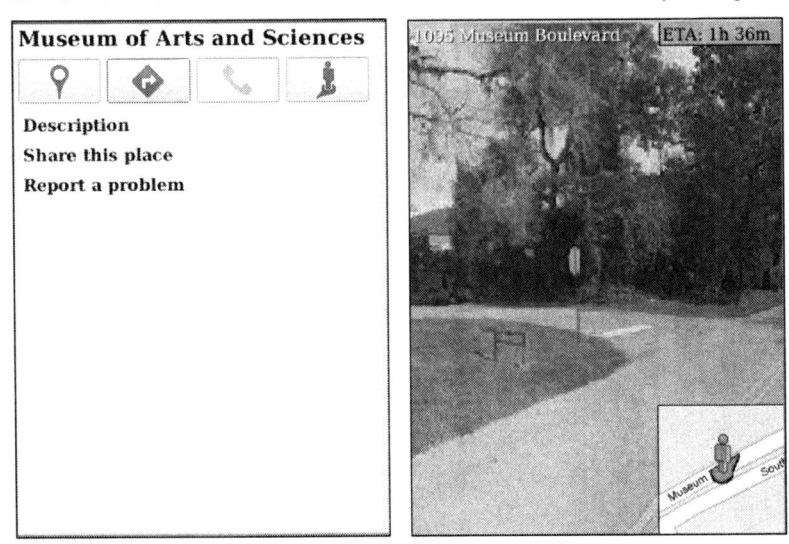

Figure 21-6. A street view of a Google Map layer

View Video Tutorials and Free Tips at www.MadeSimpleLearning.com   471

Layers – Seeing Transit Lines

Have you ever wondered where the closest subway or train station was located? You can follow these steps to find transit locations with the **Layers** feature:

1. Press the **Menu** key and select **Layers**.

2. Select **Transit Lines**. If you don't see this option listed, you will need to click the **+ More Layers** option to see it (see Figure 21-7).

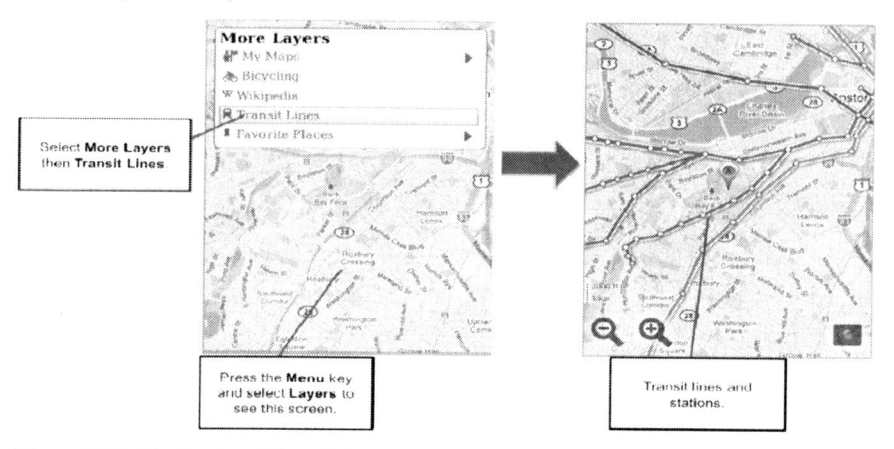

Figure 21-7. Viewing transit lines in Layers

Viewing Current Traffic

Google Maps Mobile also includes another cool feature: you can view the current traffic in major metropolitan areas or on major highways (this feature does not work everywhere). Follow these steps to do so:

1. Map the location where you want to view traffic.

2. Press the **Menu** key and select **Show Traffic,** or use the slide-out keyboard shortcut key Z/7 (see Figure 21-8).

CHAPTER 21: Traveling (Maps & More)

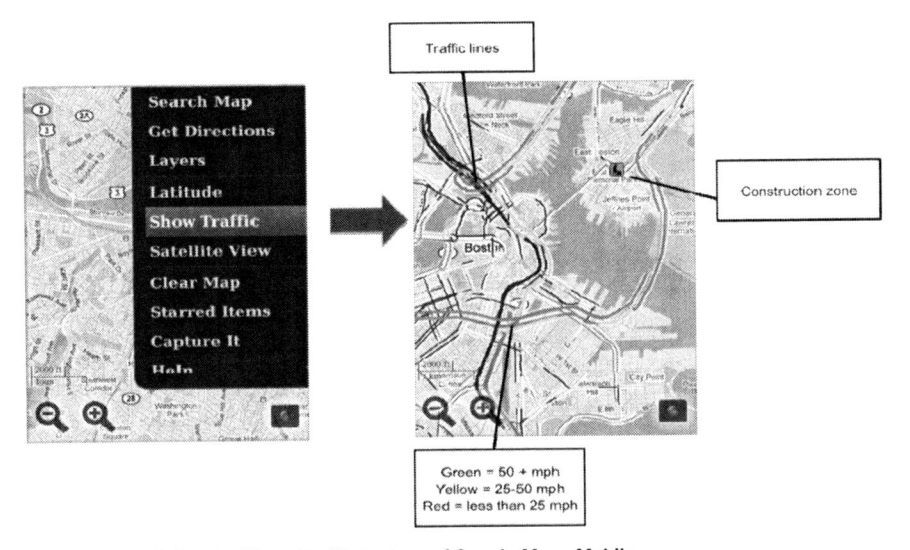

Figure 21-8. Using the Show Traffic feature of Google Maps Mobile

Chapter 22:
Other Applications

As if your BlackBerry doesn't do enough for you, RIM thoughtfully included even more utilities and programs to help you keep organized and manage your busy life. Most of these additional programs will be found in the **Applications** folder.

In this chapter, we will show you how to use your BlackBerry as a calculator and unit of measure converter, an alarm clock and timer, a voice-note recorder, a password keeper and a way to subscribe to and enjoy podcasts.

The Calculator – Tips and Unit Conversions

There are many times when having a calculator nearby is handy. I usually like to have my 15-year old "Math Genius" daughter nearby when I have a math problem to figure out, but sometimes she needs to go to school, and she isn't available to help. In such cases, I rely on the BlackBerry's built-in **Calculator** program:

1. To start the **Calculator** program, go the **Applications** folder and touch or click it. One of the icons should say **Calculator** (see Figure 22-1).

CHAPTER 22: Other Applications

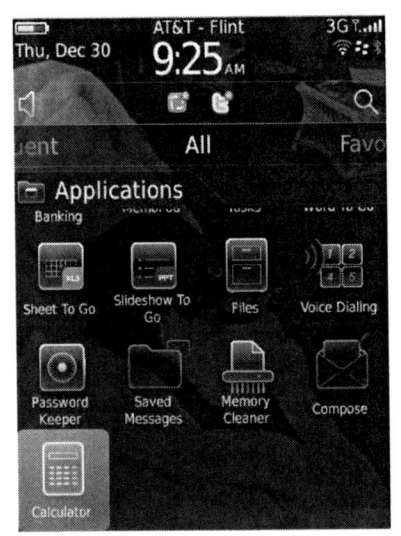

Figure 22-1. Using the Calculator app on your BlackBerry

2. Type your equation as you would in any **Calculator** program.

3. One handy tool in the **Calculator** program is that it can convert imperial values to metric ones. Just press the **Menu** button and scroll to the **To Metric** option.

4. You can then choose which metric measurement you want; from inches to centimeters to gallons to liters. You can also choose **From Metric** for reverse calculations.

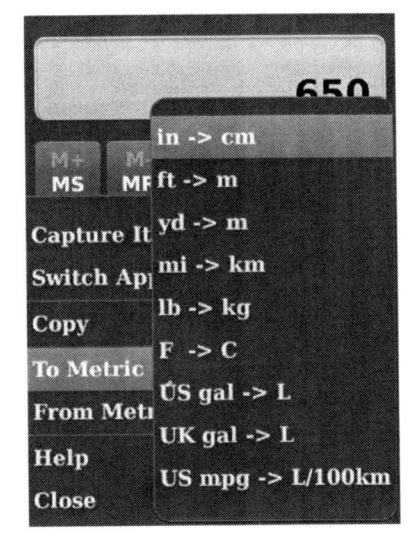

View Video Tutorials and Free Tips at www.MadeSimpleLearning.com   475

The Clock – Alarm, Stopwatch and Timer

The BlackBerry has long included a **Clock** program. However, it now includes some very cool new features and customization features that prove especially useful when the BlackBerry is on your bedside table (see Figure 22-2).

To start the **Clock** application, you can find and click the **Clock** icon; however, if you just want to set an alarm, it is easier just to tap the time in the top middle of your **Home** screen.

On some devices, the **Clock** icon might be located only in the menu, not a folder. By default, the **Clock** program displays an analog face – but that be changed.

Figure 22-2. Setting an alarm on the Clock app.

The first feature to look at is the **Alarm**. Follow these steps to bring up and manipulate the **Clock** program's **Alarm** settings:

1. Press the **Menu** button and click **Set Alarm**. The **Alarm** menu will appear in the center of the screen.

2. Press the desired field and then scroll up or down in that highlighted field to change the option's value.

3. Highlight the right-most field and scroll to select **OFF, ON,** or **Weekdays** for the alarm setting.

4. Press the **Escape** key to save your changes.

The **Clock** application also includes **Stopwatch** and **Timer** features; you can start either feature by selecting it from the menu. Finally, you give the clock facelift, selecting the following faces from the Options menu: **Analog, Digital, Flip Clock,** or **LCD Digital** (see Figure 22-3).

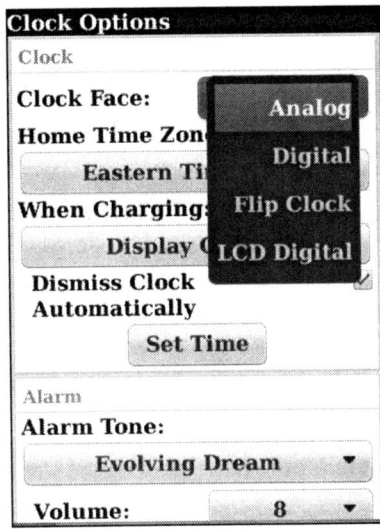

Figure 22-3. Changing the clock face

> **TIP:** If you do not want your clock to display every time you plug your BlackBerry to the charger, then change the setting for **When Charging** to **Do Nothing** in the **Clock** program's **Clock Options** screen.

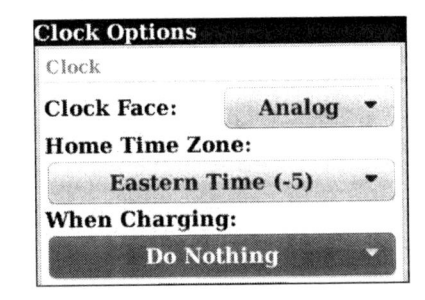

Bedside Mode:

One of the nice new features of the **Clock** application is the **Bedside Mode** setting. Many of us keep our BlackBerry by the side of the bed – we now have the option of telling the BlackBerry not to flash, ring, or buzz in the middle of the night, which is accomplished by following a handful of simple steps:

Press the **Menu** key in the **Clock** application.

Scrolling down to **Options** and click, from which you can configure the **Bedside Mode** options.

You can find the specific **Bedside Mode** options towards the bottom of the **Options** menu. From here, you can disable the LED, the radio, and dim the screen when **Bedside Mode** is set.

To enter Bedside Mode, press the **Menu** key from the **Clock** app and scroll down to **Enter Bedside Mode** and click.

> **TIP**: Bedside Mode is great for when you sleep so you won't be woken up by blinking lights, sounds or notifications.

Voice Notes Recorder

Another useful program in your **Media** folder is the **Voice Notes Recorder**. Say you need to dictate something you need to remember at a later time. Or, assume you would rather speak a note instead of composing an email. Your BlackBerry makes that very easy; simply navigate to the **Media** folder and select **Voice Notes Recorder** to launch the program.

This is a very simple program to use:

1. Launch the program, as just described.
2. Tap the screen when you are ready to record your note.
3. Tap it again when you finish recording your note.
4. You will then see the following icons along the lower part of the program: **Continue recording**, **Stop**, **Play**, , **Rename**, , or **Delete** the Voice Note. Tap the appropriate icon to perform the desired action (see Figure 22-4).

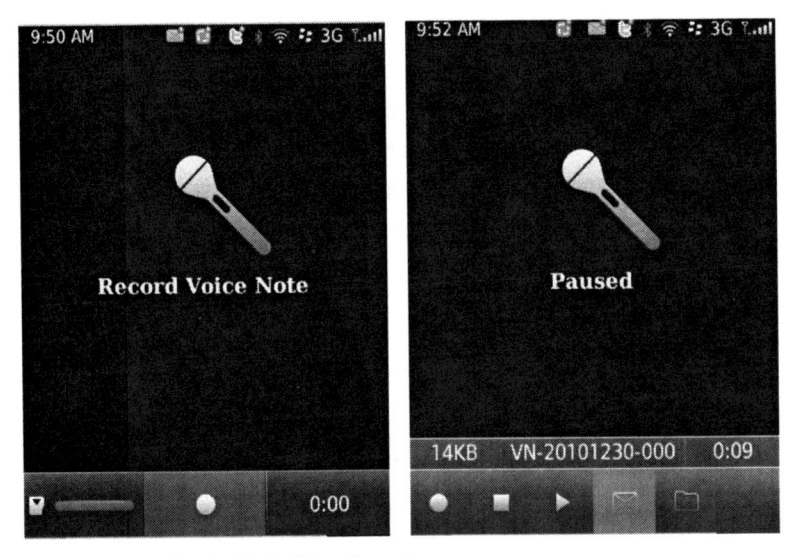

Figure 22-4. Using the Vocie Notes Recorder app

Password Keeper

It can be very hard to manage passwords in today's web-oriented world, where you need to keep track of so many sites with different passwords and different password rules. Fortunately, your BlackBerry has a very effective and safe way for you to manage all your passwords: the **Password Keeper** program (see Figure 22-5).

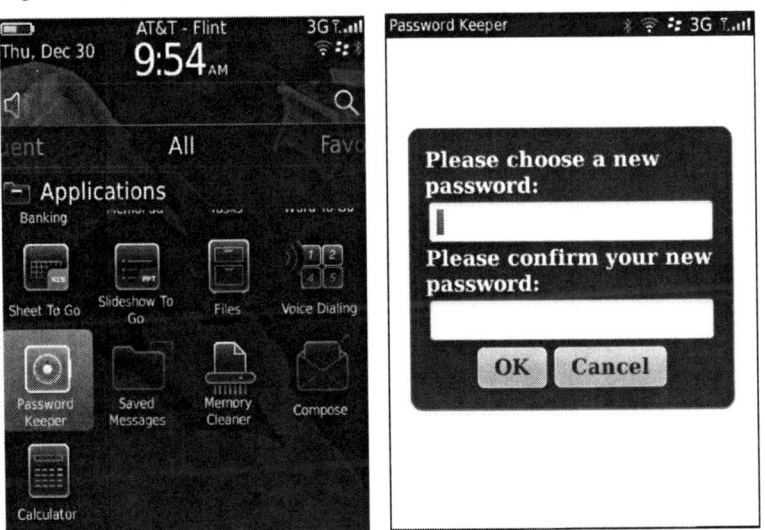

*Figure 22-5. Using the **Password Keeper** app to set a password*

Using the Password Keeper app to set a password is easy:

1. Go to your **Applications** folder, navigate to the **Password Keeper** icon, and then Tap it. The first thing you will need to do is set a program password for **Password Keeper** – pick something you will remember (see Figure 22-6)!

Figure 22-6. Using the Password Keeper app to set individual passwords

2. Once your password is set, you can add new passwords for just about anything or any web site. Press the **Menu** button and select **New** to add passwords, user names and web addresses that you want **Password Keeper** to help you remember.

Podcasts App

This is the Podcast Icon you normally see on web sites that have RSS or podcast feeds available.

Podcasts icon on your BlackBerry where you can listen to audio, text and watch video and animated podcasts.

Social Feeds icon where you can read text only RSS feeds. Learn more about **Social Feeds** in Chapter 14.

Podcasts come in many flavors: text only (RSS feeds), audio (recordings), video and even animation (like a Dilbert cartoon). As you saw in the "Social Feeds" section of Chapter 14, you can subscribe to text-only RSS (Really Simple Syndication) feeds in your **Social Feeds** app.

Now on your BlackBerry, you can subscribe to and enjoy audio, video and animated podcasts using the **Podcasts** app.

Browse and Search for Podcasts

1. Tap the **Podcasts** icon which is usually found in your **Media** folder on your **Home** screen.

2. Tap **Explore Podcasts** to browse all available podcasts.

3. Swipe left or right to view featured podcasts (or use your **Trackpad**).

4. Tap the **Top Episodes** (victory cup icon) in the lower left corner to view popular podcast episodes.

5. Tap the **Categories** (folder icon) to view all podcasts by category.

6. Use your slide-out keyboard to type a search string to find podcasts at the top of the screen. In this example, we have searched for all podcasts with the word "marketplace" in their name.

Subscribe and Unsubscribe to Podcasts

1. Once you see a podcast that interests you on the screen tap or click on it to learn more.
2. Tap the **Subscribe** button to subscribe to this podcast.
3. After you subscribe to a podcast, the episodes will be downloaded to your BlackBerry based on the settings in your **Podcasts Options** screen (see below). Typically, large podcasts will only be downloaded via Wi-Fi network connections.
4. To unsubscribe, press the **Menu** key and select **Unsubscribe**.

Enjoy Podcasts

To enjoy any podcast, from the main Podcast screen, tap the selected podcast series, then in the next screen, tap any episode that is not dimmed. (Dimmed episodes are ones that have not been downloaded to your BlackBerry).

> **TIP:** You can tap a dimmed podcast episode, then select Download from the next screen.

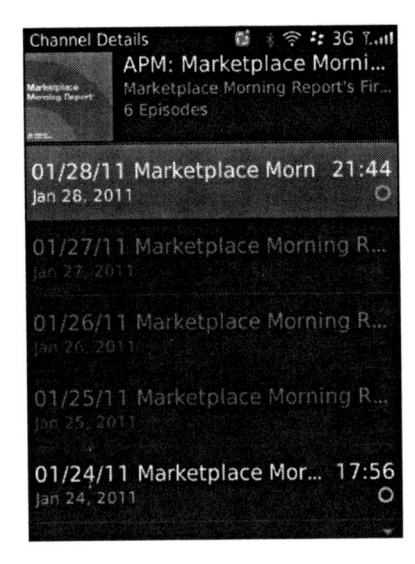

Podcast Options

In order to customize how your podcasts are downloaded (for example number of episodes, method of download), you need to adjust your Podcast Options. Start the Podcasts app, press the Menu key and select Options. Then scroll down until you see the Podcasts Options section as shown in Figure 22-7.

Figure 22-7. Podcast Options

Chapter 23: Troubleshooting

Your BlackBerry is virtually a complete computer in the palm of your hand. However, sometimes it needs a little tweaking to keep it in top running order. This chapter provides some valuable tips and tricks you can use to fix problems and keep your BlackBerry running smoothly. This chapter particularly concentrates on walking you through how to resolve problems with sending emails or connecting to the internet. It describes several possible reasons for these issues, and provides tips for how you might work around the problems described.

Solving Connection Issues

Thanks to the engineers at RIM (Research In Motion) – the company that makes BlackBerry devices – traveling away from a wireless signal for a significant amount of time (a few hours) will prompt your BlackBerry to turn off its wireless radio automatically. This feature is intended to conserve the device's battery strength; the wireless radio that makes your BlackBerry so powerful also consumes a lot of power when it is trying to find a weak or non-existent wireless signal.

The nice thing about this feature: All you need to do to turn your wireless radio back on is to hit the **Menu** key and select the **Manage Connections** icon.

You can also touch the top **Status Bar** to bring up the **Manage Connections** window.

Low Signal Strength

You may have already figured this out, but signal strength is usually stronger above ground and near the windows, if you happen to be inside a building. Signal strength is usually at its best outdoors and away from large buildings. We once worked in an office building that was on the edge of coverage, and we all had to leave our BlackBerry devices on the windowsill to get any coverage at all! We're not sure why, but sometimes turning your radio off and on will recover a stronger signal.

Managing Connections Manually (Airplane Mode)

Sometimes you might transition from a low signal to no signal at all. Many times the simple act of turning your wireless radio off and then back on again will restore your wireless connectivity. Follow these steps to manually turn your device off and then on again:

1. Go to your **Home** screen, and touch the **Status Bar** at the top of the screen.

2. Tap **Turn all Connections Off**.

3. Tap **Turn All Connections On**.

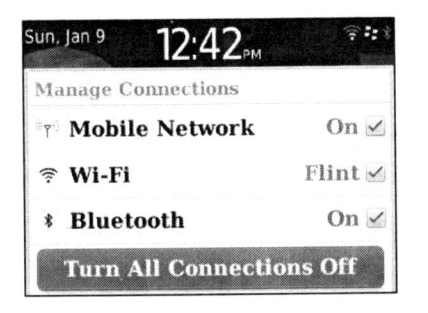

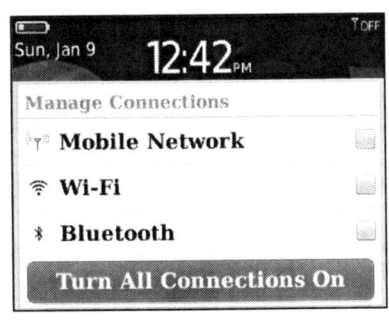

4. Look for your wireless signal meter and one of the following uppercase phrases: **3G**, **EDGE**, or **GPRS**. Check to see whether your email and internet connection are working.

Registering with the Host Routing Table

If you try all the steps we've outlined in the chapter so far, but you're still have trouble sending or receiving emails or experiencing other problems related to your wireless connection, you can try one more thing to solve the problem: registering with the Host Routing Table (see Figure 23-2). This technique lets you connect to your network provider's data connection, and it can sometimes help you resolve difficult-to-fix networking issues. Follow these steps to implement this technique:

1. From your **Home** screen, tap the Search icon and type "host."
2. Tap the **Options** icon.
3. Tap **Host Routing Table**.
4. Press the **Menu** key and select **Register Now**.

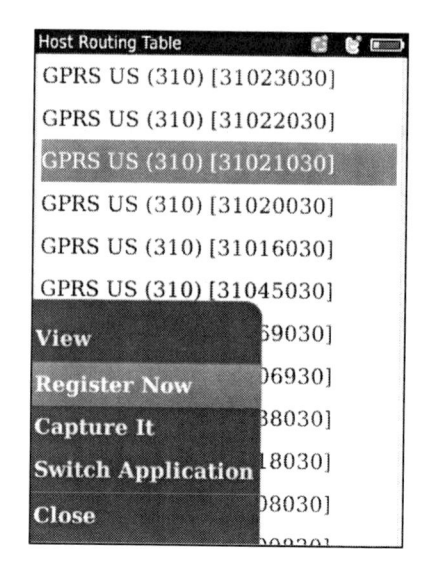

On the **Host Routing Table** screen, you will see many entries related to your BlackBerry's carrier.

Managing Your Applications

Sometimes you want to view all your installed applications, check out how much of your resources (memory, CPU (processor) power, and disk space) each one is using and possibly remove apps that are causing trouble. You can do this all from the **Application Management** screens.

1. Type the word "application" in your **Search** window on your **Home** screen.
2. Tap **Application Management.**

CHAPTER 23: Troubleshooting

3. The initial view shows you a list of everything installed on your BlackBerry.

4. You can view all required and optional apps installed by tapping the top bar and selecting a filter.

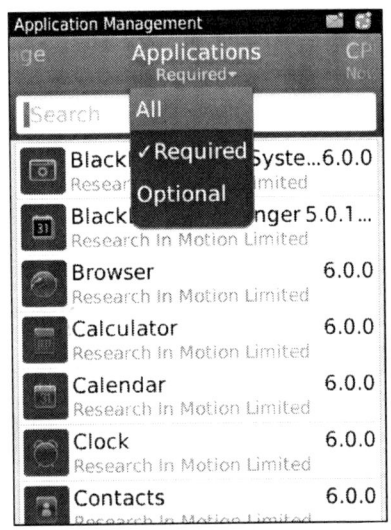

5. Swipe the top bar to the left or right to see more options **CPU** is amount of processor power used by each app. Tap the top bar to filer for specific time frames.

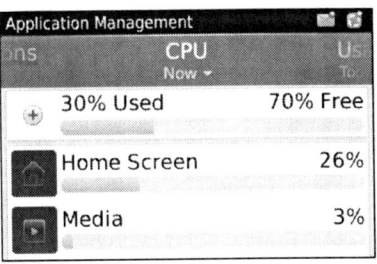

6. Usage – amount of time each application has been used on your BlackBerry. You can keep an eye on how long you have been playing games or working with your email (Messages).

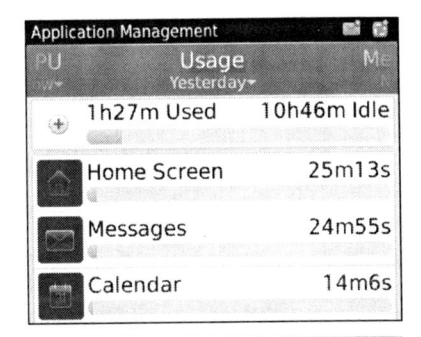

7. Amount of memory used by each application when it is running. Again, you can tap the top bar to filter the time frame for the display.

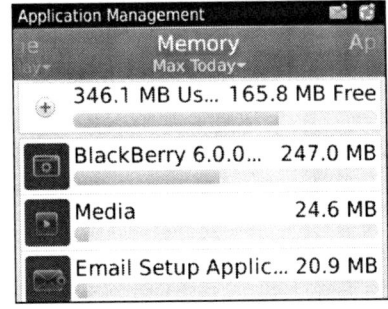

View Video Tutorials and Free Tips at www.MadeSimpleLearning.com   487

8. Amount of space used to store the applications on your BlackBerry.

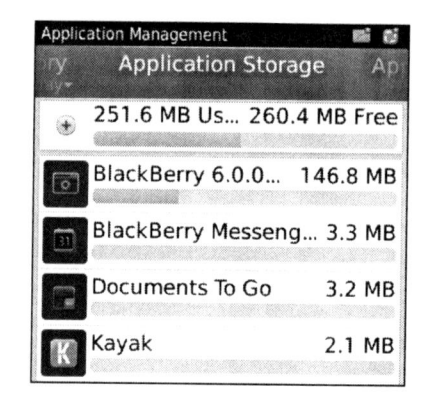

9. To view more details on any of the apps, tap them. You will be able to delete any optional apps and view specific resource usage if you scroll down the details screen (see Figure 23-1).

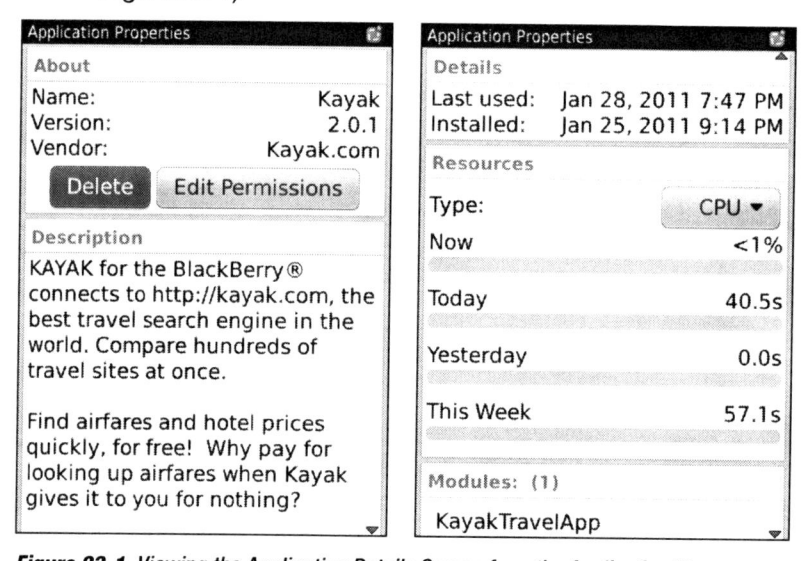

Figure 23-1. Viewing the Application Details Screen from the Application Manager

Viewing the Diagnostic Help Me! Screen

Your BlackBerry's software includes a special **Diagnostic** "Help Me!" screen that shows you detailed information about your BlackBerry. You might need this information to talk with a Help Desk or technical support person. Follow these steps to access that screen:

CHAPTER 23: Troubleshooting

1. Simultaneously press and hold three keys on your slide out keyboard: **Alt** + **Cap** + **H**.

2. At this point, you should see a **Help Me!** Status screen similar to the one shown here.

```
Help Me!
Vendor ID:                        102
Platform:                    6.4.0.105
App Version:       6.0.0.246 (695)
PIN:                          22faab9a
IMEI:            353490.04.929240.8
MAC:               40:5f:be:aa:db:87
Uptime:                 391945 secs
Signal Strength:          -70 dBm
Battery Level:                   75%
File Free:         223493077 Bytes
File Total:        420872192 Bytes
```

TIP: To view your total call minutes and other call status information, from the **Phone** app, press the **Menu** key and select **Status** to see an abbreviated status of calls made.

Clearing the Event Log

Your BlackBerry tracks absolutely everything it does. You can use this fact to help with debugging and troubleshooting in what is called an **Event Log**. Periodically clearing this log helps your BlackBerry run smoother and faster. Follow these steps to clear the **Event Log**:

1. From your **Home** screen, slide open the keyboard.

2. Press and hold down the **Alt** key while typing **LGLG** on the keyboard to display the **Event Log**.

3. Press the **Menu** key and select **Clear Log** to erase all the log entries.

4. Confirm your selection on the next screen. Once the log is cleared, press the **Escape** key to return to your **Home** screen.

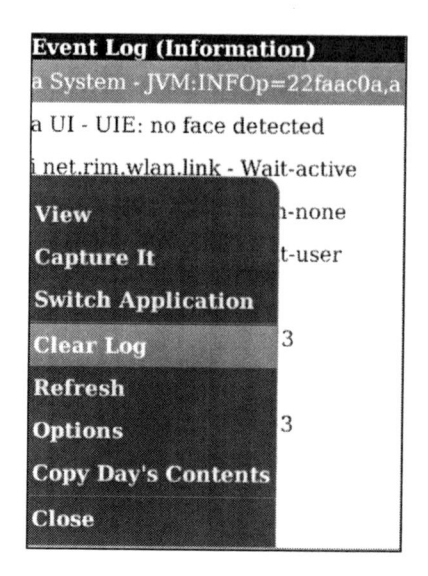

```
Event Log (Information)
a System - JVM:INFOp=22faac0a,a
a UI - UIE: no face detected
i net.rim.wlan.link - Wait-active
View                            n-none
Capture It                      t-user
Switch Application
Clear Log                       3
Refresh
Options                         3
Copy Day's Contents
Close
```

Saving Battery Life

Here's a tip that can both extend your BlackBerry's battery life and ensure that you receive the best signal possible. Specifically, you need to make sure your BlackBerry uses the optimal kind of network connection for the carrier you have your BlackBerry service with (see Figure 23-2). For example, if you're in the US, and you're using your BlackBerry on a GSM Network (such as AT&T), you will want to make sure that your device's Network Technology option is set to **3G & 2G**. This way, you will have faster coverage when available but still be covered if you roam outside of a 3G area.

1. Touch or Click on Mobile Network Options from the Manage Connections screen or touch the Status bar at the very top of the screen.

2. Make sure your Data Services are On, and then scroll down to **Network Mode**.

3. If you know you have good 3G coverage, you can set this to 3G & 2G. if you travel outside of the US, or if you are traveling through poor coverage and don't want to waste battery searching for 3G, just select 2G.

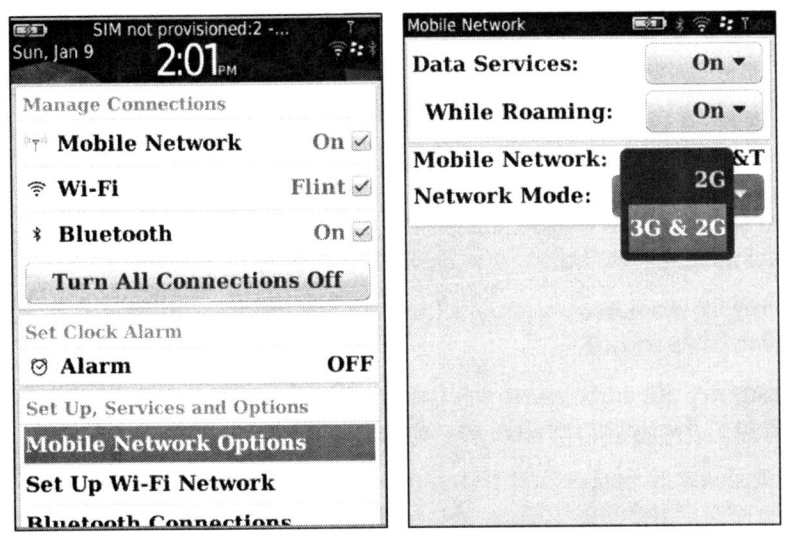

*Figure 23-2. Choose the right network under **Network Mode**.*

Resetting Your BlackBerry

There are two ways to reset or reboot your BlackBerry, a keyboard soft reset and a battery pull or hard reset.

Performing a Soft Reset

To soft reset your BlackBerry, slide out the keyboard and press and hold three keys (**ALT** + **CAP** + **DEL**) for about 15 seconds until you see the screen go black and the red LED light come on.

alt + aA⇧ + del = Soft reset

A successful soft (or hard) reset will show you the reboot screen shown to the right. As soon as the status bar on the bottom gets all the way to the right, your BlackBerry will come back on.

Performing a Hard Reset

Occasionally you will need to perform a *hard reset*; that is, you will need to remove and reinsert your battery.

Follow these steps to perform a hard reset:

1. Power down your device, if possible, by pressing and holding the power button (the **Red Phone** key).
2. Turn your BlackBerry over an slide the back cover downwards and off.
3. Gently pry out and remove the battery – look for the little indent in the upper right corner to stick in your finger tip.
4. Wait about 30 seconds, then replace the battery. Make sure you slide in the bottom of the battery first, then press down from the top.
5. Replace the battery door by aligning the tabs and then sliding the cover until it clicks into place.
6. Power-on your BlackBerry (it may come on automatically), then wait for the status bar to display. This part is similar to a soft reset.
7. Then your BlackBerry and radios should all come back on.

Let's assume you've tried all the techniques covered in this chapter so far, and your email and/or browser still aren't working. If you are using your BlackBerry with a BlackBerry Enterprise Server, please contact your organization's Help Desk or your wireless carrier's support number.

If your email and internet connection still aren't working, and you do not use your BlackBerry with a BlackBerry Enterprise Server, then proceed to the next section.

Sending Internet Account Service Books

At this point in the process, you might try sending your *service books*. BlackBerry service books are configuration files that determine how your BlackBerry will connect with the servers to enable certain features such as email and web browsing.

Without unique service books your BlackBerry will be limited in its functionality. For example, the web browser and each unique email address you have will have their own service book.

CHAPTER 23: Troubleshooting

NOTE: This troubleshooting step will work for you only if you have set up at least one Internet Email Account on your BlackBerry.

Follow these steps to send your service books:

1. Touch or click the **Setup** icon, which should be in the **All Home** screen.

2. Tap **Email Accounts**.

3. If you see this screen, tap **Internet Mail Account**.

4. Login to your Internet Email setup and tap **Continue**.

NOTE: If you have not yet logged in and created an account for your BlackBerry Internet Service with your carrier, you may now be asked to create such an account to log in for the first time.

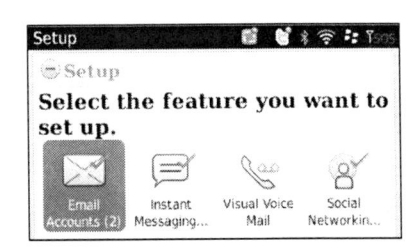

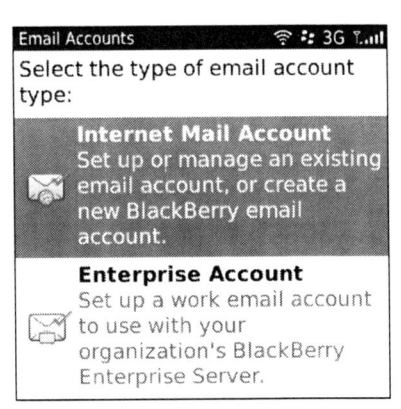

View Video Tutorials and Free Tips at www.MadeSimpleLearning.com   493

5. Press the **Menu** key and select **Service Books.**

6. Finally, on this screen, tap **Send Service Books**.

7. If everything worked, then you will see a status window saying that the **Service Books** were successfully sent.

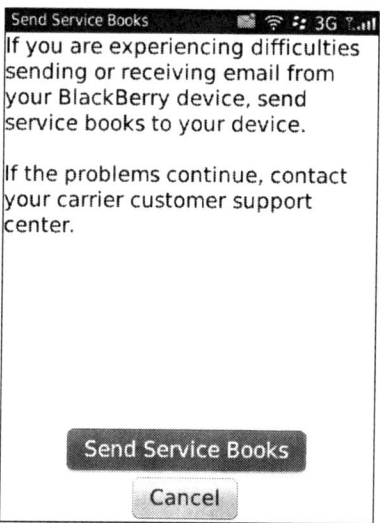

After sending the service books, you will see an **Activation Message** in your Messages email inbox for each one of your Internet email accounts set up or integrated with your BlackBerry. When opened, each message will look similar to the one shown to the right.

If the Problem Persists...

If you still cannot send email or connect to the internet, then we recommend visiting the **BlackBerry Knowledge Base** at http://www.blackberry.com/btsc. Or, you can post a question at one of the BlackBerry user forums such as www.crackberry.com, www.blackberryforums.com or www.pinstack.com.

Part 4: Keyboard Shortcuts

You have learned the basics and can now navigate around your BlackBerry, but sometimes, you just want to get somewhere even more quickly. You may not use all the following information on a daily basis, but if you take the time to learn the keyboard shortcuts – you will save time doing thing things you do most often.

Throughout this book, we have shown you how to get the most out of your BlackBerry, where to find key information, and even suggested lots of tips and tricks to help along the way.

BlackBerry's built-in **shortcut hotkeys** will quickly start an app or perform a common function within an app at the single touch of a button. For example, let's say you want to go to your **Messages** app – once your **Home** screen **Application Shortcuts** are enabled (see next few pages to learn how), just hit the letter **M**. No scrolling, searching, or fumbling. These shortcuts can be very useful.

Home Screen Hotkeys / Application Shortcuts

The Home Screen Hotkeys (**Application Shortcuts**) can help you start the most common apps on your BlackBerry with a single key click. You do sacrifice the immediate **Universal Search** by enabling these shortcuts, but you can decide – which do you use more – **Application Shortcuts** or **Universal Search**. Why not turn on the **Application Shortcuts**, and then tap the **Search** icon (magnifying glass) whenever you want to perform a search – it's not that hard.

The Switch to Select Search or Shortcuts

By default, typing on the **Home** screen will start the **Universal Search**. To change this to **Application Shortcuts**, follow these steps:

1. Press the **Menu** key from the **Home** screen and select **Options**.

2. Change the Launch by Typing to **Application Shortcuts**.

3. Press the **Menu** key and **Save**.

Once you enable your **Home Screen Hotkeys**, you will be able to launch most of the built-in applications with just one letter. This is extremely useful when you have lots of programs and you don't exactly remember where the icons are located.

List of Home Screen Hotkeys

Following are the list of one key shortcuts to start most of your major icons.

M	Messages (Email)	H	Help
L	Calendar	K	Keyboard Lock
C	Contact or A = Address Book	O	Options Icon
		U	Calculator
V	Saved Messages	S	Search Icon
N	BlackBerry Messenger	D	MemoPad
T	Tasks	P	Phone (Call Logs)
B	Web Browser		
W	WAP Browser		

If you forget the particular shortcut key, then your BlackBerry will remind you.

Once turned on, the shortcut are denoted by the underlined letter.

For example in the image to the right you can see **M**essages (M), Ca**l**ender (L), Memopa**d** **(D)**, Cont**a**cts (C) and **B**rowser (B) all have a single letter underlined.

> **TIP:** Don't press and hold any of these hotkeys, otherwise you will start Speed Dialing.

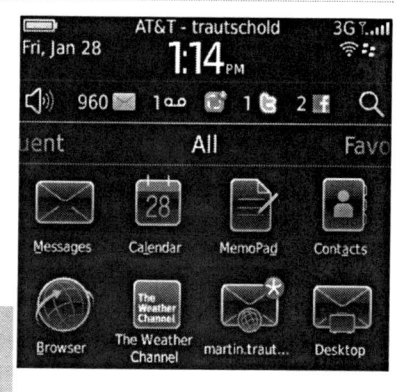

Email Messages Shortcuts

Not only are shortcut keys helpful to launch apps, they are also very useful from inside an app to take you to a specific menu command. The following are the hotkeys that can get you around your **Messages** email app.

> **NOTE:** There is nothing you need to do to activate the hotkeys in the **Mail** app, they just work from your slide-out keyboard.

T	Top of Inbox	**Space** key - Page Down	C	Compose new email
B	Bottom of Inbox	**ALT + Glide trackpad** - Page Up/Down	R	Reply to selected message
U	Go to next newest Unread Message	ENTER - Open Item (or compose new message when on date row separator)	L	Reply All to selected message
			F	Forward selected
N	Next day	E	Find Delivery Error(s)	K	Search for next message in thread (Replies, Forward, etc.)
P	Previous day	V	Go to Saved Messages Folder		
ALT + U Toggle read / unread message		Q	(When highlighting email/name) Show/Hide Email Address Or Friendly Name	S	Search
Shift + Glide trackpad - Select messages		**ALT + O** – Filter for Outgoing Messages		**ALT + M** – Filter for MMS Multimedia Messages
DEL Delete selected message(s)		**ALT + I** – Filter for Incoming Messages		**ALT + S** – Filter for SMS Text Messages
CAUTION: If wireless email synchronization is on, this will also delete email from your inbox.					

Web Browser Shortcuts

Just like in your **Messages** app, you can quickly launch certain commands or jump to specific places in your **Browser** menu by using certain hotkeys, such as the ones in the following table.

> **NOTE:** These shortcuts will work while you are viewing a web page, if your cursor is in a field on a web page such as the Google search window, then typing these letters will simply type letters on the screen and not perform the shortcut functions listed here.

T	Top of Web Page	Enter	Select (click-on) highlighted link	I	Zoom into web page
B	Bottom of Web Page	Y	View history of web pages	O	Zoom out of web page
Space key	– Page Down			**F/V**	Find text on the current web page
Shift + Space key	– Page Up	R	Refresh the current web page	A	Add new bookmark
G	Go to your Start page where you can type in a web address	P	View the address for the page you are viewing with an option to copy or send the address (via email, PIN, or SMS).	K	View your Bookmark List
H	Go to your Home page (Set in **Browser** menu → Options → Home Page)			S	View Browser Options
Escape key/**DEL** key	Go back 1 page				
D	View downloads				

> **NOTE:** These hotkeys do not work in your Bookmark List or **Start** page; only when you are actually viewing a web page. In the Bookmark List, typing letters will find your bookmarks that match the typed letters.

Calendar Shortcuts

You can turn on the **Calendar** shortcut keys listed in the following table on your BlackBerry using these steps:

1. Start your **Calendar**.
2. Press the **Menu** key.
3. Select **Options**.
4. Tap **Calendar and Display Actions** glide to **Enable Quick Entry** and change it to **No** by pressing the **Space** key.
5. Save your **Options** settings.

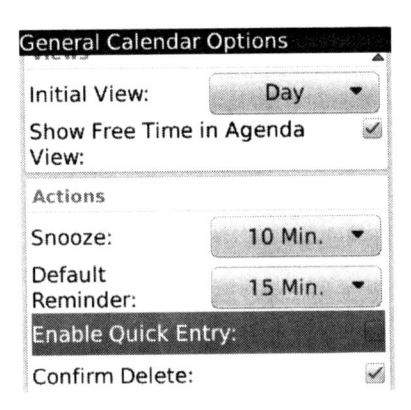

D	Day View
W	Week View
M	Month View
A	Agenda View
N	Next day
P	Previous day
G	Go To Date
T	Jump to Today (Now)
C	Schedule new event (detail view)
Space key	Next Day (day view)
DEL	Delete selected event

Shift + glide trackpad – select several hours – example if you wanted to quickly schedule a 3 hour meeting you would highlight the 3 hours and press **Enter**.

Enter – (in Day View – not on a scheduled event) Start Quick Scheduling

Enter – (in Day View – on a scheduled event) Opens it.

Quick Scheduling New Events in Day View:

Enter key -- Begin Quick Scheduling, then type the Subject of the appointment on the Day View screen.

> 4:00p Quick scheduling
> 4:15p

ALT + Glide Trackpad up/down – Change **Start** time

> 4:30p Quick scheduling
> 4:45p

Glide **Trackpad** up/down – Change **Ending** time

> 4:30p Quick scheduling
> 5:30p

Enter or click **trackpad** – Save new event

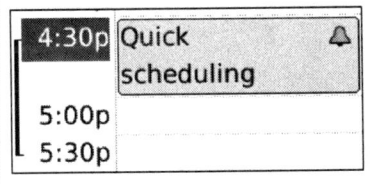

Media Player Shortcuts

The shortcuts and tips below will help you enjoy media on your BlackBerry by performing many of the common functions with a single keystroke or special key combinations.

Use your **Mute** key on the top right of your BlackBerry to play or pause any song or video.

Top Edge

Mute Key
(Also Pause/Play Music)

These keys work only inside the Media Player App	These keys work anytime (Whether inside the Media App or not)
N Next track/video	Press and hold **Volume Up** key – Next track/video
P Previous track/video	Press and hold **Volume Down** key – Previous track/video
Space key Play / pause	**Mute** Key on top – Play/pause
Enter key Selects highlighted button (same as clicking trackpad / clicking trackball)	**$/Speaker** key – Switch between speaker and handset/headset (including Bluetooth headset or Bluetooth audio speakers)
DEL Delete selected media item (song, video, and so forth, when in list or icon view)	
Jumping Out of the Media Player: Press and hold the **Menu** key to multitask or press the **Red Phone** key to jump back to the **Home** screen.	Jumping Back into the Media Player and currently playing song/video: Tap the **Menu** key and select **Now Playing**… from the top of the menu. Press and hold the **Menu** key, then select the **Media** icon from the pop-up.

Index

A

Accented Characters · 167
Accesibility
    Options · 218
Acknowledgments · xx
Address Book · *See* Contact List
    Find - Company Name · 286
    Find - First Name · 286
    Find - Last Name · 286
    Rule 1 - Add Entries Often · 279
    Rule 2 - Know how Find works · 279
AIM Messenger · 343
Airplane Mode · 456
Alarm Clock
    Overview · 26
AOL Instant Messenger · 344
App Reference Tables · *See*
App World · 437
    Archive or Delete Apps · 447
    Bill through carrier (AT&T, other) · 443
    BlackBerry ID · 438
    Categories · 440
    Change Download Folder · 450
    Changing Payment Options · 443
    Credit Card · 438, 443
    Delete Apps · 446, 451
    Direct Bill (Wireless Bill) · 438
    Download · 437
    Download and Purchase Apps · 444
    Download Themes · 439
    Downloading Apps · 443
    Featured Programs · 439
    Icons and Getting Around · 439
    Managing Your Apps · 446
    My World · 446
    My World Menu Commands · 446
    Payment Options · 438
    PayPal · 438, 443
    Purchase Apps · 444
    Search · 442
    Seeing Installed Apps · 446
    Starting · 438
    Top 25 · 441
    Using "My World" · 446
Apple Mac Setup · 132
Application Management · 486
Application Manager
    Delete Apps · 452
Application short cuts · 17
Apps · *See* App World
    Application Manager · 452
    Deleting · 451
    Finding More · 450
    Official BlackBerry Sites · 451
    Removing · 451
    Reviews · 451
    Starting and exiting · 13
Apps - Install from Web Sites · 448
Apps - Install or Delete · *See* App World
Apps - Install without App World · 448
Authors
    About · xix
AutoText
    Creating New Entries · 174
    Date, Time, Phone · 176
    Delete · 175
    Edit · 175
    Macros · 176

B

Battery

Extending Battery Life · 92
How often to charge · 90
Low Battery Warning · 91
Re-Charging · 91
Battery Life
   tips · 90
Battery Life Tips · 490
Battery Pull · 492
BBM · *See* BlackBerry Messenger
BES Express · 60
   Free Software · 60
   Microsoft Exchange · 60
   Microsoft Small Business Server · 60
BlackBerry App World · 437, *See* App World
   Purchase Apps · 445
BlackBerry Enterprise Server Express · 60
BlackBerry Maps · 459, See Mapping - BlackBerry Maps
   Directions · 462
   Directions - Turn by Turn · 463
   Directions - View on Map · 463
   Getting Directions · 460
   Icon · 460
   Menu Commands · 463
   Viewing maps · 460
BlackBerry Messenger
   Add Contacts (People) · 329
   Add Contacts by Email or PIN · 329
   Add Using Barcodes · 330
   Availability - Showing · 336
   Backup · 332
   Download Software · 328
   Emoticons (Smileys) · 333
   Emoticons / Icons · 334
   Groups · 338
   Groups - Calendars · 340
   Groups - Creating · 339
   Groups - Picture Sharing · 340
   Groups - To-Do Lists · 340
   Introduction · 328
   Menu Commands · 331
   Multi-person Chat · 337
   Options · 332
   Ping a Contact · 335
   Reply to Invitation · 331
   Restore · 332
   Send BlackBerry Contact · 335
   Send File · 335
   Send Location (GPS) · 335
   Send Messenger Contact · 335
   Send Picture · 335
   Send Voice Note · 335
   Sending Files · 335
   Set up · 329
   Smiley Faces · 334
   Start / End Conversations · 333
   Status Setting · 336
BlackBerry Technical Knowledge Base · 109
Bluetooth
   Answer and Call from Headset · 65
   Configuring · 63
   Connecting · 61
   Discoverable · 63
   Passkey (PIN) · 65
   Security Tips · 61
   Send / Receive Files · 68
   Setup Menu Commands · 67
   Supported Devices · 62
   Troubleshooting · 69
Browser · See Web Browser

C

Calculator · 474
   Unit of Measure Conversions · 474
   Units - English to Metric · 474
   Units - Metric to English · 474
   Using It · 474
Calendar
   Agenda View · 297
   Alarm · 302
   Alarm - Ring, Vibrate · 309
   All-day Event · 300
   Busy - Free - Out of Office · 302
   Conference Call - Join Now · 304
   Conference Call - Moderator and Participant Numbers · 305
   Conference Calls Scheduling · 304
   Copy & Paste · 305
   Day View · 297
   Default Day, Week, Month or Agenda Views · 304
   Default Reminder · 304
   Default View · 304
   Detailed Scheduling · 300
   Introduction · 296
   Invitation - Change Invitees · 313
   Invitation - Email Invitees · 313
   Invitation - Reply To · 312

INDEX

Invite Attendee · 311
Keyboard Shortcuts · 298
Meeting Invitations · 311
Month View · 297
Move between days · 298
Options · 303
Overview · 25
Private Events · 302
Quick Scheduling · 300
Recurring Events · 302
Reminder · 302
Scheduling Appointments · 300
Scheduling Tips · 300
Snooze Alarm · 310
Snooze Default · 304
Soft Keys · 299
Sounds - Adjusting · 309
Start and End of Day Hour · 303
Switching Views · 297
Sync with Google · 313
Tranfer or Sync to BlackBerry · 296
View Availability · 312
Week View · 297
Call Log
    Set from keyboard · 239
Call Logs
    Checking · 233
    Put in Messages (Inbox) · 236
    Using · 233
Camera · 399
    Adjusting Picture Size · 403
    Auto · 405
    Beach · 405
    Buttons · 400
    Caller ID from Pictures · 410
    Close-Up · 405
    Face Detection · 405
    Features · 400
    Geotagging · 403
    Icons · 401
    Introduction · 399
    Landscape · 405
    Night · 405
    Party · 405
    Picture Quality · 405
    Picture Size · 403
    Picture Storage · 406
    Portrait · 405
    Scene Mode · 405
    Send Pictures (Email, Text, BBM, Twitter...) · 402
    Set As / Crop · 401
    Set as Wallpaper · 402
    Snow · 405
    Sports · 405
    Starting It · 400
    Store on Device Memory · 406
    Store on Media Card · 406
    Taking Great Pictures · 399
    Text · 405
    Tips & Tricks · 411
    Tips for Great Pictures · 399
    Transferring Pictures · 410
    Viewing Pictures · 407
    Zoom · 406
Categories
    Contact List · 291
Clock · 476
    Alarm · 476
    Bedside Mode · 477
    Countdown · 476
    Timer · 476
Conference Call
    Dial from Calendar Event · 309
Connect Laptop to Internet · See Tethered Modem
Connect Mac to Internet · See Tethered Modem
Connect PC to Internet · See Tethered Modem
Contact List · 278
    Add Entry from Email Address · 283
    Add Entry from Phone Call Log · 283
    Add from phone, email or street address · 284
    Add New Contacts · 280
    Add Picture · 290
    Calling Contacts · 287
    Cannot See All Contacts · 293
    Categories · 291
    Editing Contacts · 288
    Email to a Group · 295
    Error · 293
    Filtering · 292
    Find Feature · 286
    Finding Contacts · 286
    Groups - Mailing List · 293
    Introduction · 278
    Letters in Phone Numbers · 282
    Mailing List · 293
    New Entry · 280
    Picture Add · 290

View Video Tutorials and Free Tips at www.MadeSimpleLearning.com

Picture Caller ID · 290
Recommendations for Use · 279
Remote Lookup · 284
Several Email Addresses · 282
SIM Card, Transfer from · 279
SMS Mailing List · 294
Some Contacts Missing · 293
Sort by First, Last or Company · 291
Transfer or Sync to BlackBerry · 278
Troubleshooting · 293
When it is useful · 279
Contact List XE "Contact List:Calling Contacts"
Use in Phone to Dial by Name · 287
Contact List XE "Contact List:Sort by First, Last or Company"
Sorting · 291
Contacts
Email - Sending · 257
Contacts
Unfiltering · 293
Contacts
Sync with Google · 313
Contents · viii
At a glance · vi
Contents At A Glance · vi
Controls
Showing and hiding · 15
Convenience Keys
Changing · 204
Programming · 204
Setting · 204
Converting Units · See Calculator
Copy · 18
Copy & Paste
Calendar · 305
Shortcuts · 381
Copy and Paste
PIN · 342
Corporate Address List · See Remote Contact Lookup

D

Date
Set · 197
Desktop Software
Backup data · 109
Delete data · 113

Ipdate BlackBerry system software · 115
Media Manager · 118
Restore data · 112
Switch devices · 100
Working with applications · 114
Desktop Software (Windows)
Check Your Version · 95
Download Latest · 95
Installing · 97
Overview · 98
Synchronization - Troubleshooting · 108
Synchronization Setup · 101
Desktop Software Mac
Add/Remove Applications · 142
Backup · 140
Backup options · 136
Download · 133
Restore · 141
Sync options · 137
Syncing Music · 147
Syncing Pictures · 149
Syncing Videos · 150
USB Drive Mode · 156
Wireless Music Sync · 152
Dial by Name
Tip · 287
Documents to Go
Editing · 270
Download Folder
Change · 450
Drop Down Tips
Letter Keys · 169
Typing Numbers · 169

E

Email
Adding Recipients · 257
Attach Contact Entry · 266
Attach File · 267
Attachment Viewing · 269
Attachments · 268
Attachments Supported Formats · 268
BCC · 257
Body · 257
CC · 257
Composing / Writing · 255
From Contacts · 257

INDEX

High Priority · 262
Low Priority · 262
Opening Attachments · 268
Picture View or Save · 273
Priority · 262
Priority - Setting · 262
Reply - Shortcuts · 259
Save Picture · 273
Search Recipient · 277
Search Sender · 276
Searching · 275
Selecting Recipients · 256
Send · 255
Send Picture · 267
Send Ringtone · 267
Send Song · 267
Send to a Group · 295
Sending Files · 267
Set Importance Level · 262
Shortcut keys · 260
Soft keys · 260
Spell Checking · 262
Subject · 257
TO · 256
Troubleshooting Missing Messages · 57
Urgent · 262
View Email Address (Q) · 258
View Picture · 273
View Spreadsheet · 272
Viewing Pictures · 261
Wireless reconciliation · 53
Email Accounts
   Maintaining · 48
   Setup · 43
   Troubleshooting · 54
Email Addresses - Multiple · 282
Email Icons
   Hiding · 48
Email Setup
   Activation Password · 51
   Enterprise Activation · 51
English Unit Conversion · *See* Calculator
Enterprise Activation · See Email Setup (Corporate)
   Setup · 50
Escape Key · 9

F

Facebook · 347
   Contact List Functions · 289
   Friends · 350
   Icons · 349
   Install · 345
   Introduction · 347
   Login · 346
   News Feed · 349
   Picture Upload · 351
   Setup Wizard · 348
   Tag Photo · 351
   Update Status · 349
   Uploading Pictures · 351
Features Overview
   Media Card (MicroSD Card) · 32
Flag for Follow Up · 263
   Changing Color · 264
   Create for Email · 263
   Due Date · 264
   Finding Flagged Items · 265
   Ringing Alarm · 266
   Search Hot Key · 266
Folders
   Opening · 13
   Working with · 195
Folders on BlackBerry · 194
Follow Up Flag · 263
Font
   Changing Size & Type · 199
Font Size and Type · 199

G

GAL Lookup · *See* Remote Contact Lookup
Geotagging · *See* Camera - Geotagging
Getting Started · 41
Global Address List · *See* Remote Contact Lookup
Global Address List Lookup · *See* Remote Contact Lookup
Google Contacts
   Syncing · 54
Google Latitude · *See* Google Maps - Latitude
Google Maps · 465
   Bicycle · 470
   Call Business · 467

Current Traffic · 472
Directions · 469
Downloading and Installing · 465
Find Business · 467
Finding an Address · 466
Finding Places · 436
Icon · 466
Installing · 465
Latitude · 470
Layers · 470
Menu Commands · 466
My Location · 466
Public Transit · 470
Public Transportation Maps · 472
Put Pin on Map · 467
Search / Find Address · 467
Search Results · 467
See traffic · 472
Shortcuts / Hot Keys · 469
Street View · 468
Traffic · 472
Transit Lines · 472
Walking · 470
Zoom In or Out · 466
Google Search · 435
Google Sync · 313
Address Book · 313
Calendar · 313
Contacts · 313
How it Looks · 317
Installing · 314
Google Talk Messenger · 343
Green Phone Key · 8

H

Hard Reset · 492
Help
Menus · 184
Tips and Tricks · 185
Using It · 183
Help Icon
Text Help · 184
Help Me Screen · 488
Home Screen
Downloads location · 189
Reset preferences · 190
Home Screen Background
Change or Set · 187
Home Screen Image
Change or Set · 187
Home Screens
Getting around · 11
Host Routing Table
Register Now · 326
HotSpot (WiFi) · *See* WiFi

I

Icons
Hiding · 192
Moving (between folders) · 194
Moving (in folder) · 192
Organizing · 191
What they do · 164, 165
Install Apps · *See* App World
International Travel Tips · 454
Airplane Mode · 456
Before Trip · 455
Data Roaming (Disable) · 457
Foreign SIM Card · 455
International Rate Plan · 455, 458
Re-enable Data Services · 458
Register with Network · 457
Set Time Zone · 456
Unlock BlackBerry · 455
Verify Time Zone · 458
Wi-Fi Netwoks (Free) · 458
Internet Radio · 391
Introduction · 30, 31
ISBN · ii

K

Keyboard
Full "QWERTY" Keyboard · 162
Hide and Show · 15
Multitap · 163
On screen · 19
Physical · 19
SureType™ · 163
Keyboard Shortcut Sheets · 496
Application Shortcuts · 497
Browser · 500
Calendar · 501
Email / Messages · 499
Enable Application Shortcuts · 497
Home Screen · 497

Media Player · 502
Messages · 499
Music Player · 502
Universal Search vs. Application Shortcuts (Switch) · 497
Video Player · 502
Web Browser · 500
Keyboards Three Types · 161

L

Light (LED)
    Amber · 206
    Blue · 206
    Blue - Adjusting · 207
    Green · 206
    Green - Adjusting · 208
    Red · 206
    Red - Adjusting · 206
    Repeat Notification · 206
LinkedIn
    Connections · 359
    Getting Around · 358
    Icons · 359
    Install · 345
    Introduction · 358
    Invitation · 360
    Login · 358
    Messages · 360
    Navigating · 358
    Options · 361
    Reconnect · 361
    Search · 359
Lock Key · 7
Lookup · *See* Remote Contact Lookup
Loud · *See* Profiles, *See* Profiles

M

Mac Media Sync
    Options · 145
Made Simple Learning
    Contact Information · ii
    Email tips · 35
    Video tutorials · 36
Manage Connections
    Airplane mode · 12
Managing your Contacts · 287

Mapping
    BlackBerry Maps · 459
    BlackBerry Maps - Email & Favorites · 464
    BlackBerry Maps, Google Maps · 459
    Google Maps · 465
    Introduction · 459
    TeleNav · See TeleNav
    Using BlackBerry Maps · 459
Media
    Playing Music · 383
    Playing Videos · 416
    Ringtone - Using Music As · 245
    Ringtone Folder · 245
    Shuffle Music · 386
    Supported Music Types · 390
    Transfer using USB Drive Mode · 155
    Transfer using Mass Storage Mode · 118
    Transfer using USB Drive Mode · 72
    Unique Ringtone for Callers · 247
    Viewing Videos · 416
Media (Music, Videos, Pictures)
    Introduction · 382
Media Card · 70
Media Transfer
    Pictures · 126
    USB Drive mode · 118
    Videos · 129
MemoPad · 374
    1001 Usage Ideas · 375
    Add New · 375
    Categories · 378
    Copy & Paste · 380
    Date & Time Inserting · 378
    Editing Memos · 375
    Email Memos · 380
    Finding - Tips · 377
    Forward Memo · 380
    Introduction · 374
    Menu Items Described · 380
    Ordering · 377
    Search · 377
    Tips · 377
    Transfer or Sync from Computer · 374
    Viewing · 377
Memory Card
    Checking · 71
    Free Space · 71
    Installing · 70
    Purchasing · 70

Verify Install · 71
Memos · *See* MemoPad
Menu Key · 8
Metric Conversion · *See* Calculator
MicroSD Card · *See* Memory Card
MMS · See Multi-Media Messaging
Modem (use BlackBerry as) · See Tethered Modem
Multi-Me XE "Picture Messaging" \t "*See* Multi-Media Messaging (MMS)" XE "Video Messaging" \t "*See* Multi-Media Messaging (MMS)" dia
Messaging (MMS)
  Introduction · 323
Multi-Media Messaging (MMS)
  Costs · 318
  Searching · 275
Multi-Media Messaging (MMS)
  Sending · 324
Multi-Media Messaging (MMS)
  Send from Media Player · 325
Multi-Media Messaging (MMS)
  Advanced Commands · 326
Multi-Media Messaging (MMS)
  Delivery Confirmation · 326
Multi-Media Messaging (MMS)
  Troubleshooting · 326
Multi-Media Messaging (SMS)
  Introduction · 318
Multi-task · 18
Multitasking
  Switch Application · 378
Multi-Tasking
  ALT + Escape Keys · 379
Music · 382
  Automatic Playlist · 389
  Find and Play Song · 384
  Mute (Pause / Play) key · 386
  Now Playing · 385
  Play All · 386
  Play in Background · 385
  Play Song · 384
  Playlists - Create on BlackBerry · 388
  Playlists - Using · 387
  Playlists on computer · 388
  Playlists Transfer · 382
  Ring tone - use as · 386
  Standard Playlist · 389
  Supported Song file types · 390
  Transfer or Sync from Computer · 382
  Wi-Fi Music Sync · 383

Music and Video
  Playing your Music · 383
  Transferring Media Using USB Drive Mode · 155
  Transferring Media Using Mass Storage Mode · 118
  Transferring Media Using USB Drive Mode · 72
  Using Music as Ringtones · 245
  Working with Music and Videos · 382
Music Player · 382
Mute · *See* Profiles, *See* Profiles
Mute Key · 8
MySpace
  Install · 345

N

Navigation Bar
  Using · 187
Network
  Wireless · 21
Notes · *See* MemoPad

P

Pandora · 392
  Installing · 392
  Set Up · 392
  Thumbs Up and Down · 394
  Using It · 394
Password Keeper · 479
Password Security
  Disable · 89
Paste · 18
Phone
  3-Way Calling · 250
  Add New Addresses · 234
  Answering 2nd Caller · 248
  Answering a Call · 224
  Auto-pause · 253
  Auto-wait · 253
  Call Forwarding · 248
  Call Logs · 220
  Call Logs - Names / Numbers · 234
  Call Logs in Messages (Inbox) · 236
  Call Waiting (2nd Caller) · 247
  Caller Volume · 230

INDEX

Calling · 221
Conference Calling · 250
Conference Split - Drop · 252
Contact List · 220
Dial by name · 224
Dial from call log · 223
Dial Pad · 220
Dialing Extensions Automatically (x) · 254
Dialing Letters · 252, 253
Hold · 251
Ignore & Mute Calls · 225
Important Keys · 221
Join Callers · 250
Making a Call · 221
More Time to Answer · 225
Muting a Call · 226
My Number · 230
New Call · 250
Notes while on call · 228
Pause · 253
Placing a Call · 221
Ringtone - Changing · 231
Speed Dial · 237
Split · 251
Tips & Tricks · 254
Voice Dialing · 241
Voice Mail · 232
Voice Mail Problems · 232
Wait · 253
Phone Numbers
    Dialing Letters (1800 CALL ABC) · 282
Phone Tune
    Setting your Music as · 245
Picture Caller ID · 290
Picture Messaging · *See* Multi-Media Messaging (MMS)
Pictures · 408
    Caller ID, Adding As · 410
    Save to Computer · 410
    Scrolling Through · 410
    Slide Show · 409
    Soft Keys · 409
    Transfer To-From BlackBerry · 410
    Viewing · 407
PIN Messaging
    Adding PIN to Address Book · 342
    Free · 328
    Introduction · 341
    Inviting Others · 341
    My PIN Number Trick · 341

Sending Your PIN · 341
PIN-to-PIN · *See* PIN Messaging
Playing Music · 383
Podcasts · 480
    Categories · 481
    Dimmed · 483
    Enjoy · 483
    Explore · 481
    Find · 481
    Listen · 483
    Options · 483
    Popular · 481
    Search · 481
    Subscribe · 482
    Top Episodes · 481
    Unsubscribe · 482
    Watch · 483
Profiles
    Calendar Alarm - Ring, Vibrate · 309
Profiles (Sounds)
    Basic Settings · 209
    Changing · 209
    Customize · 210
    Loud · 209
    Mute · 209
    Quiet · 209
    Ringtone - Download New · 213
    Vibrate · 209

Q

Quick Start · 1
    Getting Around Quickly · 3
    Overview of Torch · 4
Quick Status Bar · 12

R

Radio (Internet) · 391
Reboot BlackBerry · 491
Red Phone Key · 9
Register with Wireless Network · 326
Remote Contact Lookup · 284
Reset BlackBerry · 491
Ring Tone
    Changing · 245
Ring Tones
    Unique for contacts · 215

View Video Tutorials and Free Tips at www.MadeSimpleLearning.com   **511**

Ringtone
    Different · 214
    Louder Ones · 214
    Open · 214
    Save · 214
Ringtone - Using Music As · 245
Ringtones
    Setting Unique Ringtones for Callers · 247
    using music as · 245
RSS · See Social Feeds - RSS

S

Search
    Introduction · 180
    Tips & Tricks · 181
Search Email
    Assigning a Hot Key · 266
Security
    Disabling Password Security · 89
    Email and Web Browsing · 85
    Setting Owner Information · 87
    Web browsing tips · 89
Selecting Text
    Copy, cut or detete · 172
Send / Receive Files
    Bluetooth · 68
Service Boks
    Activation Message · 494
Service Books
    Sending · 492
Setup App · 42
Setup Wizard
    Overview · 23
Signal Strength Meter
    Troubleshooting · 484
SIM card · 279
Slacker Radio · 394
    Controls · 397
    Installing It · 395
    Menus and Shortcuts · 398
    Select Station · 397
    Start Up · 395
SMS · See Text Messaging
Social Feeds · 364
    Adding Feeds · 365
    Posting Updates · 367
    RSS - Adding · 365
    RSS - Managing Feeds · 368

Using · 366
Social Networking · 345
Songs · See Music
Sounds
    Ring and vibrate · 209
Space Bar
    Email Addresses · 169
Space key · 16
Speed Dial · 237
    Edit, move or delete · 240
    Set from call log · 238
Spell Checker
    Adding Custom Words · 178
    Built-In · 177
Supported Video Formats · 416
Switch Application · See Multitasking
Symbols
    Using · 165
Sync using Desktop Software · 101
Synchronization (Apple)
    Automating Daily Sync · 144
Synchronization (Windows)
    Backup & Restore · 94
    Desktop Software · 94
    Setup · 94
Syncing Music · 120
    Wi-Fi music sync · 122

T

Table of Contents · viii
Tasks · 369
    Adding New · 370
    Category · 371
    Checking Off · 373
    Completing · 373
    Filter by Category · 371
    Finding · 372
    Introduction · 369
    Options · 374
    Searching · 372
    Snooze Alarm · 310
    Sorting · 374
    Transfer or Sync from Computer · 369
    Viewing · 370
Technical Support
    Where to find · 450
Tethered Modem
    Using BlackBerry as · 83
    What is it? · 84

INDEX

Text Messaging (SMS)
    Attach · 321
    AutoText · 322
    Commands · 321
    Composing · 319
    Emoticons · 321
    Introduction · 318, 319
    Menu Commands · 321
    Open and Reply · 323
    Searching · 275
    Send from Contacts List · 321
    Send New · 319
    Smiley Faces · 321
    Switch Input Language · 321
Texting · *See* Text Messaging (SMS)
Theme
    Carrier · 200
    Download New Ones · 201
    Overview · 200
    Standard · 200
Themes · 200
Three BlackBerry Keyboards · 161
Time
    Set · 197
Time Zone
    Set · 197
Touch Screen Gestures
    Basics · 10
    Basics Graphics · 11
    Selecting Menus · 14
    Swipe for next day, email, picture · 15
Track Pad · 9
Trackpad
    Overview · 160
    Sensitivity · 161
Travel · *See* International Travel Tips
Troubleshooting · 484
    App Manager · 486
    Battery Life · 490
    Battery Pull · 492
    Clear Event Log · 489
    Connection Issues · 484
    CPU · 486
    Event Log · 489
    Hard Reset · 492
    Help Me Screen · 488
    Internet Email Problems · 492
    Manage Apps · 486
    Memory Usage · 486
    More Resources · 495
    Poor / Low Wireless · 484
    Radio Not Working · 484
    Register with Host Routing Table · 485
    Reset BlackBerry · 491
    Send Service Books · 492
    Soft Reset BlackBerry · 491
    Storage (Apps) · 486
    Web Browsing Problems · 492
    Wi-Fi Network · 484
Twitter
    Create Account · 352
    Direct Messages · 355
    Find People · 355
    Icons · 354
    Install · 345
    Introduction · 352
    Login · 353
    Mentions · 354
    My Lists · 354
    My Profile · 355
    Options · 357
    Popular Topics · 356
    Search · 356
    Send out Tweet · 354
    Tweet · 354
    What's happening? · 354
Typing
    Accented Characters · 167
    Editing Text · 168
Typing Tips
    Auto Period and Uppercase · 165
    Automatic Capitalization · 164
    Editing Text · 168
    Space Bar · 168

U

UberTwitter
    Install · 345
Unit of Measure · *See* Calculator
Universal Search · 16

V

Vibrate · *See* Profiles, *See* Profiles
Video
    Supported Formats · 416
Video Camera · 412
    Format · 414

Light · 414
MMS Mode or Normal Mode · 415
Options · 414
Options - Scene Modes · 414
Small Size · 414
Video Light · 414
Video Messaging · *See* Multi-Media Messaging (MMS)
Video Supported Formats · 416
Videos · 412
    Controls - Show or Hide · 418
    Converting DVDs to Play · 415
    Playing Videos · 416
    Transfer from Computer · 412
    Transfer To Computer · 412
Voice Dialing
    Basics · 241
    Battery & Signal · 242
    Call Extension · 242
    Calling a Number · 242
    Nicknames · 244
    Options · 242
    Prompts On / Off · 242
    Tips · 243
Voice Notes
    Recorder · 478

W

Wallpaper
    Change or Set · 187
Web Browser · 420
    Add Bookmark · 430
    Address Bar · 425
    Bookmark Naming Tips · 430
    Bookmarks · 430
    Bookmarks - Using · 433
    Copy or Email Web Page · 429
    Editing Web Addresses · 428
    Globe Shortcut Menu · 428
    Go To... · 425
    Google Search · 435
    Google Search for Location · 436
    Icon · 420
    Introduction · 420

    Keyboard Shortcuts · 422
    Menu · 423
    New Bookmark · 430
    Open New Tab (Page) · 421
    Selecting Your Starting View (Start vs. Home) · 432
    Start Page · 420, 425
    Tabs or Pages · 421
    Tips · 422
    Using Tabs (Pages) · 421
    Zoom In or Out · 428
Web Site
    mobile.blackberry.com · 214
WiFi
    Advantages · 76
    Changing Your Network · 82
    Connecting to Networks · 76
    Diagnostics · 83
    HotSpot Connection · 81
    Introduction · 75
    Prioritizing Networks · 82
    Setup · 76
    Troubleshooting · 83
Wi-Fi
    Connecting · 75
Wi-Fi
    Push-button setup · 78
Wi-Fi
    Manually connect · 79
Wi-Fi Music Sync
    Using · 125
Windows Live Messenger · 344
Windows PC Setup · 94
Wireless Network
    Troubleshooting · 326

Y

Yahoo Messenger · 343
YouTube · 362
    App · 362
    Favorites · 364
    In Browser · 363
    On the Web · 363

INDEX